Image
Restoration

Fundamentals
and Advances

Digital Imaging and Computer Vision Series

Series Editor

Rastislav Lukac
Foveon, Inc./Sigma Corporation
San Jose, California, U.S.A.

Image Restoration

Fundamentals and Advances

EDITED BY

Bahadir K. Gunturk • Xin Li

CRC Press
Taylor & Francis Group
Boca Raton   London   New York

CRC Press is an imprint of the
Taylor & Francis Group, an **informa** business

MATLAB® is a trademark of The MathWorks, Inc. and is used with permission. The MathWorks does not warrant the accuracy of the text or exercises in this book. This book's use or discussion of MATLAB® software or related products does not constitute endorsement or sponsorship by The MathWorks of a particular pedagogical approach or particular use of the MATLAB® software.

CRC Press
Taylor & Francis Group
6000 Broken Sound Parkway NW, Suite 300
Boca Raton, FL 33487-2742

First issued in paperback 2017

© 2013 by Taylor & Francis Group, LLC
CRC Press is an imprint of Taylor & Francis Group, an Informa business

No claim to original U.S. Government works

Version Date: 20120711

ISBN 13: 978-1-4398-6955-0 (hbk)
ISBN 13: 978-1-138-70177-3 (pbk)

Library of Congress Cataloging-in-Publication Data

Image restoration : fundamentals and advances / editors, Bahadir K. Gunturk, Xin Li.
     p. cm. -- (Digital imaging and computer vision)
    Summary: "Digital Imaging and Computer Vision"-- Provided by publisher.
    Includes bibliographical references and index.
    ISBN 978-1-4398-6955-0 (hardback)
    1. Image reconstruction. I. Gunturk, Bahadir K. (Bahadir Kursat) II. Li, Xin, 1974-

TA1632.I4828 2012
006.6--dc23
                                        2012021570

Visit the Taylor & Francis Web site at
http://www.taylorandfrancis.com

and the CRC Press Web site at
http://www.crcpress.com

Contents

11 Correction of Spatially Varying Image and Video Motion Blur Using a Hybrid Camera   311

Yu-Wing Tai and Michael S. Brown

Preface

Image restoration refers to the recovery of an unknown *true* image from its degraded measurement. The degradation may occur during image formation, transmission, and storage; and it may be in a number of forms, including additive noise, space invariant or variant blur, aliasing, and compression artifact. With the advances in imaging, computing, and communication technologies over the past decades, image restoration has evolved into a field at the intersection of image processing, computer vision, and computational imaging. Its derivatives include image denoising (also known as noise removal/reduction), image deblurring/deconvolution (including optical/motion deblurring), image inpainting (also called image completion), image interpolation (including super resolution and color demosaicking), image reconstruction (including computed tomography and compressed sensing), and image deblocking/deringing (also referred to as compression artifact removal). Apparently, image restoration techniques have become a fundamental tool to support low-level vision tasks arising from various scientific and engineering fields.

As two mid-career researchers in the field of image processing, it occurred to us that many reference books devoted to image restoration were published over twenty years ago, and more recent works on image restoration have been scattered around in the literature. There is a significant gap between what we can learn from standard image processing textbooks and what the current state-of-the-art is in image restoration. This book was conceived to fill in this gap, at least to some extent. We understand there are already monographs on similar topics such as sparse representation and super resolution. Therefore, we have chosen to edit a book featured by (1) focusing on algorithms rather than theories or applications, and (2) striking a good balance between fundamentals and advances.

Image restoration algorithms are important not only because they serve a wide range of real-world applications (e.g., astronomical imaging, photo editing, medical imaging, and so on), but also due to their intrinsic connection with underlying image models/representations. Breakthroughs in algorithm development often bring novel insights into fundamental properties of image sources—for example, Shapiro's [36] embedded zerotree wavelet (EZW) coding reshaped our thinking about the importance of modeling location uncertainty for images, Portilla et al.'s [23] Gaussian scalar mixture (GSM) denoising polished our understanding about variance estimation for wavelet coefficients, and Dabov et al.'s [27] block-matching 3D (BM3D) denoising challenged the conventional wisdom of modeling image signals from a local view. Meantime, the reproducibility of published works on algorithms makes it easier for other researchers to build upon each other's work, which often benefits the vitality of the technical community as a whole. For this reason, we have attempted to make this book as experimentally reproducible as possible. The source codes accompanying many chapters of this book can be downloaded from its homepage: http://www.csee.wvu.edu/~xinl/IRFA.html.

This book is neither a textbook nor a monograph, but it attempts to connect with a wider range of audience. For young minds entering the field of image processing, we recommend the first two chapters as a starting point. For image processing veterans, any individual chapter and its associated bibliography can serve as a quick reference. As with any edited book, we do acknowledge that our contributors have varying styles of writing and reasoning, but hopefully they also reflect the intellectual understanding of similar topics from diverse perspectives. Please feel free to challenge the claims and models contained in this book and use the released research codes to jump-start your own research.

To facilitate readers, we have organized the chapters as follows. The first three chapters (Chapters 1–3) serve as introductory chapters presenting the fundamentals of image denoising and blurring. Chapter 1 provides an overview of the image denoising field with an intuitive development of ideas and methods; Chapter 2 provides comprehensive coverage of image deconvolution methods, focusing on linear space-invariant systems; and Chapter 3 goes beyond linear space-invariant systems and discusses blind image restoration under space-varying blur. Chapters 4 through 6 concentrate on two important ideas that have been developed recently: nonlocality and sparsity. Chapter 4 reviews the image restoration methods that use the nonlocality idea; Chapter 5 focuses on the idea of sparsity and its extension from local to nonlocal representations; and Chapter 6 focuses on a specific prior driven by the sparsity idea. Chapters 7 and 8 are on super resolution image restoration. Chapter 7 briefly surveys the super resolution methods and presents a learning-based method; Chapter 8 demonstrates super resolution restoration in multispectral imaging with a new Bayesian approach. The final three chapters (Chapters 9–11) extend the treatment of the topic further. Chapter 9 deals with restoration of color images; Chapter 10 exemplifies the importance and variety of image formation modeling with the restoration of document images; and finally, Chapter 11 demonstrates that hybrid imaging systems may bring new possibilities in image restoration.

Last but not least, we want to thank CRC Press/Taylor & Francis for endorsing this book project. We are also grateful to all our colleagues and their collaborators for contributing their work to this book.

MATLAB® is a registered trademark of The Mathworks, Inc. For product information, please contact:
3 Apple Hill Drive
Natick, MA 01760-2098 USA
Tel: 508-647-7000
Fax: 508-647-7001
E-mail: info@mathworks.com
Web: www.mathworks.com

BAHADIR K. GUNTURK
XIN LI

Editors

BAHADIR K. GUNTURK

Bahadir K. Gunturk received his B.S. degree from Bilkent University, Turkey, and his Ph.D. degree from the Georgia Institute of Technology in 1999 and 2003, respectively, both in electrical engineering. Since 2003, he has been with the Department of Electrical and Computer Engineering at Louisiana State University, where he is an associate professor. His research interests are in image processing and computer vision. Dr. Gunturk was a visiting scholar at the Air Force Research Lab in Dayton, Ohio, and at Columbia University in New York City. He is the recipient of the Outstanding Research Award at the Center of Signal and Image Processing at Georgia Tech in 2001, the Air Force Summer Faculty Fellowship Program (SFFP) Award in 2011 and 2012, and named as a Flagship Faculty at Louisiana State University in 2009.

XIN LI

Xin Li received his B.S. degree with highest honors in electronic engineering and information science from the University of Science and Technology of China, Hefei, in 1996, and his Ph.D. degree in electrical engineering from Princeton University, Princeton, New Jersey, in 2000. He was a member of the technical staff with Sharp Laboratories of America, Camas, Washington, from August 2000 to December 2002. Since January 2003, he has been a faculty member in the Lane Department of Computer Science and Electrical Engineering at West Virginia University. He is currently a tenured associate professor at West Virginia University. His research interests include image/video coding and processing. Dr. Li received a Best Student Paper Award at the Visual Communications and Image Processing Conference in 2001; a runner-up prize of Best Student Paper Award at the IEEE Asilomar Conference on Signals, Systems and Computers in 2006; and a Best Paper Award at the Visual Communications and Image Processing Conference in 2010.

Contributors

PETER VAN BEEK
Sharp Laboratories of America, Camas, Washington
pvanbeek@sharplabs.com

MICHAEL S. BROWN
School of Computing, National University of Singapore
brown@comp.nus.edu.sg

CHARLES L. BYRNE
Department of Mathematical Sciences, University of Massachusetts, Lowell, Massachusetts
charles_byrne@uml.edu

WEISHENG DONG
School of Electronic Engineering, Xidian University, China
wsdong@mail.xidian.edu.cn

MICHAEL A. FIDDY
Center for Optoelectronics and Optical Communications, University of North Carolina, Charlotte, North Carolina
mafiddy@uncc.edu

DONALD FRASER
School of Engineering & Information Technology, University of New South Wales @ ADFA, Australia
don.fraser@adfa.edu.au

PRAKASH P. GAJJAR
Dhirubhai Ambani - Institute of Information and Communication Technology, Gandhinagar, Gujarat, India
prakash_gajjar@daiict.ac.in

IVAN GERACE
Consiglio Nazionale delle Ricerche, Istituto di Scienza e Tecnologie dell'Informazione, Pisa, Italy
ivan.gerace@isti.cnr.it
Dipartimento di Matematica e Informatica, Università degli Studi di Perugia, Perugia, Italy
gerace@dmi.unipg.it

BAHADIR K. GUNTURK
Department of Electrical and Computer Engineering, Louisiana State University, Baton Rouge, Louisiana
bahadir@ece.lsu.edu

XIUPING JIA
School of Engineering & Information Technology, University of New South Wales @ ADFA, Australia
x.jia@adfa.edu.au

MANJUNATH V. JOSHI
Dhirubhai Ambani - Institute of Information and Communication Technology, Gandhinagar, Gujarat, India
mv_joshi@daiict.ac.in

ANDREW LAMBERT
School of Engineering & Information Technology, University of New South Wales @ ADFA, Australia
a.lambert@adfa.edu.au

FENG LI
Academy of Opto-Electronics, Chinese Academy of Sciences, China
lifeng@aoe.ac.cn

XIN LI
Lane Department of Computer Science and Electrical Engineering, West Virginia University, Morgantown, West Virginia
xin.li@ieee.org

RASTISLAV LUKAC
Foveon, Inc. / Sigma Corp., San Jose, California
lukacr@colorimageprocessing.com

FRANCESCA MARTINELLI
Consiglio Nazionale delle Ricerche, Istituto di Scienza e Tecnologie dell'Informazione, Pisa, Italy
francesca.martinelli@isti.cnr.it

HSIN M. SHIEH
Department of Electrical Engineering, Feng Chia University, Taichung, Taiwan
hmshieh@fcu.edu.tw

MICHAL ŠOREL
Department of Image Processing, Institute of Information Theory and Automation, Academy of Sciences of the Czech Republic, Prague, Czech Republic
sorel@utia.cas.cz

FILIP ŠROUBEK
Department of Image Processing, Institute of Information Theory and Automation, Academy of Sciences of the Czech Republic, Prague, Czech Republic
sroubekf@utia.cas.cz

YEPING SU
Apple Inc., Cupertino, California
yeping_su@apple.com

YU-WING TAI
Department of Computer Science, Korea Advanced Institute of Science and Technology, South Korea
yuwing@cs.kaist.ac.kr

ANNA TONAZZINI
Consiglio Nazionale delle Ricerche, Istituto di Scienza e Tecnologie dell'Informazione, Pisa, Italy
anna.tonazzini@isti.cnr.it

KISHOR P. UPLA
Electronics and Communication Engineering Department, Sardar Vallabhbhai National Institute of Technology, Surat, Gujarat, India
kishorupla@gmail.com

JUNLAN YANG
Marseille Inc., Santa Clara, California
julia.jyang@gmail.com

FRANCESCA MARTINELLI
Consiglio Nazionale delle Ricerche, Istituto di Scienza e Tecnologie dell'Informazione, Pisa, Italy

CHI M. SHIEH
Department of Electrical Engineering, Feng Chia University, Taichung, Taiwan

XUE SHEN
Department of Image Processing, Institute of Information Theory and Automation, Academy of Sciences of the Czech Republic, Prague, Czech Republic

FILIP SROUBEK
Department of Image Processing, Institute of Information Theory and Automation, Academy of Sciences of the Czech Republic, Prague, Czech Republic

YEPING SU
Apple Inc, Cupertino, California

YU WON TAI
Department of Computer Science, Korea Advanced Institute of Science and Technology, South Korea

ANNA TONAZZINI
Consiglio Nazionale delle Ricerche, Istituto di Scienza e Tecnologie dell'Informazione, Pisa, Italy

Chapter 1

Image Denoising: Past, Present, and Future

XIN LI
West Virginia University

1.1 Introduction

Image denoising refers to the restoration of an image contaminated by additive white Gaussian noise (AWGN). Just like AWGN has served as the simplest situation in modeling channel degradation in digital communication, image denoising represents the simplest task in image restoration and therefore has been extensively studied by several technical communities. It should be noted that the study of the more general problem of signal denoising dates back to at least Norbert Wiener in the 1940s. The celebrated Wiener filter provides the optimal solution to the recovery of Gaussian signals contaminated by AWGN. The derivation of Wiener filtering, based on the so-called orthogonality principle, represents an elegant solution and the only known situation where constraining to linear solutions does not render any sacrifice on the performance. Therefore, at least in theory the problem of image denoising can be solved if we can reduce it to a problem that satisfies the assumptions behind the Wiener filtering theory. The challenge of image denoising ultimately boils down to the art of modeling images.

As George Box once said, "All models are wrong; but some are useful." Under the context of image denoising, the usefulness of models heavily depends on the class of images of interest. The class of photographic images (a.k.a. natural images) are likely to be the most studied in the literature of image coding and denoising. Even though denoising research has been co-evolving with coding research, image models developed for one do not lend themselves directly to the other. The bit rate constraint and accessibility to the original image define the boundary of image coding differently from that of image denoising. Taking an analogy, image denoising behaves more like a source decoding instead of an encoding one — for example, the role played by the redundancy of signal representation is diametrically different in denoising and coding scenarios. An overcomplete representation — often undesirable and deemed "wrong" in image coding — turns out to be a lot more "useful" in image denoising.

Image models underlying all existing image denoising algorithms, no matter explicitly or implicitly stated, can be classified into two categories: deterministic and statistical. Deterministic models include those studied in functional analysis (e.g., Sobolov and Besov-space functions) and partial differential equations (PDE); statistical models include Markov Random Field (MRF), conditional random field (CRF), Gaussian scalar mixture (GSM) and so on. Despite the apparent difference at the surface, deterministic and statistical models have intrinsic connections (e.g., the equivalence between wavelet shrinkage and total variation diffusion). The subtle difference between deterministic and statistical models is highlighted by Von Neumann's famous quote on randomness, "Anyone who considers arithmetical methods of producing random digits is, of course, in a state of sin." Indeed, a theoretically optimal denoising algorithm (though of little practical value) is to recognize the deterministic procedure of simulating AWGN on digital computers. By reverse-engineering the noise simulation process, one can always perfectly remove it and reach zero errors!

The above reasoning raises another issue that has not received as much attention from the image processing community as image modeling — mathematical modeling of *noise*. Even though computer simulation of AWGN has become the gold standard of image denoising, there is little justification that the contaminating noise in real-world images satisfies the AWGN assumption. In fact, noise sources in the physical world are often non-additive (e.g., multiplicative) and non-Gaussian (e.g., Poisson). Nevertheless, algorithms developed for AWGN can often be twisted to match other types of noise in more general restoration tasks (e.g., involving motion or optical blur). As regularization strategies aim at incorporating a priori knowledge about either the image or noise source into the solution algorithms, we expect that mathematical modeling of the noise source is going to play a more important role in the recovery of images contaminated by real-world noise in the future.

The rest of this chapter is organized as follows. We first provide a historical review of image denoising in Section 1.2, especially its revival in the past decade. Due to space limitation, our review is concise and attempts to complement existing ones (e.g., [1]). Then we will work with a pair of popular test images — *lena* and *barbara* — and walk through a series of representative denoising algorithms in Sections 1.3 through 1.5. These two images — one abundant with regular edges and the other regular textures — serve to illustrate the effectiveness of incorporating complementary priori knowledge such as local smoothness and nonlocal similarity. Fully reproducible experimental results will be reported to help young minds entering the field get acquainted with the current state-of-the-art algorithms yet maintain a healthy skepticism toward authoritative models. We make some concluding remarks and discuss future research directions in Section 1.6.

1.2 Historical Review of Image Denoising

Signal denoising dates back to the pioneering work of Wiener and Kolmogorov in the 1940s. The Wiener–Kolmogorov filtering theory was the first rigorous result of designing statistically optimal filters for the class of stationary Gaussian processes. Its long-lasting impact has been witnessed in the past six decades, as we will elaborate next. In the 1950s, Peter Swerling — one of the most influential radar theoreticians — made significant contributions to the optimal estimation orbits and trajectories of satellites and missiles at the RAND Corporation, while the Soviet mathematician Ruslan Stratonovich solved the problem of optimal nonlinear filtering based on his theory of conditional Markov processes in 1959–1960. The next milestone was marked by Rudolf Kalman's adaptive filtering, which extends the Wiener–Kolmogorov theory from a stationary to a nonstationary process. The capability of tracking changes of local statistics by Kalman filtering has led to a wide range of applications in space and military technology.

In the 1970s, two-dimensional signals such as digital imagery started to attract more attention. To the best of our knowledge, image denoising was first studied as a problem of statistical image enhancement by Nasser Nahi and Ali Habibi of the University of Southern California in [2, 3]. Test images used in their study are apparently oversimplified from today's standard, but given the limited computing power and memory resources, those early works were still visionary and it is not surprising that the USC image database is likely the most popular since then. By contrast, theoretic extension of Kalman filtering from 1D to 2D (e.g., [4]) had received relatively less attention partially due to the practical limitations at that time. The full potential of 2D Kalman filtering had to wait until advances in computing technology caught up in 1980s to make its implementation more feasible. The highly cited work of Jong-Sen Lee [5] on image enhancement/noise filtering by local statistics is a standard implementation of 2D Kalman filtering — namely, through the estimation of local mean/variance from a centralized window (the origin of image patches). Nevertheless, [5] was the first algorithmic achievement of applying local Wiener filtering to image denoising, and its conceptual simplicity (in contrast to mathematically more demanding state-space formulation in 2D Kalman filtering) greatly contributed to its impact on engineering applications.

The history of image denoising took an interesting turn in the late 1980s as wavelet theory was established independently by applied mathematicians, computer scientists, and electrical engineers [54]. Wavelet transforms rapidly became the favorite tool for various image processing tasks from compression to denoising. Simple ideas such as wavelet shrinkage/thresholding [7] became the new fashion; while orthodox approaches of applying local Wiener filtering in the wavelet domain (e.g., [8–12]) found themselves in an awkward position — they had to prove they work better than ad-hoc shrinkage techniques (e.g., [7, 13–16]). Not to mention that some more sophisticated models in the wavelet domain (e.g., hidden Markov model [17,18] and Markov random field [19–21]) often achieve modest performance gain over local Wiener filtering while at the price of prohibitive complexity. The rapid growth of wavelet-based image denoising algorithms from the late 1990s to early 2000s might be the consequence of a bandwagon effect (unfortunately this author was also caught during his Ph.D. study). Hindsight reveals that what is more important than the invention of a tool (e.g., wavelet transform [22]) are the novel insights it could bear to a fundamental understanding of the problem. Good localization property of wavelet bases does indicate a good fit with the strategy of local Wiener filtering (even its more sophisticated extension such as Gaussian scalar mixture [23]), but what makes it a success is often what blinds its vision from seeing further.

At the turn of the century, two influential works related to texture synthesis appeared: Efros and Leung's nonparametric resampling in the spatial domain [24] and Portilla and Simoncelli's parametric models in the wavelet domain [25]. Experimental results clearly show that the nonparametric approach is more favored, which for the first time suggests that clustering (e.g., nearest-neighbor in the patch space) might play a more fundamental role than transform. The ripple of nonparametric resampling initiated by the community of texture synthesis took five years to reach the community of image denoising. In the summer of 2005, when I was attending the Computer Vision and Pattern Recognition (CVPR) conference for the first time, I was intrigued by a talk on nonlocal means denoising [17], which received the Best Paper Award Honorable Mention. While I was reasoning with myself about this new idea of nonlocal and the existing fashion of transform, I accidentally ran into a conference version of the now-celebrated BM3D denoising that was first published at a SPIE conference in the winter of 2005–2006 [27]. The reported experimental results were so impressive that I immediately recognized the potential impact of nonlocal sparsity. The rest of the story is easy to tell; since the publication of the journal version of BM3D [18], there has been increasing interest in not only image denoising (please refer to a plot of citation record in Figure 1.1), but also other restoration tasks where nonlocal sparse representations could benefit (please refer to the chapter on sparse representation in this book).

Next, we will break the history of image denoising into three episodes and re-run them in fast-forward mode. Due to space limitation, we will only review the most representative

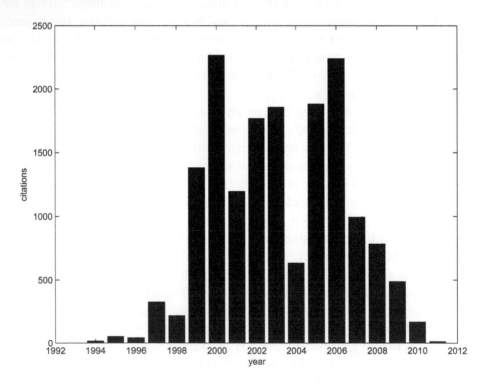

Figure 1.1 The evolutionary history of image denoising as measured by the total number of citations each year (as of May 2011 and based on *Publish or Perish*).

algorithms for each time period and report fully reproducible results for a pair of test images (*lena* and *barbara*) and noise levels ($\sigma_w = 15, 50$). These two images from the University of Southern California (USC) data set — despite being less perfect — do contain the mixture of edges and textures and have been widely used in the literature on image denoising. The general message we deliver through these experiments is: as our understanding about image signals improves, we can achieve better denoising results, though at the price of increased computational complexity. Some idea might be rediscovered years later purely for the reason of waiting for computing technology to catch up. And not all bright ideas or new theories will pay off when applied to experimental validation.

1.3 First Episode: Local Wiener Filtering

What makes an image different from Gaussian noise? There are many possible lines of intuitive reasoning: any linear combination of two Gaussian noise processes is still Gaussian while the average of two photographic images does not produce a new legitimate one; random permutation of Gaussian noise does not change its statistical property (still iid Gaussian) while the random permutation of pixels usually destroys the image content for sure; AWGN is translation invariant or strongly stationary while the local statistics within an image often vary from region to region. There are two generic ways of turning intuitive reasoning into deeper understanding: mentally reproducible (i.e., through the use of mathematical models) and experimentally reproducible (i.e., through the design of computer algorithms). The boundary between these two lines is often vague: any mathematical model must be verified by experiments and the design of any image denoising algorithm inevitably involves some kind of mathematical model — explicitly or implicitly. We have opted to emphasize the line of experimentally reproducible attack here but the connections between an exemplar denoising algorithm and various theories/models will also be analyzed at least, at a conceptual level.

We first study the local Wiener filtering technique developed in [29]. Despite being published over thirty years ago, the basic idea behind the local estimation of mean and variance is still relevant and has been reused extensively in the literature of wavelet-based image denoising. Therefore, we deem it an appropriate baseline scheme to start with. As the formula of classical Wiener filtering suggests, an optimal filter for a Gaussian random variable contaminated by AWGN $y = x + w$ is given by

$$\hat{x} = \frac{\sigma_x^2}{\sigma_x^2 + \sigma_w^2} y, \qquad (1.1)$$

where σ_x^2, σ_w^2 denotes the variance of signal and noise, respectively. In the presence of n noisy samples $y_1, ..., y_n$, a maximum-likelihood estimation of signal variance can be obtained by

$$\hat{\sigma}_x^2 = max[0, \frac{1}{n} \sum_{k=1}^{n} (y_k - \hat{m}_x)^2 - \sigma_w^2], \qquad (1.2)$$

where $\hat{m}_x = \hat{m}_y = \frac{1}{n} \sum_{k=1}^{n} y_k$ is the mean of signal (under the zero-mean assumption of AWGN). Note that the optimality of Equation (1.1) is conditioned on the iid assumption

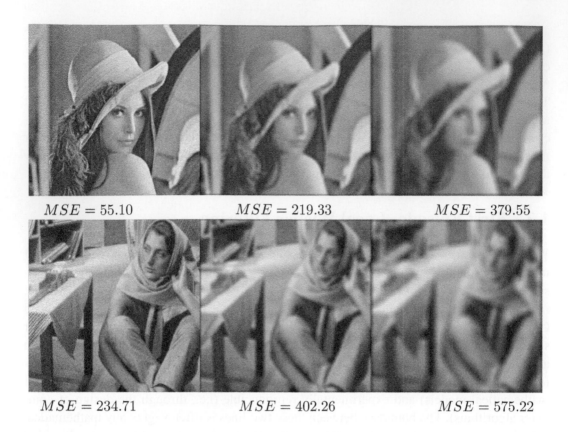

$MSE = 55.10$ $MSE = 219.33$ $MSE = 379.55$

$MSE = 234.71$ $MSE = 402.26$ $MSE = 575.22$

Figure 1.2 Comparison of denoised *lena* (top) and *barbara* (bottom) images at $\sigma_w = 15$ by local Wiener filtering (implemented by MATLAB® function *wiener*2) with different window sizes: left-$[3, 3]$, middle-$[11, 11]$, right-$[19, 19]$.

about $x_1, ..., x_n$. Because such an assumption is seldom satisfied by real-world images, the design of image denoising algorithms is intrinsically related to the modeling of image models, that is, to relax the strict assumption about $x_1, ..., x_n$ such that the Wiener filtering solution can better match the situation of real-world image data.

The key motivation behind local Wiener filtering is to recursively apply Equation (1.2) on a sliding-window basis. Therefore, the only user-defined parameter is the size of the local window $[T_1, T_2]$, decreasing/increasing the window size would reflect the assumption that local statistics change faster/slower from region to region. The algorithm of local Wiener filtering has been well documented in standard textbooks (e.g., [30]) and implemented by the *wiener*2 function in the MATLAB® image processing toolbox. Figures 1.2 and 1.3 include the comparison of denoised images at two different noise levels by local Wiener filtering with varying window sizes. It can be observed that (1) as the window size increases (i.e., the stationarity assumption underlying the image model goes from local to global), noise suppression is more effective but the denoised images appear more blurred; (2) as the noise level increases, larger window size is desirable for the purpose of obtaining a more accurate estimation of signal variance; (3) between *lena* and *barbara*, the latter is a

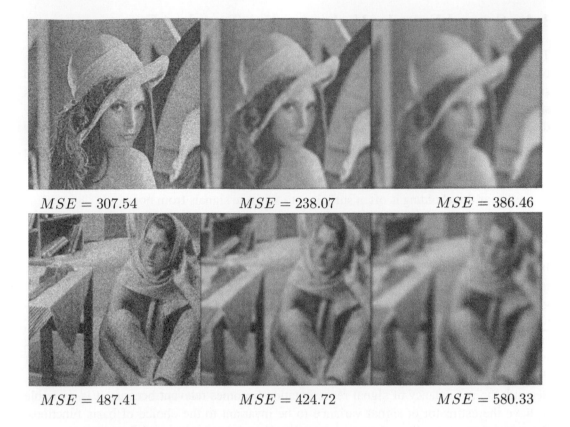

$MSE = 307.54$ $MSE = 238.07$ $MSE = 386.46$

$MSE = 487.41$ $MSE = 424.72$ $MSE = 580.33$

Figure 1.3 Comparison of denoised *lena* (top) and *barbara* (bottom) images at $\sigma_w = 50$ by local Wiener filtering (implemented by MATLAB function *wiener2*) with different window sizes: left-$[3, 3]$, middle-$[11, 11]$, right-$[19, 19]$.

worse match for local Wiener filtering because its abundant textures tend to overshoot the estimated signal variance.

Just like the same physical law could admit seemingly different but deeply equivalent mathematical formulations, the intuitive idea of images being locally smooth can be characterized by different mathematical theories/tools. For example, the concept of local signal variance in Wiener filtering is connected with the smoothness of analytical functions in Sobolev space [31] and the definition of stabilizing operator in Tikhonov regularization [32]. Even though their rigorous connections in the sense of mentally reproducibility have remained elusive, we argue that what is more important and fundamental than tools themselves is what novel insights a new tool can bring about. Under the context of image denoising, a novel insight — if we put ourselves in the situation of the late 1980s — would be the characterization of local changes or transient events [5]. In other words, how to preserve singularities (edges, lines, corners, etc.) in photographic images would reshape our understanding of local Wiener filtering and its related tools.

1.4   Second Episode: Understanding Transient Events

The importance of preserving singularities such as edges and textures has been long recognized but it is the construction of wavelet theory in later 1980s that offers a principled solution. Intuitively, both singularities and noise involve changes but what distinguishes singularities from noise? Wavelet transforms are change-of-coordinates; they are carefully designed in such a way that signals of our interest (singularities) would be characterized by so-called heavy tails in the transform domain. The whole argument has a statistical flavor; it is possible that AWGN could produce some significant coefficient in the wavelet space; but the probability of such a rare event is so small that a conceptually simple strategy such as nonlinear thresholding is often sufficient to separate signals from noise.

1.4.1   Local Wiener Filtering in the Wavelet Space

In a nutshell, wavelet-based image denoising is nothing but the coupling of local Wiener filtering with wavelet transform. However, there are several technical subtleties that could guide us toward a deeper understanding of wavelet-based image denoising. First, the connection between nonlinear thresholding and local Wiener filtering, as pointed out in [8], suggests that wavelet thresholding can be viewed as a simplified version of Wiener filtering with reduced computational complexity. So it is often less fruitful to refine the strategy of thresholding than to improve the statistical model underlying signal variance estimation. Second, the redundancy of signal representation becomes relevant because it is desirable to have the estimator of signal variance to be invariant to the choice of basis functions. We argue that such a line of reasoning makes it easier to understand the idea behind so-called translation-invariant denoising [34]; the cycle-spin technique should be cast into the same category as the sliding windowing technique used in local Wiener filtering. Third, maximum-likelihood (ML) estimation of signal variance in Equation 1.2 represents an empirical Bayesian approach; there are plenty of tools developed by the statistical community to improve upon it. For example, there exists an iterative expectation-maximization (EM) based approach toward ML estimation of signal variance [35] and a fully Bayesian approach where signal variance is modeled by a hidden variable (e.g., in Gaussian scalar mixture [23]).

   To illustrate how those ideas work, we continue our experimental studies as follows. Three wavelet-based image denoising algorithms are compared in this new experiment: TI-denoising [34] available from Wavelab tool box, local Wiener filtering[1] in the mrdwt domain (the implementation of mrdwt is taken from Rice Wavelet toolbox), and the famous BLS-GSM algorithm. Figures 1.4 and 1.5 include the subjective quality comparison of denoised images along with their objective measures in terms of MSE. It can be observed that (1) local Wiener filtering often offers a more principled approach than ad-hoc thresholding (to say the least, the theoretic formula for choosing the optimal threshold does not always match our empirical findings with practical data); (2) GSM offers further improvement over empirical Bayesian estimation of signal variance. The gain for *barbara* is more

---

[1]In our released implementation, we have adopted an iterative ML estimation of signal variance as presented in [35] in contrast to the noniterative solution of Equation (1.2).

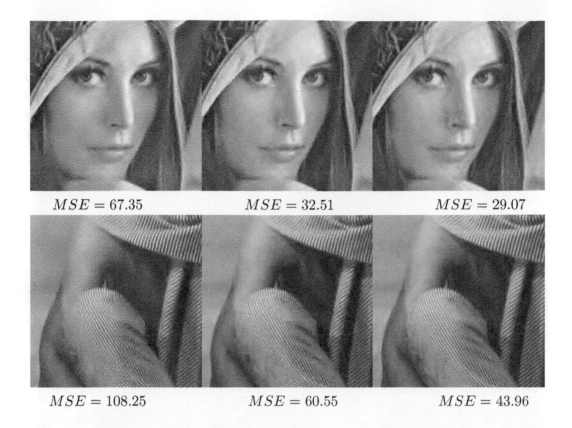

<table>
<tr><td>$MSE = 67.35$</td><td>$MSE = 32.51$</td><td>$MSE = 29.07$</td></tr>
<tr><td>$MSE = 108.25$</td><td>$MSE = 60.55$</td><td>$MSE = 43.96$</td></tr>
</table>

Figure 1.4 Comparison of denoised *lena* (top) and *barbara* (bottom) images at $\sigma_w = 15$ by three wavelet-based denoising algorithms: left - TI thresholding [34] (from Wavelab8.50), middle - local Wiener filtering in mrdwt domain [10] (from Rice Wavelet toolbox), right - BLS-GSM denoising [23].

significant than that for *lena*, which suggests that texture regions are where signal variance estimation falls apart in local Wiener filtering (we will revisit this issue in Section 1.5).

1.4.2 Wavelet vs. DCT Denoising

For a long time, there was debate about wavelet transform (as adopted by JPEG2000) and DCT (as adopted by JPEG) within the community of image coding. Despite the popularity and impact of wavelet-based image coders (e.g., EZW [32], SPIHT [37], and EBCOT [38]), their success seems to attribute more to the statistical modeling of wavelet coefficients than the wavelet transform itself. A comparative study [39] has clearly shown that the embedding coding strategy could also significantly boost DCT-based image coders. In fact, the choice between wavelet transform and DCT is also relevant to the task of image denoising. To the best of my knowledge, a comparative study between wavelet-based and DCT-based image denoising has not been undertaken in the open literature. Therefore, it seems a proper contribution for this review chapter to experimentally conduct such a comparison, which we hope will shed some insights into our understanding. As we elaborate next, the choice

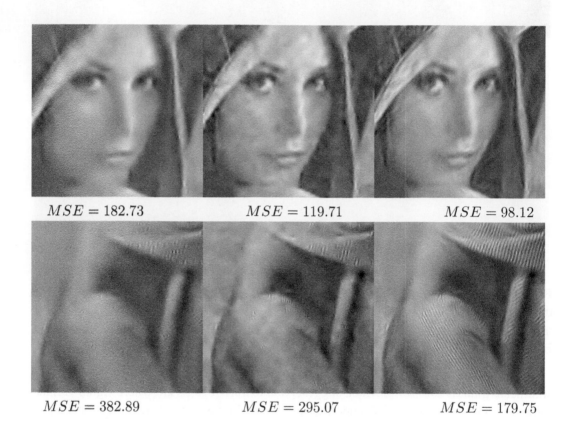

$MSE = 182.73$ $MSE = 119.71$ $MSE = 98.12$

$MSE = 382.89$ $MSE = 295.07$ $MSE = 179.75$

Figure 1.5 Comparison of denoised *lena* (top) and *barbara* (bottom) images at $\sigma_w = 50$ by three wavelet-based denoising algorithms: left - TI thresholding [34] (from Wavelab8.50), middle - local Wiener filtering in mrdwt domain [10] (from Rice Wavelet toolbox), right - BLS-GSM denoising [23].

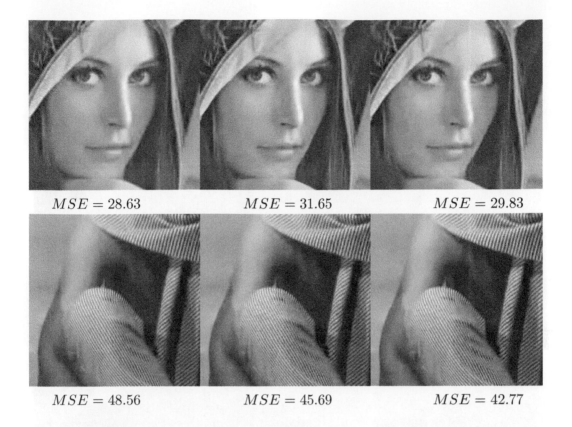

$MSE = 28.63$ $MSE = 31.65$ $MSE = 29.83$

$MSE = 48.56$ $MSE = 45.69$ $MSE = 42.77$

Figure 1.6 Comparison of denoised *lena* (top) and *barbara* (bottom) images at $\sigma_w = 15$ by three dictionary-based denoising algorithms: left - overcomplete DCT-based [41] (from KSVD toolbox), middle - Shape-Adaptive DCT [78] (from SA-DCT toolbox), right - weighted overcomplete-DCT denoising [42].

of transform or a fixed dictionary is secondary to the learning of dictionary and more fundamental issues related to the statistical modeling of photographic images such as locality.

We have tested three exemplar DCT-based denoising algorithms: (1) overcomplete-DCT denoising from KSVD toolbox [41]; (2) shape-adaptive DCT denoising [78]; (3) weighted overcomplete-DCT denoising [42]. Figures 1.6 and 1.7 include the comparison of denoising results for three DCT-based denoising techniques. Comparing them against Figures 1.4 and 1.5, we can observe that the best of DCT-based is indeed highly comparable to the best of wavelet based. Probably it is fair to say that DCT-based does not fall behind wavelet-based on *barbara*, an image with abundant textures (note that a similar observation was also made for the image coding task in [39]). Meanwhile, within the category of DCT-based denoising, we can see that sophisticated strategies such as shape adaptation [78] and weighed estimation [42] have their own merits but the gain is often modest.

In addition to DCT, we want to mention at least two other classes of attack on image denoising: geometric wavelets (e.g., curvelet [43] and contourlet [37]) and diffusion-based (e.g., total-variation diffusion [1, 46] and nonlinear diffusion [89]). Geometric wavelets have been shown to be particularly effective for a certain class of images such as finger-

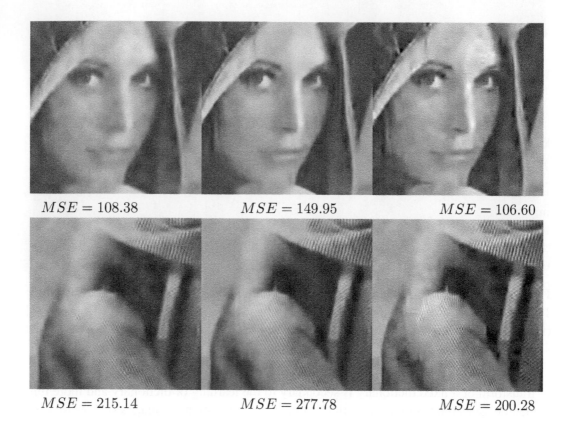

$MSE = 108.38$ $MSE = 149.95$ $MSE = 106.60$

$MSE = 215.14$ $MSE = 277.78$ $MSE = 200.28$

Figure 1.7 Comparison of denoised *lena* (top) and *barbara* (bottom) images at $\sigma_w = 50$ by three dictionary-based denoising algorithms: left - overcomplete DCT-based [41] (from KSVD toolbox), middle - Shape-Adaptive DCT [78] (from SA-DCT toolbox), right - weighted overcomplete-DCT denoising [42].

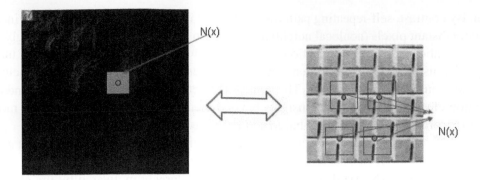

Figure 1.8 Local versus nonlocal neighborhood of a pixel x due to the relativity of defining $N(x)$.

prints with abundant ridge patterns, and diffusion-based models are often a good match with images containing piecewise constant content (e.g., cartoons). However, their overall performance on generic photographic images has not shown convincing improvement over wavelet or DCT-based approaches. It has also been shown that total-variation diffusion is mathematically equivalent to wavelet shrinkage with the simplest Haar filters [48]. Therefore we argue that those more sophisticated technical tools — nevertheless useful — fail to deliver new insights into the modeling of photographic images. What is more fundamental than discovering or polishing a tool is to gain a novel insight; can we think outside the box of transform-based image models?

1.5   Third Generation: Understanding Nonlocal Similarity

Second-generation image models based on wavelet transform or DCT attempt to characterize the a priori knowledge about photographic images by their local smoothness. For modeling transient events or singularities, such a local point of view is indeed appropriate. However, it is important to recognize that local variation and global invariance are two sides of the same coin. To define any change, we must first articulate the frame of reference for measuring such a change. For a pixel of interest within an image, it is often a default to speak of the change with respect to its local neighborhood but that does not imply that spatially adjacent pixels are the only possible frame of reference. For example, a texture image is often decomposed of self-repeating patterns originating from the joint interaction between local reaction and global diffusion [49]. Even for the class of regular edge structures, their geometric constraint implies the relativity of defining local variations — that is, the intensity field is homogeneous along the edge orientation.

It is more enlightening to understand the breakdown of locality assumption from a Wiener filtering perspective. The fundamental assumption underlying Equations (1.1) and (1.2) is that $\{y_1, ..., y_n\}$ belong to the same class (or associated with the same uncorrupted x). The locality principle assumes that $N(x) = \{y_1, ..., y_n\}$ are spatially adjacent pixels of x, no matter if it is in the pixel (first generation) or transform (second generation) domain. By contrast, self-repeating patterns in a texture image often dictate that $N(x)$ include spatially distant pixels (nonlocal neighbors) as shown in Figure 1.8. Such a seemingly simple observation has deep implications for the way we understand image signals, that is, image denoising is intrinsically connected with other higher-level vision tasks, including segmentation or even recognition. The connection between regression/denoising and classification/clustering offers novel insights beyond the reach of conventional image models in the Hilbert space (e.g., wavelet-based and DCT-based).

The idea of using data clustering techniques to solve the denoising problem has gained increased attention in the past five years. One of the earliest nonlocal denoising algorithms — nonlocal means (NLM) denoising [17, 50] — was largely motivated by the effectiveness of nonparametric sampling for texture synthesis [13] and adopted a weighted filtering strategy similar to spectral clustering [95]. It has inspired a flurry of patch-based nonlocal denoising algorithms (e.g., [52–54] and Total-Least-Square denoising [55]). Another pioneering work is KSVD denoising [41]; it generalizes the kmeans clustering algorithm and adaptive PCA denoising [56] by making a connection with matching pursuit-based sparsity optimization [57]. Various follow-up work includes K-LLD [43], learned simultaneous sparse coding (LSSC) [21], and stochastic approximation [60]. The breakthrough made by Finnish researchers — namely, BM3D denoising [18] — was based on a variation of k-Nearest-Neighbor clustering and a two-stage simplification of the EM-based estimation of signal variance as described in Equation (1.2). Despite the conceptual simplicity of BM3D, its outstanding performance (especially in terms of the trade-off between computational complexity and visual quality) has inspired renewed interest in the problem of image denoising (e.g., exemplar-based EM denoising [61], LPG-PCA denoising [62,67]). According to Google Scholar, [18] was the most often cited paper published by *IEEE Transactions on Image Processing* in 2007.

In our experimental study, we selected three exemplar nonlocal denoising algorithms whose implementations appear most robust and efficient: KSVD [41], nonlocal extension of MRF [64], and BM3D [18]. Figures 1.9 and 1.10 include the comparison of both subjective and objective quality comparisons for the two test images with low and high noise contaminations (all results are based on the authors' original source codes release without further parameter tuning). It can be observed that BM3D still outperforms others in all test scenarios and have also been confirmed by other experimental studies such as [64]. When compared with their local counterparts, we can see that nonlocal denoising algorithms can achieve at least comparable and often smaller MSE results.

As the field advances rapidly, one cannot help wondering: Is there any order in the jungle of nonlocal image denoising? We feel that the following two points are important, especially to those young minds entering the field. First, some clustering is more appropriate for denoising than others, if we extrapolate George Box's quoted above, "all clustering tools are wrong; some are useful." The usefulness of any data clustering technique depends on the task it serves. For image denoising, our experience suggests that outliers often have a negative impact on the sparsity of a signal representation and therefore it is desirable to use a clustering tool with minimum risk of outlier (e.g., kNN is preferred over kmeans). We suggest that denoising might be used as an evaluation tool for benchmarking clustering techniques; that is, an optimized clustering result should produce the most accurate estimation of signal variance and therefore the best denoising result. Second, the gap be-

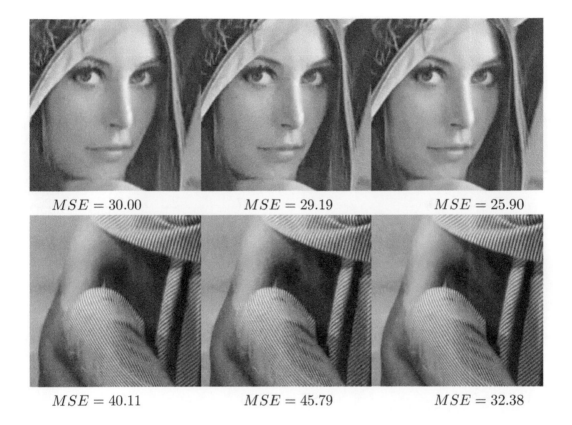

$$MSE = 30.00 \qquad MSE = 29.19 \qquad MSE = 25.90$$

$$MSE = 40.11 \qquad MSE = 45.79 \qquad MSE = 32.38$$

Figure 1.9 Comparison of denoised *lena* (top) and *barbara* (bottom) images at $\sigma_w = 15$ by three nonlocal denoising algorithms: left - KSVD denoising [41] (from KSVD toolbox), middle - nonlocal regularization with GSM [64] (from NLR-GSM toolbox), right - BM3D denoising [18] (from BM3D toolbox).

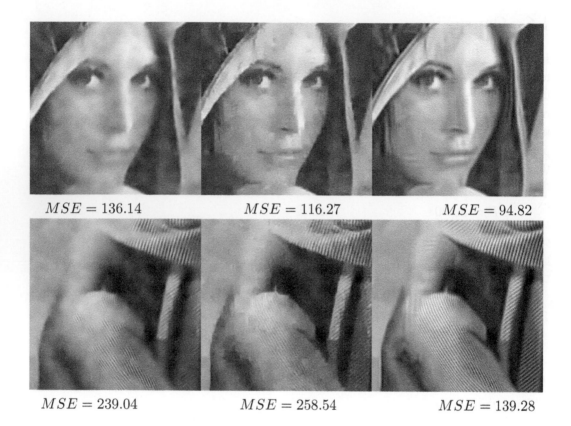

$MSE = 136.14$ $MSE = 116.27$ $MSE = 94.82$

$MSE = 239.04$ $MSE = 258.54$ $MSE = 139.28$

Figure 1.10 Comparison of denoised *lena* (top) and *barbara* (bottom) images at $\sigma_w = 50$ by three nonlocal denoising algorithms: left - KSVD denoising [41] (from KSVD toolbox), middle - nonlocal regularization with GSM [64] (from NLR-GSM toolbox), right - BM3D denoising [18] (from BM3D toolbox).

tween mentally reproducible and experimentally reproducible research is often the biggest challenge in image denoising research. A good and solid theory is mentally reproducible but does not always lead to better experimental results, while experimental breakthrough (e.g., BM3D) often suggests there is something missing in the existing theoretic framework, which calls for deeper logical reasoning. Our own recent work [61] attempts to fill in such a gap by deriving a nonlocal extension of Equation (1.2), but there is still a lot to explore (e.g., the extension from translational invariance to more generic invariant properties).

1.6   Conclusions and Perspectives

In this chapter, we have used extensive experiments to compare various image denoising techniques across three generations: local Wiener filtering, transform based, and nonlocal techniques. A cautious warning for readers who are eager to interpret the reported experimental results in an inductive fashion: we have only reported comparison results for two images at two different noise levels; it is likely that readers could easily find counterexamples for which KSVD outperforms BM3D or NLM-GSM falls behind GSM. We suggest that a sound interpretation of our reported experimental results here is that they have shown the general tendency or typical events (events with a high probability). There will always be exceptions at the "microscopic" level; but what matters more to both theoretic understanding and practical applications related to image denoising is the "macroscopic" behavior of different denoising algorithms. We conclude this chapter by making several comments about the role of representation and optimization in developing denoising algorithms and our own perspective about the evolutionary path of the field.

1.6.1   Representation versus Optimization

Representation is as important as optimization. Again taking BM3D as the example, its effectiveness is largely due to its right intuition and efficient implementation (the authors did not make serious claims about the optimization). In fact, it is possible to interpret the thresholding step and Wiener-filtering step in BM3D (they share lots of similarity) as the first two iterations of an EM-like algorithm. However, our experience has shown that more iterations do not always achieve further gain. Apparently, optimization is more successful in the non-blind scenario of lossy image coding [65] (where the original image is given) than the blind situation such as image denoising. A theoretically powerful tool such as Stein's unbiased risk estimator (SURE) has only found limited success in wavelet-based [66] and nonlocal means [67] denoising. A deeper reason seems to be connected with the definition of randomness or determinism of Turing machines.

One might argue that it is possible to attack the denoising problem without the necessity of addressing the representational issue (e.g., learning-based [68–72]). On the surface, learning-based does appear to be an appealing framework as the alternative to model-based. As of today, learning-based image denoising has been less explored than model-based. However, we believe that the additional assumption about the availability of training data does not necessarily make the problem easier to solve. At least for Bayesians, training data simply "transforms" prior to posterior, so nothing has fundamentally changed. What is more, a pitfall with a learning-based approach is that it could be mathematically equivalent to many existing approaches unless we seriously attack learning as a separate problem on its own merit (i.e., from a neural computation perspective [73]).

1.6.2   Is Image Denoising Dead?

How much more room is left for image denoising research? Recent studies (e.g., [74]) have argued that there is often still plenty of room to improve for a wide range of generic images at certain noise levels. In fact, as long as we believe in the intrinsic connection between denoising and segmentation/recognition, the denoising problem will not be completely dead until others are solved. It is reasonable to expect that advances in image segmentation and object recognition could offer new insights into the way we define patch similarity and therefore really push the art of Wiener filtering to the next level. We also anticipate that the hierarchical representation in the Hilbert space (i.e., wavelet-based) could be generalized into nonlocal image representation (likely outside Hilbert space, e.g., metric space [75]).

Why should we still care about image denoising? Based on our own experience, a better denoising algorithm offers more than just a new tool. The new insights brought by BM3D have inspired many researchers (including this author) to revisit other conventional image processing tasks, including deblurring [22], interpolation [77], inpainting [78], and so on. Such leverage to other low-level vision tasks is almost a free lunch because they share the common objective of building a better image prior model. It might be more fruitful to leverage advances in image denoising to high-level tasks like the feedback connections in the human brain. Nevertheless, a higher SNR often implies more likelihood of object detection [79]; so it might be more rewarding to treat the problem of image denoising not as an isolated engineering problem but as one component in the bigger scientific picture of visual perception [80].

Acknowledgment

This work was supported in part by NSF Award CCF-0914353.

Bibliography

[1] A. Buades, B. Coll, and J. M. Morel, "A review of image denoising algorithms, with a new one," *Multiscale Modeling and Simulation*, vol. 4, no. 2, pp. 490–530, 2005.

[2] N. Nahi, "Role of recursive estimation in statistical image enhancement," *Proceedings of the IEEE*, vol. 60, no. 7, pp. 872–877, 1972.

[3] A. Habibi, "Two-dimensional Bayesian estimate of images," *Proceedings of the IEEE*, vol. 60, no. 7, pp. 878–883, 1972.

[4] J. Woods and C. Radewan, "Kalman filtering in two dimensions," *IEEE Transactions on Information Theory*, vol. 23, pp. 473–482, July 1977.

[5] J.-S. Lee, "Digital image enhancement and noise filtering by use of local statistics," *IEEE Transactions on Pattern Analysis and Machine Intelligence*, vol. 2, pp. 165–168, Mar. 1980.

[6] I. Daubechies, "Where do wavelets come from? A personal point of view," *Proceedings of the IEEE*, vol. 84, no. 4, pp. 510–513, 1996.

[7] D. Donoho and I. Johnstone, "Ideal spatial adaptation by wavelet shrinkage," *Biometrika*, vol. 81, pp. 425–455, 1994.

[8] E. P. Simoncelli and E. H. Adelson, "Noise removal via Bayesian wavelet coring," in *IEEE International Conference on Image Processing*, pp. 379–382, 1996.

[9] P. Moulin and J. Liu, "Analysis of multiresolution image denoising schemes using generalized Gaussian and complexity priors," *IEEE Transactions on Information Theory*, vol. 45, pp. 909–919, Apr. 1999.

[10] I. K. M.K. Mihcak and K. Ramchandran, "Local statistical modeling of wavelet image coefficients and its application to denoising," in *IEEE International Conference on Acoust. Speech Signal Processing*, pp. 3253–3256, 1999.

[11] X. Li and M. Orchard, "Spatially adaptive image denoising under overcomplete expansion," in *IEEE International Conference on Image Processing*, pp. 300–303, 2000.

[12] L. Zhang, P. Bao, and X. Wu, "Multiscale lMMSE-based image denoising with optimal wavelet selection," *IEEE Transactions on Circuits and Systems for Video Technology*, vol. 15, no. 4, pp. 469–481, 2005.

[13] N. Weyrich and G. Warhola, "Wavelet shrinkage and generalized cross validation for image denoising," *IEEE Transactions on Image Processing*, vol. 7, no. 1, pp. 82–90, 1998.

[14] S. G. Chang, B. Yu, and M. Vetterli, "Adaptive wavelet thresholding for image denoising and compression," *IEEE Transactions on Image Processing*, vol. 9, no. 9, pp. 1532–1546, 2000.

[15] S. Chang, B. Yu, and M. Vetterli, "Spatially adaptive wavelet thresholding with context modeling for image denoising," *IEEE Transactions on Image Processing*, vol. 9, no. 9, pp. 1522–1531, 2000.

[16] L. Sendur and I. W. Selesnick, "Bivariate shrinkage functions for wavelet-based denoising exploiting interscale dependency," *IEEE Transactions on Signal Processing*, vol. 50, pp. 2744–2756, 2002.

[17] M. Crouse, R. Nowak, and R. Baraniuk, "Wavelet-based statistical signal-processing using hidden Markov-models," *IEEE Transactions on Signal Processing*, vol. 46, pp. 886–902, 1998.

[18] G. Fan and X.-G. Xia, "Image denoising using a local contextual hidden Markov model in the wavelet domain," *IEEE Signal Processing Letters*, vol. 8, pp. 125–128, May 2001.

[19] M. Malfait and D. Roose, "Wavelet-based image denoising using a Markov random field a priorimodel," *IEEE Transactions on Image Processing*, vol. 6, pp. 549–565, April 1997.

[20] A. Pizurica, W. Philips, I. Lemahieu, and M. Acheroy, "A joint inter- and intrascale statistical model for Bayesian wavelet based image denoising," *IEEE Transactions on Image Processing*, vol. 11, pp. 545–557, May 2002.

[21] A. Pizurica and W. Philips, "Estimating the probability of the presence of a signal of interest in multiresolution single-and multiband image denoising," *IEEE Transactions on Image Processing*, vol. 15, no. 3, pp. 654–665, 2006.

[22] I. Daubechies, "Orthonormal bases of compactly supported bases," *Communications on Pure and Applied Mathematics*, vol. 41, pp. 909–996, 1988.

[23] J. Portilla, V. Strela, M. Wainwright, and E. Simoncelli, "Image denoising using scale mixtures of Gaussians in the wavelet domain," *IEEE Transactions on Image Processing*, vol. 12, pp. 1338–1351, Nov 2003.

[24] A. Efros and T. Leung, "Texture synthesis by non-parametric sampling," in *International Conference on Computer Vision*, pp. 1033–1038, 1999.

[25] J. Portilla and E. Simoncelli, "A parametric texture model based on joint statistics of complex wavelet coefficients," *International Journal of Computer Vision*, vol. 40, pp. 49–71, 2000.

[26] A. Buades, B. Coll, and J.-M. Morel, "A non-local algorithm for image denoising," *IEEE Conference on Computer Vision and Pattern Recognition*, vol. 2, pp. 60–65, 2005.

[27] K. Dabov, A. Foi, V. Katkovnik, and K. Egiazarian, "Image denoising with block-matching and 3D filtering," in *Proc. SPIE Electronic Imaging: Algorithms and Systems V*, vol. 6064, (San Jose, CA), January 2006.

[28] K. Dabov, A. Foi, V. Katkovnik, and K. Egiazarian, "Image denoising by sparse 3-D transform-domain collaborative filtering," *IEEE Transactions on Image Processing*, vol. 16, pp. 2080–2095, Aug. 2007.

[29] J. Lee, "Digital image enhancement and noise filtering by use of local statistics," *IEEE Transactions on Pattern Analysis and Machine Intelligence*, vol. PAMI-2 no. 2, pp. 165–168, 1980.

[30] J. Lim, *Two-Dimensional Signal and Image Processing*, Englewood Cliffs, NJ, Prentice Hall, 1990, 710 p., vol. 1, 1990.

[31] S. Sobolev, "On a theorem of functional analysis," *Mat. Sbornik*, vol. 4, pp. 471–497, 1938.

[32] A. Tikhonov and V. Arsenin, *Solutions of Ill-Posed Problems*. New York: Wiley, 1977.

[33] S. Mallat, *A Wavelet Tour of Signal Processing*. San Diego: CA, Academic Press, 2nd ed., 1999.

[34] D. Donoho and R. Coifman, "Translation invariant denoising," Tech. Rep., 1995. Stanford Statistics Dept.

[35] R. E. Blahut, *Theory of Remote Image Formation*. New York: Cambridge University Press, Jan. 2005.

[36] J. M. Shapiro, "Embedded image coding using zerotrees of wavelet coefficients," *IEEE Transactions on Acoustic Speech and Signal Processing*, vol. 41, no. 12, pp. 3445–3462, 1993.

[37] A. Said and W. A. Pearlman, "A new fast and efficient image codec based on set partitioning in hierarchical trees," *IEEE Transactions on Circuits and Systems for Video Technology*, vol. 6, pp. 243–250, 1996.

[38] D. Taubman, "High-performance scalable image compression with EBCOT," *IEEE Transactions on Image Processing*, vol. 7, pp. 1158–1170, 2000.

[39] Z. Xiong, K. Ramchandran, M. Orchard, and Y. Zhang, "A comparative study of DCT- and wavelet-based image coding," *IEEE Transactions on Circuits and Systems for Video Technology*, vol. 9, no. 5, pp. 692–695, 1999.

[40] M. Elad and M. Aharon, "Image denoising via sparse and redundant representations over learned dictionaries," *IEEE Transactions on Image Processing*, vol. 15, pp. 3736–3745, Dec. 2006.

[41] A. Foi, V. Katkovnik, and K. Egiazarian, "Pointwise shape-adaptive DCT for high-quality denoising and deblocking of grayscale and color images," *IEEE Transactions on Image Processing*, vol. 16, pp. 1395–1411, May 2007.

[42] O. G. Guleryuz, "Weighted averaging for denoising over overcomplete dictionaries," *IEEE Transactions on Image Processing*, vol. 16, no. 12, pp. 3020–3034, 2007.

[43] J. Starck, D. L. Donoho, and E. J. Candes, "The curvelet transform for image denoising," *IEEE Transactions on Image Processing*, vol. II, no. 6, pp. 670–684, 2002.

[44] M. N. Do and M. Vetterli, "The contourlet transform: An efficient directional multiresolution image representation," *IEEE Transactions on Image Processing*, vol. 14, pp. 2091–2106, Dec. 2005.

[45] L. Rudin, S. Osher, and E. Fatemi, "Nonlinear total variation based noise removal algorithms," *Physica D*, vol. 60, pp. 259–268, 1992.

[46] G. Aubert and L. Vese, "A variational method in image recovery," *SIAM Journal on Numerical Analysis*, vol. 34, no. 5, pp. 1948–1979, 1997.

[47] P. Perona and J. Malik, "Scale space and edge detection using anisotropic diffusion," *IEEE Transactions on Pattern Analysis and Machine Intelligence*, vol. 12, no. 7, pp. 629–639, 1990.

[48] G. Steidl, J. Weickert, T. Brox, P. Mrázek, and M. Welk, "On the equivalence of soft wavelet shrinkage, total variation diffusion, total variation regularization, and sides," *SIAM Journal on Numerical Analysis*, vol. 42, no. 2, pp. 686–713, 2004.

[49] A. Turing, "The chemical basis of morphogenesis," *Philosophical Transactions of the Royal Society of London. Series B, Biological Sciences*, vol. 237, no. 641, pp. 37–72, 1952.

[50] M. Mahmoudi and G. Sapiro, "Fast image and video denoising via nonlocal means of similar neighborhoods," *IEEE Signal Processing Letters*, vol. 12, pp. 839–842, Dec. 2005.

[51] A. Ng, M. Jordan, and Y. Weiss, "On spectral clustering: Analysis and an algorithm," *Advances in Neural Information Processing Systems (NIPS)*, vol. 2, pp. 849–856, 2002.

[52] C. Kervrann and J. Boulanger, "Unsupervised patch-based image regularization and representation," in *Proc. of European Conference on Computer Vision (ECCV06)*, pp. IV: 555–567, 2006.

[53] C. Kervrann and J. Boulanger, "Optimal spatial adaptation for patch-based image denoising," *IEEE Transactions on Image Processing*, vol. 15, pp. 2866–2878, Oct. 2006.

[54] C. Kervrann and J. Boulanger, "Local adaptivity to variable smoothness for exemplar-based image regularization and representation," *International Journal of Computer Vision*, vol. 79, no. 1, pp. 45–69, 2008.

[55] K. Hirakawa and T. Parks, "Image denoising using total least squares," *IEEE Transactions on Image Processing*, vol. 15, pp. 2730– 2742, Sept. 2006.

[56] D. Muresan and T. Parks, "Adaptive principal components and image denoising," in *IEEE International Conference on Image Processing*, vol. 1, pp. 101–104, 2003.

[57] F. Bergeaud and S. Mallat, "Matching pursuit: Adaptive representations of images," *Computational and Applied Mathematics*, vol. 15, no. 2, pp. 97–109, 1996.

[58] P. Chatterjee and P. Milanfar, "Clustering-based denoising with locally learned dictionaries," *IEEE Transactions on Image Processing*, vol. 18, no. 7, pp. 1438–1451, 2009.

[59] J. Mairal, F. Bach, J. Ponce, G. Sapiro, and A. Zisserman, "Non-local sparse models for image restoration," in *2009 IEEE 12th International Conference on Computer Vision*, pp. 2272–2279, 2009.

[60] J. Mairal, F. Bach, J. Ponce, and G. Sapiro, "Online dictionary learning for sparse coding," in *Proceedings of the 26th Annual International Conference on Machine Learning*, pp. 689–696, 2009.

[61] X. Li, "Exemplar-based EM-like image denoising via manifold reconstruction," in *IEEE International Conference on Image Processing*, pp. 73–76, 2010.

[62] L. Zhang, W. Dong, D. Zhang, and G. Shi, "Two-stage image denoising by principal component analysis with local pixel grouping," *Pattern Recognition*, vol. 43, no. 4, pp. 1531–1549, 2010.

[63] W. Dong, X. Li, L. Zhang, and G. Shi, "Sparsity-based image via dictionary learning and structural clustering," *IEEE Conference on Computer Vision and Pattern Recognition*, 2011.

[64] J. Sun and M. F. Tappen, "Learning non-local range markov random field for image restoration," *IEEE Conference on Computer Vision and Pattern Recognition*, 2011.

[65] A. Ortega and K. Ramchandran, "Rate-distortion methods for image and video compression," *IEEE Signal Processing Magazine*, vol. 15, no. 6, pp. 23–50, 1998.

[66] F. Luisier, T. Blu, and M. Unser, "A new SURE approach to image denoising: Interscale orthonormal wavelet thresholding," *IEEE Transactions on Image Processing*, vol. 16, no. 3, pp. 593–606, 2007.

[67] D. Van De Ville and M. Kocher, "SURE-based non-local means," *IEEE Signal Processing Letters*, vol. 16, no. 11, pp. 973–976, 2009.

[68] S. Roth and M. J. Black, "Fields of experts: A framework for learning image priors," *IEEE Conference on Computer Vision and Pattern Recognition*, vol. 2, pp. 860–867, 2005.

[69] K. Kim, M. Franz, and B. Schölkopf, "Iterative kernel principal component analysis for image modeling," *IEEE Transactions on Pattern Analysis and Machine Intelligence*, vol. 27, no. 9, pp. 1351–1366, 2005.

[70] P. V. Gehler and M. Welling, "Product of "edge-perts"," *Neural Information Processing System*, vol. 18, Aug. 2005.

[71] Y. Weiss and W. Freeman, "What makes a good model of natural images?" in *IEEE Conference on Computer Vision and Pattern Recognition*, pp. 1–8, 2007.

[72] A. Barbu, "Training an active random field for real-time image denoising," *IEEE Transactions on Image Processing*, vol. 18, no. 11, pp. 2451–2462, 2009.

[73] J. Hertz, A. Krogh, and R. G. Palmer, *Introduction to the Theory of Neural Computation*. Boston, MA: Addison-Wesley Longman Publishing Co., Inc., 1991.

[74] P. Chatterjee and P. Milanfar, "Is denoising dead?" *IEEE Transactions on Image Processing*, vol. 19, no. 4, pp. 895–911, 2010.

[75] X. Li, "Collective Sensing: A Fixed-Point Approach in the Metric Space," in *Proceedings of the SPIE Conference on Visual Communication and Image Processing*, pp. 7744–7746, 2010.

[76] X. Li, "Fine-granularity and spatially-adaptive regularization for projection-based image deblurring," *IEEE Transactions on Image Processing*, vol. 20, no. 4, pp. 971–983, 2011.

[77] A. Danielyan, R. Foi, V. Katkovnik, and K. Egiazarian, "Image and video super-resolution via spatially adaptive blockmatching filtering," in *Proceedings of International Workshop on Local and Non-Local Approximation in Image Processing (LNLA)*, 2008.

[78] X. Li, "Image recovery from hybrid sparse representation: A determinisitc annealing approach," *IEEE Journal of Selected Topics in Signal Processing*, vol. 5, no. 5, pp. 953–962, 2011.

[79] H. Barlow, "Redundancy reduction revisited," *Network: Computation in Neural Systems*, vol. 12, pp. 241–253(13), 1 March 2001.

[80] E. Simoncelli and B. Olshausen, "Natural image statistics and neural representation," *AnnNeuro*, vol. 24, pp. 1193–1216, May 2001.

Chapter 2

Fundamentals of Image Restoration

BAHADIR K. GUNTURK
Louisiana State University

2.1 Introduction

In many imaging applications, the measured image is a degraded version of the *true* (or *original*) image that ideally represents the scene. The degradation may be due to (1) atmospheric distortions (including turbulence and aerosol scattering), (2) optical aberrations (such as diffraction and out-of-focus blur), (3) sensor blur (resulting from spatial averaging at photosites), (4) motion blur (resulting from camera shake or the movements of the objects in the scene), and (5) noise (such as shot noise and quantization). Image restoration algorithms aim to recover the true image from degraded measurements. This inverse problem is typically ill-posed, meaning that the solution does not satisfy at least one of the following: existence, uniqueness, or stability. Regularization techniques are often adopted to obtain a solution with desired properties, indicating a knowledge of prior information about the true image.

Image restoration is widely used in almost all technical areas involving images; astronomy, remote sensing, microscopy, medical imaging, photography, surveillance, and HDTV systems are just a few. For example, license plates may appear illegible due to motion blur; photographs captured under low-light conditions may suffer from noise; out-of-focus photographs may look blurry; standard TV signals may not be sufficiently sharp for high-definition TV sets; archived movies may be corrupted by artifacts and noise; atmospheric distortions may degrade the quality of images in remote sensing. In these examples and in many other scenarios, the importance of image restoration ranges from beneficial to essential.

The term *deblurring* is commonly used to refer to restoration of images degraded by blur. Although the degradation process is in general nonlinear and space varying, a large number of problems could be addressed with a linear and shift-invariant (LSI) model. Because the output of an LSI system is the convolution of the true image with the impulse

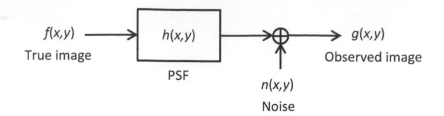

Figure 2.1 Linear shift-invariant image degradation model.

response of the system, the point spread function (PSF), image restoration in LSI systems is called image *deconvolution*. When the impulse response of the system is a delta function and there is only noise, the restoration process becomes image *denoising*. If the PSF is unknown, then the problem is referred to as *blind* image restoration. When there are multiple images that overlap but are not identical, it is possible to recover spatial details that are not visible in any of the input images; this type of restoration is known as *super resolution* image restoration.

Deblurring is the subject matter of this chapter, with the focus on LSI imaging systems. We provide mathematical modeling, and discuss basic image deblurring techniques and recent significant developments. We briefly review blind image restoration techniques and also provide an introduction to super resolution image restoration.

2.2   Linear Shift-Invariant Degradation Model

Suppose that $f(x, y)$ is the true image that we would like to recover from the degraded measurement $g(x, y)$, where (x, y) are spatial coordinates. For a linear shift-invariant system, the imaging process can be formulated as

$$g(x, y) = h(x, y) * f(x, y) + n(x, y), \qquad (2.1)$$

where "$*$" is the convolution operation, $h(x, y)$ is the PSF of the imaging system, and $n(x, y)$ is the additive noise. (The block diagram of this process is given in Figure 2.1.) The imaging formulation can also be done in matrix-vector form or in frequency domain. Defining $\mathbf{g}$, $\mathbf{f}$, and $\mathbf{n}$ as the vectorized versions of $g(x, y)$, $f(x, y)$, and $n(x, y)$, respectively, the matrix-vector formulation is

$$\mathbf{g} = \mathbf{H}\mathbf{f} + \mathbf{n}, \qquad (2.2)$$

where $\mathbf{H}$ is a two-dimensional sparse matrix with elements taken from $h(x, y)$ to have the proper mapping from $\mathbf{f}$ to $\mathbf{g}$. The vectorization could be done in a number of ways, including raster-scan, row-major, and column-major order.

On the other hand, the Fourier-domain version of the imaging model is

$$G(u, v) = H(u, v)F(u, v) + N(u, v), \qquad (2.3)$$

where $G(u, v)$, $H(u, v)$, $F(u, v)$, and $N(u, v)$ are the Fourier transforms of $g(x, y)$, $h(x, y)$, $f(x, y)$, and $n(x, y)$, respectively.

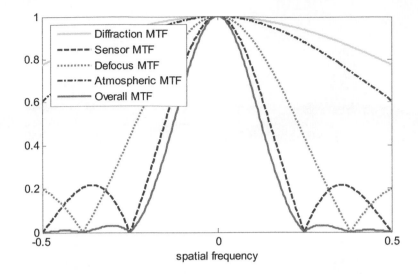

Figure 2.2 An example illustrating that the overall MTF is the product of the individual MTFs in an imaging system.

Because recorded images are of finite spatial extent, the support of the PSF goes beyond the borders of the degraded image; this boundary value problem may create ringing artifacts affecting the entire restored image [1]. The simplest way of extending an image is zero padding, assuming all pixels outside measured region have zero values. With this extension, the blur matrix **H** has a Block Toeplitz with Toeplitz Blocks (BTTB) structure [2]. Another common method is periodic extension, in which case the blur matrix **H** is Block Circulant with Circulant Blocks (BCCB). Because a BCCB matrix can be diagonalized by discrete Fourier transform, periodic extension is the implicit method in Fourier domain solutions. As the periodic extension is obviously not natural, there may be severe artifacts if no additional precautions are taken. In [3, 4], it is shown that a BTTB matrix can be approximated better with a block-Toeplitz-plus-Hankel with Toeplitz-plus-Hankel-block (BTHTHB) matrix, which can be diagonalized with a discrete cosine transform (DCT), leading to reduced artifacts. Other methods of extending images include pixel repetition, mirroring, and extrapolation by polynomial fitting.

The PSF $h(x, y)$ is the impulse response of the imaging system; it is the image of a point from the scene and essentially tells the smallest image detail an imaging system can form. Its Fourier transform $H(u, v)$ is the frequency response of the imaging system and is called the optical transfer function (OTF). The magnitude and phase of the OTF are called the modulation transfer function (MTF) and phase transfer function (PTF), respectively. Unless $h(x, y)$ satisfies certain symmetry conditions, the PTF is nonzero, causing phase distortions during the imaging process. On the other hand, the MTF yields the relative transmittance of an imaging system for different frequency components and is very useful in analyzing system performance, including the resolution of an optical system. In general, frequency domain analysis of an imaging system provides additional insight about the behavior of the system. Also, when there are multiple subsystems, we can multiply the individual transfer functions to get the overall transfer function, as illustrated in Figure 2.2. This is typically

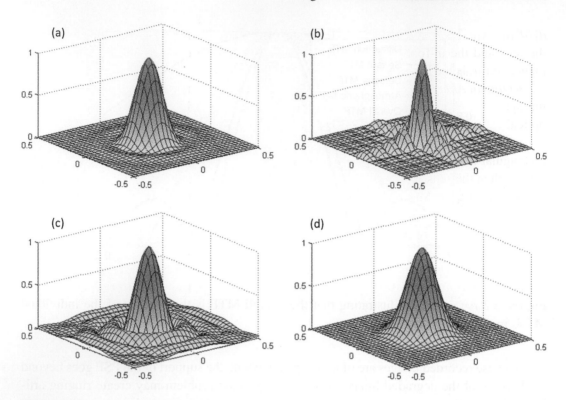

Figure 2.3 (a) Airy disk; (b) MTF of a rectangular PSF; (c) MTF of a disk PSF; (d) Gaussian PSF/MTF.

easier than the repeated convolutions of the impulse responses and allows easy visualization and understanding of the performance of the overall system.

The major factors contributing to image degradation are described below, and some of the PSF/MTF shapes are presented in Figure 2.3.

• *Diffraction:* Because of the wave nature of the light, an optical system can never form a point image. Even when there is no other degradation (such a system is called a diffraction-limited system), an optical system will have a characteristic minimum blur, which is determined by the aperture of the system [5]. For a circular aperture of diameter D, the diffraction PSF has a specific shape, known as the Airy disk . The well-known Rayleigh resolution limit to discern two discrete points, $1.22\lambda f/D$, is approximately the distance between the maximum and the first minimum of the Airy disk.

• *Atmospheric Blur:* Atmospheric distortions are caused mainly by turbulence and aerosol scattering/absorption. A turbulent medium causes wavefront tilt, resulting in local or global image shifts. The MTF of atmospheric turbulence typically has an exponential decay, with dependence on various parameters, including medium characteristics, path length, and exposure time [6]. Aerosol effects, on the other hand, cause diffusion and are modeled well with a Gaussian MTF [7]. It is possible to estimate these parameters and restore images [8].

• *Out-of-Focus Blur:* The out-of-focus PSF takes the shape of the camera aperture. For a circular aperture, the PSF is a disk, which is sometimes referred to as the circle of confusion. For a thin-lens model, it can be shown that the diameter of the disk is $D(|d' -$

$d|/d')m$, where D is the diameter of the disk, f is the focal length, d is the distance between the lens and the in-focus plane, d' is the distance between the lens and the out-of-focus plane, and m is the lens magnification.

• *Optical Aberrations:* An optical system may introduce additional aberrations, including spherical aberration, chromatic aberration, and geometric distortion. Blur caused by optical aberrations could be space varying and color channel dependent.

• *Sensor Blur:* A sensor integrates light over a photosensitive area, which is typically a rectangular area forming the PSF of the sensor. If there is a micro-lens array in front of the sensor, then the sensor PSF takes the shape of the micro-lens.

• *Motion Blur:* Motion blur occurs when the scene is not static and the exposure time is not small enough with respect to the motion in the scene or the camera. In such a case, the projected imagery is smeared over the sensor according to the motion.

• *Anti-Aliasing Filter:* Aliasing happens if the spatial sampling rate is less than the Nyquist rate. To prevent aliasing, an anti-aliasing filter should be applied before the image plane. The anti-aliasing filter should be designed according to the photosite pitch and the color filter array (if used).

• *Noise:* In addition to the blurring effects, the image could also be corrupted by noise. There are different sources of noise, including dark current, shot noise, read noise, and quantization.

2.3   Image Restoration Methods

2.3.1   Least Squares Estimation

The inverse problem of estimating $\mathbf{f}$ from the observation $\mathbf{g} = \mathbf{Hf}$ is typically not well-posed because of nonexistence, nonuniqueness, or instability of the solution. The standard approach to solve such a problem is to introduce regularization terms in a least squares estimation framework. The least squares estimator minimizes the sum of squared differences between the real observation $g(x, y)$ and the predicted observation $h(x, y) * f(x, y)$. The cost function to be minimized can be equivalently written as $\sum_{(x,y)} |g(x, y) - h(x, y) * f(x, y)|^2$ in spatial domain, $\sum_{(u,v)} |G(u, v) - H(u, v)F(u, v)|^2$ in (discrete) Fourier domain, and $||\mathbf{g} - \mathbf{Hf}||^2$ in matrix-vector notation. The solution in matrix-vector notation is

$$\hat{\mathbf{f}} = \arg\min_{\mathbf{f}} ||\mathbf{g} - \mathbf{Hf}||^2 = \mathbf{H}^+\mathbf{g}, \qquad (2.4)$$

where $\mathbf{H}^+$ is known as the pseudo-inverse of $\mathbf{H}$. For an over-determined full-rank system, the pseudo-inverse is $\mathbf{H}^+ = \left(\mathbf{H}^T\mathbf{H}\right)^{-1}\mathbf{H}^T$, which can be derived by setting the derivative of the cost function to zero: $\frac{\partial}{\partial \mathbf{f}}||\mathbf{g} - \mathbf{Hf}||^2 = 2\mathbf{H}^T\mathbf{Hf} - 2\mathbf{H}^T\mathbf{g} = 0$. On the other hand, for an under-determined system, there are an infinite number of solutions, and the problem can be restated as minimizing $||\mathbf{f}||$ subject to the constraint $\mathbf{g} = \mathbf{Hf}$. The solution, also known as the least (or minimum) norm estimate, is $\mathbf{H}^+ = \mathbf{H}^T\left(\mathbf{HH}^T\right)^{-1}$, which is obtained using the method of Lagrange multipliers. (The derivation for a slightly more general case will be done shortly.)

The least squares and least norm solutions can also be written using the singular value decomposition $\mathbf{H} = \mathbf{U}\mathbf{\Lambda}\mathbf{V}^T$, where, for an $\mathbf{H}$ of dimensions $m \times n$, $\mathbf{U}$ is an $m \times m$ unitary matrix, $\mathbf{V}$ is an $n \times n$ unitary matrix, and $\mathbf{\Lambda}$ is an $m \times n$ diagonal matrix with singular values of $\mathbf{H}$ on the diagonal. The pseudo-inverse $\mathbf{H}^+$ can be shown to be equal to $\mathbf{H}^+ = \mathbf{V}\mathbf{\Lambda}^+\mathbf{U}^T$, where $\mathbf{\Lambda}^+$ is the pseudo-inverse of $\mathbf{\Lambda}$, which is obtained by transposing and taking the inverse of every nonzero entry in $\mathbf{\Lambda}$. This implies that the singular value decomposition (SVD) approach provides an estimate even under rank-deficient situations. Suppose that $\mathbf{u}_i$ and $\mathbf{v}_i$ are the ith columns of $\mathbf{U}$ and $\mathbf{V}$, and s_i is ith singular value along the diagonal of $\mathbf{\Lambda}$; then $\mathbf{H}$ can be written as $\mathbf{H} = \sum_{i=1}^r s_i \mathbf{u}_i \mathbf{v}_i^T$, where r is the rank of $\mathbf{H}$. Therefore, the SVD-based generalized inverse becomes

$$\hat{\mathbf{f}} = \mathbf{H}^+\mathbf{g} = \mathbf{V}\mathbf{\Lambda}^+\mathbf{U}^T\mathbf{g} = \sum_{i=1}^r \frac{1}{s_i}(\mathbf{u}_i^T\mathbf{g})\mathbf{v}_i. \tag{2.5}$$

As seen in this equation, a system with small singular values s_i is not stable: A small perturbation in $\mathbf{g}$ would lead to a large change in the solution $\hat{\mathbf{f}} = \mathbf{H}^+\mathbf{g}$. Such a system is called an *ill-conditioned* system. (The *condition number* of a matrix is defined as the ratio of the largest singular value to the smallest one. A system with large condition number, that is, with relatively small singular values, is called ill-conditioned.)

One method to improve the stability is to set the singular values of $\mathbf{\Lambda}$ that are less than a threshold to zero:

$$\hat{\mathbf{f}} = \sum_{i=1}^r w(s_i)\frac{1}{s_i}(\mathbf{u}_i^T\mathbf{g})\mathbf{v}_i, \tag{2.6}$$

where $w(s_i)$ is equal to 1 if s_i is larger than the threshold, 0 otherwise. This method is known as the truncated SVD method.

An alternative way to improve the conditioning of the system and also to give preference to a solution with desirable properties is to incorporate regularization terms into the problem. Tikhonov regularization [9, 10] is a commonly used technique in least squares estimation: the cost function to be minimized is modified to $\|\mathbf{g} - \mathbf{H}\mathbf{f}\|^2 + \lambda\|\mathbf{L}\mathbf{f}\|^2$, where λ is a nonnegative regularization parameter controlling the trade-off between the fidelity and regularization terms, and $\mathbf{L}$ is an identity matrix or a high-pass filter matrix used to impose smoothness on the solution. With Tikhonov regularization, the solution (in matrix-vector notation) is

$$\hat{\mathbf{f}} = \left(\mathbf{H}^T\mathbf{H} + \lambda\mathbf{L}^T\mathbf{L}\right)^{-1}\mathbf{H}^T\mathbf{g}. \tag{2.7}$$

This direct solution in matrix-vector form is not computationally feasible because of the dimensions involved in computing the inverse of $\left(\mathbf{H}^T\mathbf{H} + \lambda\mathbf{L}^T\mathbf{L}\right)$; instead, iterative matrix inversion methods [11] could be used. On the other hand, defining $L(u, v)$ as the Fourier domain version of $\mathbf{L}$, the direct solution in Fourier domain is

$$\hat{F}(u, v) = \frac{H^*(u, v)G(u, v)}{|H(u, v)|^2 + \lambda|L(u, v)|^2}, \tag{2.8}$$

which can be computed very efficiently using fast Fourier transform. This solution extends the inverse filter solution $\hat{F}(u, v) = G(u, v)/H(u, v)$, and with proper choices of λ and $L(u, v)$, it does not suffer from noise amplification when $H(u, v)$ is zero or close to zero.

(See Figure 2.4.) For further discussion, consider the regularizer $L(u, v) = 1$, which corresponds to $\mathbf{L} = \mathbf{I}$ in matrix-vector domain. From the Fourier domain perspective, the denominator of Equation (2.8) becomes $|H(u, v)|^2 + \lambda$; therefore, division by zero is prevented with nonzero λ. From the singular value decomposition perspective, the solution (2.7) becomes $\hat{\mathbf{f}} = \mathbf{V}(\mathbf{\Lambda}^T\mathbf{\Lambda} + \lambda\mathbf{I})^{-1}\mathbf{\Lambda}^T\mathbf{U}^T\mathbf{g}$ if we substitute $\mathbf{H} = \mathbf{U}\mathbf{\Lambda}\mathbf{V}^T$ into it. Explicitly,

$$\hat{\mathbf{f}} = \sum_{i=1}^{r} \frac{s_i}{s_i^2 + \lambda}(\mathbf{u}_i^T\mathbf{g})\mathbf{v}_i. \tag{2.9}$$

In other words, the inverse matrix has singular values $s_i/(s_i^2 + \lambda)$, which demonstrates that λ prevents large singular values in the inverse matrix even when s_i is small, improving the conditioning of the system to small perturbations. Also note that the Tikhonov solution (2.9) can be interpreted as weighting the singular values as in the truncated SVD solution given in Equation (2.6), but this time the weighting function is a smooth one, $s_i^2/(s_i^2 + \lambda)$, instead of the step function in Equation (2.6).

In general, the regularized least squares estimation has the following form:

$$\hat{\mathbf{f}} = \arg\min_{\mathbf{f}} \left\{ \|\mathbf{g} - \mathbf{H}\mathbf{f}\|^2 + \lambda\phi(\mathbf{f}) \right\}, \tag{2.10}$$

where $\phi(\mathbf{f})$ is the regularization function and λ is the regularization parameter. We discuss different regularization functions and how regularization parameter is selected later in the chapter.

As an alternative to the problem statement in Equation (2.10), the optimization problem can be stated as

$$\hat{\mathbf{f}} = \arg\min_{\mathbf{f}} \left\{ \phi(\mathbf{f}) \right\} \text{ subject to } \mathbf{g} = \mathbf{H}\mathbf{f}, \tag{2.11}$$

which is a constrained optimization problem and more strict than Equation (2.10) in terms of data fidelity. The solution to Equation (2.11) could be obtained using the Lagrange multipliers technique, which minimizes the Lagrangian $\mathcal{L}(\mathbf{f}) = \phi(\mathbf{f}) + \beta^T(\mathbf{g} - \mathbf{H}\mathbf{f})$, where β is the vector of Lagrange multipliers. When $\phi(\mathbf{f}) = \|\mathbf{L}\mathbf{f}\|^2$ as in the Tikhonov regularization, the closed-form solution to Equation (2.11) is obtained by taking the derivative of the Lagrangian with respect to $\mathbf{f}$ and setting it to zero:

$$\frac{\partial\mathcal{L}(\mathbf{f})}{\partial\mathbf{f}} = 2\mathbf{L}^T\mathbf{L}\mathbf{f} - \mathbf{H}^T\beta = 0 \Rightarrow \hat{\mathbf{f}} = 0.5(\mathbf{L}^T\mathbf{L})^{-1}\mathbf{H}^T\beta, \tag{2.12}$$

and using the constraint $\mathbf{g} = \mathbf{H}\hat{\mathbf{f}}$, one can find that $\beta = 2[\mathbf{H}(\mathbf{L}^T\mathbf{L})^{-1}\mathbf{H}^T]^{-1}\mathbf{g}$, and plugging β into Equation (2.12), the constrained solution turns out to be

$$\hat{\mathbf{f}} = (\mathbf{L}^T\mathbf{L})^{-1}\mathbf{H}^T[\mathbf{H}(\mathbf{L}^T\mathbf{L})^{-1}\mathbf{H}^T]^{-1}\mathbf{g}. \tag{2.13}$$

For $\mathbf{L} = \mathbf{I}$, the constrained solution becomes the least norm solution that we have already provided for the under-determined case. In practice, the constraint $\mathbf{g} = \mathbf{H}\mathbf{f}$ might be too strict because of noise and inaccuracies in the PSF estimate, and therefore either the unconstrained approach (2.10) is taken or the constraint is relaxed to $\|\mathbf{g} - \mathbf{H}\mathbf{f}\|^2 \leq \epsilon$, for some small value ϵ that possibly reflects the noise variance. The latter approach may also be converted to an unconstrained optimization problem using the method of Lagrange multipliers.

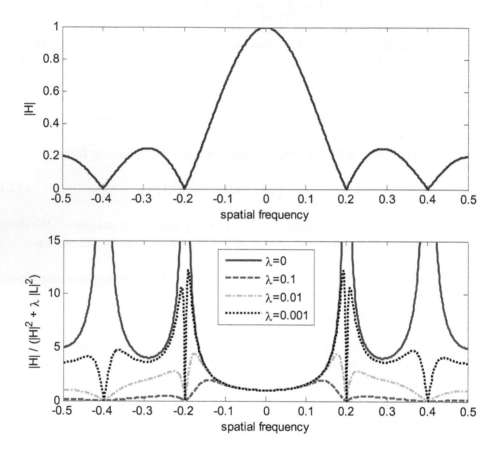

Figure 2.4 Top: The frequency response (MTF) of an averaging filter. Bottom: The frequency response of the inverse filter regularized with the high-pass filter [1 -2 1]. The regularization parameter λ controls the behavior of the inverse filter. When λ is close to zero, noise could be amplified; when λ is large, high-frequency information could be lost, resulting in an overly smooth solution.

2.3.2 Steepest Descent Approach

In cases where there is no direct solution or a direct solution is not computationally feasible, an iterative scheme is adopted. The steepest descent method updates an initial estimate iteratively in the reverse direction of the gradient of the cost function $C(\mathbf{f})$. An iteration of the steepest descent method is

$$\mathbf{f}^{(i+1)} = \mathbf{f}^{(i)} - \alpha \left. \frac{\partial C(\mathbf{f})}{\partial \mathbf{f}} \right|_{\mathbf{f}^{(i)}}, \tag{2.14}$$

where α is the step size and $\mathbf{f}^{(i)}$ is the ith estimate. In the case of a least squares problem without regularization, the cost function can be defined as $C(\mathbf{f}) = (1/2)\|\mathbf{g} - \mathbf{Hf}\|^2$, which results in the iteration

$$\mathbf{f}^{(i+1)} = \mathbf{f}^{(i)} - \alpha \left. \frac{\partial C(\mathbf{f})}{\partial \mathbf{f}} \right|_{\mathbf{f}^{(i)}} = \mathbf{f}^{(i)} + \alpha \mathbf{H}^T \left(\mathbf{g} - \mathbf{Hf}^{(i)} \right), \tag{2.15}$$

which is also known as the Landweber iteration [12,13]. The algorithm suggested by Equation (2.15) can be taken literally by converting images to vectors and obtaining the matrix form of blurring. The alternative implementation requires observing that $\mathbf{H}^T$ corresponds to convolution with the flipped PSF kernel [14]; that is, the iteration (2.15) can be implemented as

$$f^{(i+1)}(x,y) = f^{(i)}(x,y) + \alpha h^T(x,y) * \left(g(x,y) - h(x,y) * f^{(i)}(x,y) \right). \tag{2.16}$$

In [15], it is discussed that the back-propagation operator $h^T(x,y)$ in Equation (2.16) can indeed be replaced by other kernels and still achieve convergence. (For instance, in the case of the Van Cittert algorithm [16], the back-propagation operator is a delta function.) The choice of back-propagation operator, however, affects the characteristics of the solution.

The iterations are repeated until a convergence criterion is reached, which could be the maximum number of iterations, or the rate of change in the estimated signal $\|\mathbf{f}^{(i+1)} - \mathbf{f}^{(i)}\|/\|\mathbf{f}^{(i)}\|$ or in the cost $|C(\mathbf{f}^{(i+1)}) - C(\mathbf{f}^{(i)})|/C(\mathbf{f}^{(i)})$ between two successive iterations. The step size α should be small enough to guarantee convergence; on the other hand, the convergence rate would be slow if it is too small [17]. We could determine the convergence condition on α by rewriting Equation (2.15) as $\mathbf{f}^{(i+1)} = (\mathbf{I} - \alpha \mathbf{H}^T \mathbf{H}) \mathbf{f}^{(i)} + \alpha \mathbf{H}^T \mathbf{g}$, which implies that the amplification of $(\mathbf{I} - \alpha \mathbf{H}^T \mathbf{H})$ should be less than 1 in order for the iterations to converge. In other words, for an arbitrary nonzero vector $\mathbf{v}$, the supremum $sup\{\|(\mathbf{I} - \alpha \mathbf{H}^T \mathbf{H})\mathbf{v}\|/\|\mathbf{v}\|\}$ should be less than one to guarantee convergence. This condition is satisfied when $|1 - \alpha s_{max}| < 1 \Rightarrow 0 < \alpha < 2/s_{max}$, where s_{max} is the maximum singular value of $\mathbf{H}^T \mathbf{H}$.

The step size α could also be updated at each iteration. A commonly used method to choose the step size is the *exact line search* method. Defining $\mathbf{d}^{(i)} = \left. \frac{\partial C(\mathbf{f})}{\partial \mathbf{f}} \right|_{\mathbf{f}^{(i)}}$, the step size (at ith iteration) that minimizes the next cost $C(\mathbf{f}^{(i+1)}) = C(\mathbf{f}^{(i)} - \alpha \mathbf{d}^{(i)})$ is

$$\alpha = \frac{\mathbf{d}^{(i)T} \mathbf{d}^{(i)}}{\mathbf{d}^{(i)T} (\mathbf{H}^T \mathbf{H}) \mathbf{d}^{(i)}}, \tag{2.17}$$

which can be found by differentiating the cost $C\left(\mathbf{f}^{(i)} - \alpha\mathbf{d}^{(i)}\right)$ with respect to α.

It is possible to achieve faster convergence by utilizing the conjugate gradient method [18]. In the steepest descent method, the estimate is updated in the direction of the residual; the conjugate gradient method, on the other hand, ensures that the update direction is conjugate to previous directions. Another method to minimize $C(\mathbf{f})$ is the Newton's method [18], where at each iteration the image is updated as $\mathbf{f}^{(i+1)} = \mathbf{f}^{(i)} - \left(\nabla^2 C(\mathbf{f})\right)^{-1} \nabla C(\mathbf{f})\Big|_{\mathbf{f}^{(i)}}$, where $\nabla^2 C((\mathbf{f}))$ is the Hessian matrix. When applicable, Newton's method converges very fast. There are also quasi-Newton methods that approximate the inverse of the Hessian matrix [18].

One advantage of the iterative algorithms is the possibility of imposing additional constraints on the solution during the iterations. For example, one may impose nonnegativity on the pixel intensities by adding the projection operator

$$P\left[f^{(i)}(x,y)\right] = \left\{ \begin{array}{l} f^{(i)}(x,y), \text{ for } f^{(i)}(x,y) \geq 0 \\ 0, \text{ for } f^{(i)}(x,y) < 0 \end{array} \right. \tag{2.18}$$

to each iteration step.

2.3.3   Regularization Models

Regularization is essential in most image restoration problems, which has motivated research and development of a wide variety of regularization functions $\phi(\mathbf{f})$. The classical Tikhonov regularization uses ℓ_2 norm: either the energy of the signal $\|\mathbf{f}\|$ or the energy of its gradient field $\|\nabla\mathbf{f}\|$ is used as the regularization term. Over the years, it has been observed that regularization functions leading to sparse solutions are a better fit for natural images than the ones using ℓ_2 norm. Some regularization functions are as follows.

- $\phi(\mathbf{f}) = \|\mathbf{f}\|_1$: Compared to the ℓ_2 norm $\|\mathbf{f}\|$, the ℓ_1 norm is known to have a greater tendency to allow occasional large-amplitude samples. This can be understood from the Bayesian perspective. The ℓ_2 norm corresponds to a Gaussian prior in MAP estimation; on the other hand, the ℓ_1 norm corresponds to Laplace distribution, which is more heavy-tailed than the Gaussian distribution. In other words, it is more likely to have large-amplitude samples with Laplace distribution compared to Gaussian distribution. While $\|\mathbf{f}\|_1$ could be a good regularization function in certain applications, such as astronomical and medical imaging where spike-like features occur, it may not represent the true image well in some other applications.

- $\phi(\mathbf{f}) = \|\nabla\mathbf{f}\|_1$: Instead of intensity distribution, modeling of spatial gradient distribution is a better choice for natural images. Because natural images tend to be piecewise smooth, the gradient field shows a sparse distribution: most gradient values are zero (or very close to zero), and few are large due to discontinuities. The regularization function $\|\nabla\mathbf{f}\|_1$ has been shown to preserve discontinuities better than $\|\nabla\mathbf{f}\|$, which leads to solutions with blurred edges [19, 20]. The regularization $\|\nabla\mathbf{f}\|_1$ or approximately equal $\sum_{x,y} \sqrt{(\nabla_x f(x,y))^2 + (\nabla_y f(x,y))^2}$ is known as *total variation* regularization [21].

- $\phi(\mathbf{f}) = \|\nabla\mathbf{f}\|_0$: We promote more and more sparsity in the solution by pushing $p \to 0$ in ℓ_p norm. When $p = 0$, the regularization term essentially counts the number of nonzero elements. In other words, the regularization promotes the solution with the least number of nonzero terms. While this looks like an attractive direction, computational complexity of the resulting optimization problem does not lead to practical algorithms.

- $\phi(\mathbf{f}) = \mathbf{1}^T \rho(\mathbf{\Phi f})$: A large number of regularization functions in the literature can be written in the form $\phi(\mathbf{f}) = \mathbf{1}^T \rho(\mathbf{\Phi f})$, where $\mathbf{1}$ is a column vector of 1s, $\rho(\cdot)$ is a function operating element-wise on the input vector, and $\mathbf{\Phi}$ is a 2D matrix. The $\mathbf{\Phi}$ matrix could represent operations such as gradient, Laplacian, discrete cosine transform (DCT), and wavelet decomposition. From the sparsity perspective, it is known that a sparser representation of a signal using a *dictionary* $\mathbf{\Phi}$; that is, $\mathbf{\Phi f}$ is sparser than $\mathbf{f}$. For instance, the DCT coefficients or the wavelet coefficients are sparser than the signal itself, which is, by the way, the motivation behind the popular image compression schemes. Therefore, a good choice of $\mathbf{\Phi}$ could lead to a successful restoration algorithm by promoting sparsity of the solution. We should note that instead of choosing the dictionary, it is possible to *learn* the dictionary [22].

Depending on $\rho(\cdot)$, a direct closed-form solution is feasible, as in the case of Tikhonov regularization, where we have the quadratic function $\rho(z) = z^2$. If the derivative of $\rho(\cdot)$ exists, then one can apply a gradient descent type of iterative algorithm; at each iteration, the current estimate is updated in the reverse direction of the gradient, which is $\nabla \phi(\mathbf{f}) = \mathbf{\Phi}^T \rho'(\mathbf{\Phi f})$. This type of regularization is more interesting when nonquadratic robust functions are used. For example, the Huber function (given in Table 2.1) does not penalize large values as much as $\rho(z) = z^2$, and therefore preserves edges better.

- $\phi(\mathbf{f}) = \sum_{x,y} f(x,y) log(f(x,y))$: An alternative way to promote sparsity is to maximize entropy, which reaches its smallest value under uniform distribution and its largest value under delta distribution. This type of deconvolution is known as maximum entropy deconvolution [9,23]. The resulting optimization problem is nonlinear and requires iterative methods. There are different definitions of entropy, and the input to the entropy function could be the intensities or the gradients.

- $\phi(\mathbf{f}) = \sum_{c \in \mathcal{C}} V_c(\mathbf{f})$: Markov random field (MRF) models provide an easy way of defining local interactions. They are defined in terms of a potential function $V_c(\cdot)$ of a *clique c*, which is a single site or a set of sites in a neighborhood, and $\mathcal{C}$ is the set of all cliques. By summing over all possible cliques, the *energy* of the image is obtained. The way the cliques and the clique potentials are defined, different structural properties can be modeled. If quadratic potentials are chosen, for example $V_c(\mathbf{f}) = \left(\mathbf{d}_c^T \mathbf{f}\right)^2$ with $\mathbf{d}_c$ representing a finite difference operator, then the energy can be written as $\sum_{c \in \mathcal{C}} V_c(\mathbf{f}) = \mathbf{f}^T \mathbf{\Theta f}$, where $\mathbf{\Theta}$ is obtained from $\mathbf{d}_c$. Instead of quadratic potentials, it is also possible to use an arbitrary (usually convex) function $\rho(\cdot)$ and define a clique potential as $V_c(\mathbf{f}) = \rho\left(\mathbf{d}_c^T \mathbf{f}\right)$. When the Huber function is used for $\rho(\cdot)$, the resulting MRF (also known as the Huber Markov random field) preserves edges well by replacing the quadratic term after a threshold with a linear term.

2.3.4 Robust Estimation

Robust functions could be used not only in the regularization terms, but also to handle outliers that do not fit the assumed observation model. For example, the difference vector

Table 2.1 Some Functions Used in Regularization and Robust Estimation

Function	Derivative	Notes
$\rho(z) = z^2$	$\rho'(z) = 2z$	Square function
$\rho(z) = \|z\|$	$\rho'(z) = sign(z)$	Modulus (absolute value) function
$\rho(z) = \|z\| - c \log(1 + \|z\|/c)$	$\rho'(z) = \|z\| sign(z)/(\|z\| + c)$	Approximates modulus as $c \to 0$.
$\rho(z) = \begin{cases} z^2, & \text{for } \|z\| \le c \\ 2c\|z\| - c^2, & \text{for } \|z\| > c \end{cases}$	$\rho'(z) = \begin{cases} 2z, & \text{for } \|z\| \le c \\ 2c\,sign(z), & \text{for } \|z\| > c \end{cases}$	Huber function
$\rho(z) = \log\left(1 + \frac{z^2}{2c^2}\right)$	$\rho'(z) = \frac{2z}{z^2 + 2c^2}$	Lorentz function
$\rho(z) = \begin{cases} \frac{z^6}{3c^6} - \frac{z^4}{c^4} + \frac{z^2}{c^2}, & \text{for } \|z\| \le c \\ \frac{1}{3}, & \text{for } \|z\| > c \end{cases}$	$\rho'(z) = \begin{cases} z\left(1 - \frac{z^2}{c^2}\right)^2, & \text{for } \|z\| \le c \\ 0, & \text{for } \|z\| > c \end{cases}$	Tukey's biweight function

$\mathbf{g} - \mathbf{Hf}$ could be large for some entries due to noise in imaging process or errors in modeling. To address such situations, robust functions are utilized; and the cost function to be minimized is defined as

$$C(\mathbf{f}) = \mathbf{1}^T \rho_1 (\mathbf{g} - \mathbf{Hf}) + \lambda \mathbf{1}^T \rho_2 (\mathbf{Lf}), \tag{2.19}$$

where $\mathbf{1}$ is a column vector of 1s, $\rho_1(\cdot)$ and $\rho_2(\cdot)$ are robust functions operating element-wise on the input vector, and $\mathbf{L}$ is a 2D matrix, which may be an identity matrix or may output, for instance, horizontal and vertical gradients, wavelet coefficients, etc. (Note that $\mathbf{L}$ is not necessarily a square matrix.) Some well-known estimators can be considered special cases of the robust estimator. For example, the least squares estimation with Tikhonov regularization is a special case of this formulation, with $\rho_1(z) = \rho_2(z) = z^2$. When $\rho_1(z) = z^2$, $\rho_2(z) = |z|$, and $\mathbf{L}$ is the gradient operator, the robust estimator becomes the total variation estimator.

If the robust functions are differentiable, one may apply iterative techniques, such as the steepest descent method, and determine the solution iteratively using the gradient of the cost function:

$$\frac{\partial C(\mathbf{f})}{\partial \mathbf{f}} = -\mathbf{H}^T \rho_1' (\mathbf{g} - \mathbf{Hf}) + \lambda \mathbf{L}^T \rho_2' (\mathbf{Lf}), \tag{2.20}$$

where $\rho_1'(\cdot)$ and $\rho_2'(\cdot)$ are the derivatives of the robust functions. Some choices of $\rho(\cdot)$ and their derivatives are given in Table 2.1 and are plotted in Figure 2.5. More on robust statistics and the behavior of robust functions can be found in [24].

2.3.5 Regularization with ℓ_p Norm, $0 < p \le 1$

To obtain a sparse solution with the fewest possible number of nonzero elements, ℓ_0 norm is used, and the constrained optimization problem is stated as minimizing $||\mathbf{f}||_0$ subject to $\mathbf{g} = \mathbf{Hf}$. An exact solution requires exhaustive search, which is not computationally feasible in general; therefore, either algorithms approximating the solution are used or the optimization problem is relaxed to a manageable one. Greedy algorithms are popular approximation methods. Because $\mathbf{g}$ should be a linear combination of columns of $\mathbf{H}$, greedy algorithms aim to determine the correct set of columns iteratively, starting with an empty

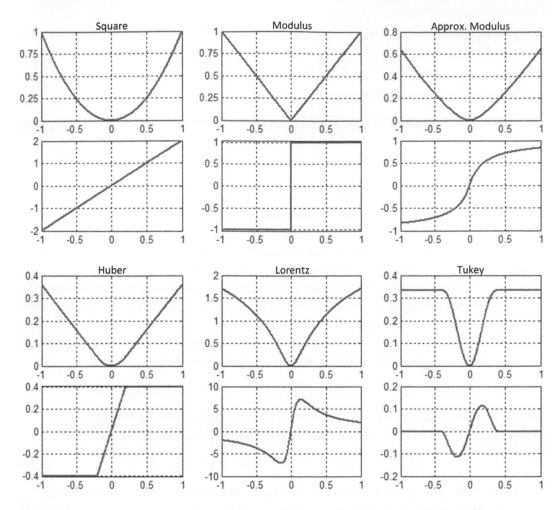

Figure 2.5 The functions shown in Table 2.1 and their derivatives are plotted. Note that the modulus function is not strictly differentiable at the origin; it is often approximated by a differentiable function.

set. At each iteration, the support (the set of columns) is incremented by adding a column, which is selected to minimize the residual $||\mathbf{g} - \mathbf{Hf}||$. The iterations are stopped when the residual is less than a predetermined threshold. The variations of the greedy algorithms include matching pursuit [25], orthogonal-matching-pursuit [26], and orthogonal least squares [27]. It is also possible to relax the problem to a nonconvex optimization problem [28, 29] or a convex one [31, 32]. A detailed survey of methods for obtaining sparse solutions is given in [33]. In this chapter, we describe two methods.

• *Iterative Reweighted Least Squares (IRLS) Method:* The IRLS method [29] is designed to minimize ℓ_p norms with $0 < p \leq 1$. The underlying idea is to rewrite the ℓ_p norm as a weighted ℓ_2 norm. Suppose that the current estimate is $\mathbf{f}^{(i)}$; if we choose a diagonal matrix $\mathbf{D}^{(i)}$, whose diagonal is $|\mathbf{f}^{(i)}|^{1-p/2}$, and define $\mathbf{D}^{(i)^+}$ as the pseudo-inverse $\mathbf{D}^{(i)}$ where all nonzero entries are inverted and the others are kept as zeros, then $||\mathbf{D}^{(i)^+}\mathbf{f}^{(i)}||_2^2$ becomes equal to the ℓ_p norm of the image $||\mathbf{f}^{(i)}||_p^p$. In other words, the ℓ_p norm minimization problem becomes an iterative quadratic optimization problem, where at each iteration

$$\mathbf{f}^{(i+1)} = \arg\min_{\mathbf{f}} \left\{ \mathbf{D}^{(i)+}\mathbf{f} \right\} \text{ subject to } \mathbf{g} = \mathbf{Hf} \qquad (2.21)$$

is minimized for constrained optimization, or

$$\mathbf{f}^{(i+1)} = \arg\min_{\mathbf{f}} \left\{ ||\mathbf{g} - \mathbf{Hf}||_2^2 + \lambda ||\mathbf{D}^{(i)+}\mathbf{f}||^2 \right\} \qquad (2.22)$$

is minimized for unconstrained optimization. The same idea has also been used in total variation-based restoration [30], where the regularizer is $||\mathbf{D}^{(i)+}\nabla\mathbf{f}||^2$.

• *Iterative Shrinkage/Thresholding (IST) Method:* The problem of minimizing $(1/2)||\mathbf{g} - \mathbf{Hf}||^2 + \lambda\mathbf{1}^T\rho(\mathbf{f})$ could be easily addressed when the regularizer is a quadratic function. On the other hand, traditional optimization algorithms are not efficient when the regularizer is nonquadratic. IST algorithms have been recently developed to do ℓ_1 minimization. There are different ways of developing IST algorithms, including the proximal (or surrogate) function technique of [34] and the expectation-maximization technique of [35]. According to the surrogate function technique, the solution is determined iteratively minimizing a modified cost function that includes a distance term promoting the proximity between subsequent estimates. The distance term is quadratic and convex, and leads to a simple and effective algorithm. The modified cost function is

$$C(\mathbf{f}) = (1/2)||\mathbf{g} - \mathbf{Hf}||^2 + \lambda\mathbf{1}^T\rho(\mathbf{f}) + (\mu/2)||\mathbf{f} - \mathbf{f}^{(i)}||^2 - (1/2)||\mathbf{Hf} - \mathbf{Hf}^{(i)}||^2, \quad (2.23)$$

where μ is chosen such that the distance term is strictly convex [34,36,37]. Defining $\mathbf{v}^{(i)} = \mathbf{f}^{(i)} + (1/\mu)\mathbf{H}^T(\mathbf{g} - \mathbf{Hf}^{(i)})$ and reorganizing the terms, Equation (2.23) becomes

$$C(\mathbf{f}) = (1/2)||\mathbf{v}^{(i)} - \mathbf{f}||^2 + \lambda/\mu\mathbf{1}^T\rho(\mathbf{f}) + constant. \qquad (2.24)$$

In other words, the problem becomes finding $\mathbf{f}$ that minimizes $(1/2)||\mathbf{v}^{(i)} - \mathbf{f}||^2 + (\lambda/\mu)\mathbf{1}^T\rho(\mathbf{f})$, which can also be written as $\sum_{(x,y)} \left[(1/2)\left(v^{(i)}(x,y) - f(x,y)\right)^2 + (\lambda/\mu)\rho(f(x,y)) \right]$. As a result, the IST algorithms have the following basic form [34,38,39]:

$$\mathbf{f}^{(i+1)} = \Psi_{\lambda/\mu}\left(\mathbf{v}^{(i)}\right) = \Psi_{\lambda/\mu}\left(\mathbf{f}^{(i)} + (1/\mu)\mathbf{H}^T(\mathbf{g} - \mathbf{Hf}^{(i)})\right), \qquad (2.25)$$

where $\Psi_{\lambda/\mu}$ is a function that minimizes $(1/2)(a - z)^2 + \lambda\rho(z)$ with respect to z. Depending on the choice of $\rho(\cdot)$, a closed-form solution for $\Psi_{\lambda/\mu}$ could be found.

Consider now the ℓ_1 norm regularization $||\mathbf{f}||_1$, in which case $\rho(z) = |z|$. Because $\rho(z)$ is not differentiable at $z = 0$, the method of setting the derivative to zero is replaced by making sure that the subgradient set includes zero. When $z > 0$, $\rho'(z) = 1$, and the gradient of the cost function is $(z - a) + \lambda$, which leads to $z = a - \lambda$. The condition $z > 0$ corresponds to $a > \lambda$. When $z < 0$, $\rho'(z) = -1$, and the gradient of the cost function is $(z - a) - \lambda$, which leads to $z = a + \lambda$; and the condition $z < 0$ corresponds to $a < -\lambda$. When $z = 0$, $\rho'(z) \in [-1, 1]$, and this corresponds to $a \in [-\lambda, \lambda]$. To summarize, we have the following solution:

$$\hat{z} = \begin{cases} a - \lambda, & \text{for } a > \lambda \\ 0, & \text{for } |a| \leq \lambda \\ a + \lambda, & \text{for } a < -\lambda \end{cases}, \qquad (2.26)$$

which corresponds to soft-thresholding operation. That is, the optimization problem $(1/2)||\mathbf{g} - \mathbf{Hf}||^2 + \lambda||\mathbf{f}||_1$ can be solved iteratively as in Equation (2.25) with $\Psi_{\lambda/\mu}$ being the soft-thresholding operator. Similarly, when the regularizer is ℓ_1 norm of the wavelet coefficients: $\rho(\mathbf{f}) = ||\mathbf{\Phi_w f}||_1$ with $\mathbf{\Phi_w}$ producing the detail wavelet coefficients, then $\Psi_{\lambda/\mu}$ becomes wavelet denoising operation.

2.3.6 Wiener Filter

The Wiener estimator looks for an estimate in the form $\hat{\mathbf{f}} = \mathbf{Wg} + \mathbf{b}$ that minimizes the mean square error between the true image and the estimated image, $E\{(\mathbf{f} - \hat{\mathbf{f}})^T(\mathbf{f} - \hat{\mathbf{f}})\}$, where $E\{\cdot\}$ is the expectation operator. The optimal $\mathbf{W}$ and $\mathbf{b}$ values can be obtained using the orthogonality principle [40], which specifies two conditions: (1) the expected value of the estimate must be equal to the expected value of the true image, that is, $E\{\hat{\mathbf{f}}\} = E\{\mathbf{f}\}$; and (2) the restoration error must be orthogonal to the observation about its mean, that is, $E\{(\mathbf{f} - \hat{\mathbf{f}})(\mathbf{g} - E\{\mathbf{g}\})^T\} = \mathbf{0}$. From the first condition, the bias term can be written as

$$\mathbf{b} = E\{\mathbf{f}\} - \mathbf{W}E\{\mathbf{g}\}, \tag{2.27}$$

and by substituting this bias term and $\hat{\mathbf{f}} = \mathbf{Wg} + \mathbf{b}$ into the second condition, the restoration matrix is found to be

$$\mathbf{W} = \mathbf{Q_{fg}Q_g}^{-1}, \tag{2.28}$$

where $\mathbf{Q_{fg}} = E\{(\mathbf{f} - E\{\mathbf{f}\})(\mathbf{g} - E\{\mathbf{g}\})^T\}$ is the cross-covariance matrix between the true image and the observation, and $\mathbf{Q_g} = E\{(\mathbf{g} - E\{\mathbf{g}\})(\mathbf{g} - E\{\mathbf{g}\})^T\}$ is the covariance matrix of the observation. While this solution applies to space-varying blur as well, it can be simplified further for a linear space-invariant system. Using the assumption that true image and noise are uncorrelated, the Wiener filter (2.28) for a linear space-invariant system becomes

$$\mathbf{W} = \mathbf{Q_f H}^T \left(\mathbf{HQ_f H}^T + \mathbf{Q_n}\right)^{-1}, \tag{2.29}$$

where $\mathbf{Q_f}$ and $\mathbf{Q_n}$ are the covariance matrices of the true image and noise. Unless the expected values of true and observed images are zero, the bias term $\mathbf{b}$ would be nonzero and should be taken into account in the estimation, but in practice, it is possible to avoid dealing with the bias term by estimating the means of the observed and true images, subtracting the observed image mean from $\mathbf{g}$ before applying the filter (2.28), and finally adding the true image mean to obtain the solution.

We can obtain additional insight by examining the Fourier domain version of the Wiener filter (2.29), which is

$$W(u, v) = \frac{H^*(u, v)}{|H(u, v)|^2 + \frac{S_n(u,v)}{S_f(u,v)}}, \tag{2.30}$$

where $S_n(u, v) = E\{|N(u, v)|^2\}$ and $S_f(u, v) = E\{|F(u, v)|^2\}$ are the power spectral densities of the noise and true image, and the estimated image is $\hat{F}(u, v) = W(u, v)G(u, v)$. At frequencies where the signal power is much larger than the noise power, the term $\frac{S_n(u,v)}{S_f(u,v)}$ is close to zero, and the Wiener filter $W(u, v)$ becomes the inverse filter $\frac{1}{H(u,v)}$. On the other hand, at frequencies where the noise power is much larger

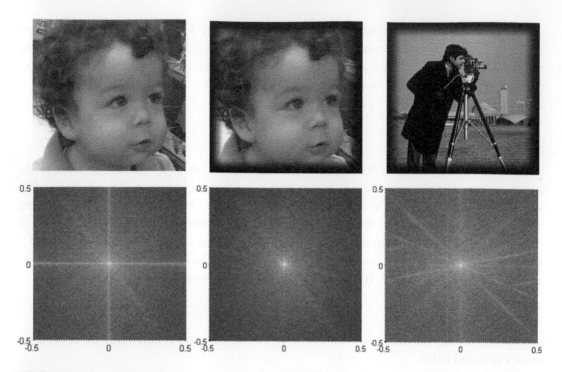

Figure 2.6 Left: Power spectral density of an image. The spectral rays passing through the origin along the horizontal and vertical directions are due to the discontinuities at the borders. Middle: The artificial spectral rays could be eliminated by tapering the image borders. Right: Power spectral densities could deviate well from isotropic model, depending on the image.

than the signal power, the Wiener filter will approach zero, preventing the noise from being amplified and dominating the solution. To implement the Wiener filter in practice, one needs to estimate the power spectral densities $S_n(u, v)$ and $S_f(u, v)$. For additive white noise, $S_n(u, v)$ is equal to the noise variance, which can be estimated in a number of ways, such as the robust median estimator [41]. To estimate $S_f(u, v)$, one may first estimate the power spectral density of the measurement $S_g(u, v) \approx |G(u, v)|^2$ and use the relation $S_g(u, v) = |H(u, v)|^2 S_f(u, v) + S_n(u, v)$ to estimate $S_f(u, v)$ from the already estimated $S_n(u, v)$ and $H(u, v)$. A second way to estimate $S_f(u, v)$ is to simply use the power spectral density obtained from a (set of) representative image(s) with similar content. A third way of estimating $S_f(u, v)$ is to use a parametric model, which is typically an exponentially decaying function, and estimate the model parameters from the measured data [42, 43]. As illustrated in Figure 2.6, the power spectral densities could be nonisotropic, and the parametric model should be chosen accordingly; also, the boundary effects due to the discontinuity at the borders should be taken care of in fitting the parameters.

2.3.7   Bayesian Approach

Bayesian estimation provides an elegant statistical perspective to the image restoration problem. The unknown image, noise, and PSF (in case of blind deconvolution) are viewed

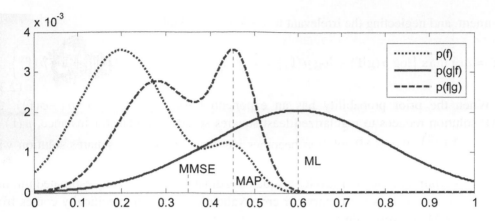

Figure 2.7 An illustration of ML, MAP, and MMSE estimators.

as random variables. The maximum likelihood (ML) estimator seeks the solution that maximizes the probability $p(\mathbf{g}|\mathbf{f})$ while the maximum *a posteriori* (MAP) estimator maximizes $p(\mathbf{f}|\mathbf{g})$. Using the Bayes rule, the MAP estimator can be written in terms of the conditional probability $p(\mathbf{g}|\mathbf{f})$ and the prior probability of $p(\mathbf{f})$:

$$\hat{\mathbf{f}} = \arg\max_{\mathbf{f}} p(\mathbf{f}|\mathbf{g}) = \arg\max_{\mathbf{f}} \frac{p(\mathbf{g}|\mathbf{f})\,p(\mathbf{f})}{p(\mathbf{g})} = \arg\max_{\mathbf{f}} p(\mathbf{g}|\mathbf{f})\,p(\mathbf{f}), \qquad (2.31)$$

where $p(\mathbf{g})$ is dropped in the last term as it is constant with respect to the argument $\mathbf{f}$. And the last term reveals that the MAP solution indeed reduces to the ML solution when there is image prior does not favor a specific solution, in other words, when $p(\mathbf{f})$ is a uniform distribution. (Note that while the MAP estimator aims to find $\mathbf{f}$ that maximizes the posterior $p(\mathbf{f}|\mathbf{g})$, the minimum mean squared error (MMSE) estimator aims to find the expected value of $\mathbf{f}$, defined as $E[\mathbf{f}|\mathbf{g}] = \int \mathbf{f}p(\mathbf{f}|\mathbf{g})df$. An illustration of ML, MAP, and MMSE estimators are given in Figure 2.7.)

In most image restoration problems, image noise is modeled to be a zero-mean independent identically distributed (iid) Gaussian random variable: $p(n(x,y)) = \frac{1}{\sqrt{2\pi}\sigma_n} \exp\left(-\frac{1}{2\sigma_n^2}(n(x,y))^2\right)$. Then, the conditional probability of the observed image is

$$
\begin{aligned}
p(\mathbf{g}|\mathbf{f}) &= \prod_{x,y} \frac{1}{\sqrt{2\pi}\sigma_n} \exp\left(-\frac{1}{2\sigma_n^2}(g(x,y) - h(x,y) * f(x,y))^2\right) \qquad (2.32) \\
&= \frac{1}{\left(\sqrt{2\pi}\sigma_n\right)^M} \exp\left(-\frac{1}{2\sigma_n^2}\sum_{x,y}(g(x,y) - h(x,y) * f(x,y))^2\right) \\
&= \frac{1}{\left(\sqrt{2\pi}\sigma_n\right)^M} \exp\left(-\frac{1}{2\sigma_n^2}\|\mathbf{g} - \mathbf{Hf}\|_2^2\right),
\end{aligned}
$$

where σ_n is the noise standard deviation and M is the total number of pixels in the observed image. Substituting Equation (2.32) into Equation (2.31) after taking the logarithm of the argument, and neglecting the irrelevant terms, one will obtain

$$\hat{\mathbf{f}} = \arg\max_{\mathbf{f}} \left\{ \log p\left(\mathbf{g}|\mathbf{f}\right) + \log p\left(\mathbf{f}\right) \right\} = \arg\max_{\mathbf{f}} \left\{ -\frac{1}{2\sigma_n^2} \|\mathbf{g} - \mathbf{Hf}\|^2 + \log p\left(\mathbf{f}\right) \right\}.$$

(2.33)

When the prior probability has an exponential form, $p\left(\mathbf{f}\right) \propto \exp\left(-\phi(\mathbf{f})\right)$, the MAP solution reduces to regularized least squares solution. When, for instance, $p\left(\mathbf{f}\right) \propto \exp\left(-\|\mathbf{Lf}\|^2\right)$, the MAP solution becomes identical to the least squares solution with Tikhonov regularization with proper value of λ. (Note that the regularization parameter λ is somewhat arbitrary in regularized least squares estimation and typically estimated through some technique such as the L-curve or cross-validation, while it specifically comes from the PDFs in MAP estimation.)

The Gaussian model is sometimes used for both the prior and the noise. Assuming zero mean, the distributions are $p\left(\mathbf{f}\right) \propto \exp\left(-\frac{1}{2}\mathbf{f}^T \mathbf{Q}_f^{-1} \mathbf{f}\right)$ and $p\left(\mathbf{n}\right) \propto \exp\left(-\frac{1}{2}\mathbf{n}^T \mathbf{Q}_n^{-1} \mathbf{n}\right)$, where $\mathbf{Q}_f$ and $\mathbf{Q}_n$ are the covariance matrices of the image prior and noise. This would lead to the solution

$$\begin{aligned} \hat{\mathbf{f}} &= \arg\min_{\mathbf{f}} \left\{ (\mathbf{g} - \mathbf{Hf})^T \mathbf{Q}_n^{-1} (\mathbf{g} - \mathbf{Hf}) + \mathbf{f}^T \mathbf{Q}_f^{-1} \mathbf{f} \right\} \\ &= \left(\mathbf{H}^T \mathbf{Q}_n^{-1} \mathbf{H} + \mathbf{Q}_f^{-1} \right)^{-1} \mathbf{H}^T \mathbf{Q}_n^{-1} \mathbf{g}, \end{aligned}$$

(2.34)

which is identical to the solution based on the Wiener filter in Equation (2.29). (The equivalence of these two solutions can be shown using the *matrix inversion lemma*.) Under the assumption of uncorrelated distributions $\mathbf{Q}_f = \sigma_f^2 \mathbf{I}$ and $\mathbf{Q}_n = \sigma_n^2 \mathbf{I}$, the solution becomes

$\hat{\mathbf{f}} = \left(\mathbf{H}^T \mathbf{H} + \frac{\sigma_n^2}{\sigma_f^2} \mathbf{I} \right)^{-1} \mathbf{H}^T \mathbf{g}$ in matrix-vector notation or $\hat{F}(u,v) = \frac{H^*(u,v)G(u,v)}{|H(u,v)|^2 + \frac{\sigma_n^2}{\sigma_f^2}}$ in Fourier domain.

It should be noted that noise models other than the Gaussian model have also been considered in the literature. For instance, when the noise model is Poisson, which is the case for data-dependent photon noise, the conditional probability of the observed image becomes

$$p\left(\mathbf{g}|\mathbf{f}\right) = \prod_{x,y} \frac{(h(x,y) * f(x,y))^{g(x,y)} \exp(-(h(x,y) * f(x,y)))}{g(x,y)!},$$

(2.35)

which leads to solutions different than the Gaussian scenario. For instance, the ML approach results in the well-known iterative Richardson-Lucy algorithm [44, 45]:

$$f^{(i+1)}(x,y) = \left[\frac{g(x,y)}{h(x,y) * f^{(i)}(x,y)} * h^T(x,y) \right] f^{(i)}(x,y).$$

(2.36)

2.3.8   Projection onto Convex Sets

The projection onto convex sets (POCS) technique [46], which consists of iterative projection of an initial estimate onto predefined constraint sets, is a well-known and successful

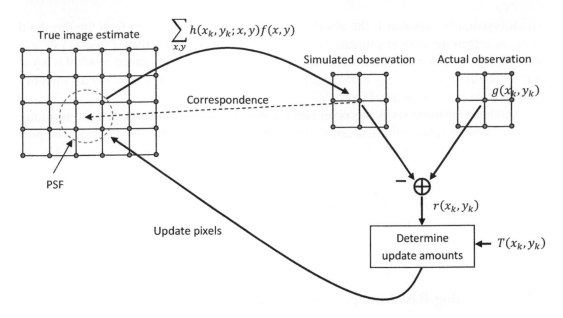

Figure 2.8 An illustration of the projection onto data fidelity constraint set.

approach in image restoration. When the constraint sets are convex and not disjoint, the technique guarantees convergence to a solution that is consistent with all the constraint sets. On the other hand, the solution is not necessarily unique; depending on the initial estimate and the order of the projections, the iterations may lead to different solutions. One advantage of the POCS technique is the ease of incorporating space-variant PSF into the restoration [47].

Data fidelity is commonly used as a constraint set to ensure consistency with observations. The set is defined at each observation pixel (x_k, y_k) and constrains the difference between the predicted pixel value and the observed pixel value:

$$C_D = \{f : |r(x_k, y_k)| \le T(x_k, y_k)\}, \tag{2.37}$$

where $r(x_k, y_k) = g(x_k, y_k) - \sum_{x,y \in S_k} h_k(x_k, y_k; x, y) f(x, y)$ is the residual, S_k is the set of pixels in the $f(x, y)$ that contributes to the pixel (x_k, y_k), $h(x_k, y_k; x, y)$ is the contribution of the pixel $f(x, y)$, and $T(x_k, y_k)$ is the threshold value that controls the data fidelity constraint. The threshold should be chosen considering the noise power [48]. If the noise power is small, the threshold could be chosen small; otherwise it should be large to allow space for disturbances caused by the noise. The projection operation onto the data fidelity constraint set is formulated as [47, 48]:

$$f^{(i+1)}(x, y) = \begin{cases} f^{(i)}(x, y) + \frac{(r(x_k, y_k) - T(x_k, y_k)) h(x_k, y_k; x, y)}{\sum\limits_{x,y \in S_k} h^2(x_k, y_k; x, y)}, & r(x_k, y_k) > T(x_k, y_k) \\ f^{(i)}(x, y), & |r(x_k, y_k)| \le T(x_k, y_k) \\ f^{(i)}(x, y) + \frac{(r(x_k, y_k) + T(x_k, y_k)) h(x_k, y_k; x, y)}{\sum\limits_{x,y \in S_k} h^2(x_k, y_k; x, y)}, & r(x_k, y_k) < -T(x_k, y_k) \end{cases},$$

$$\tag{2.38}$$

which essentially says that if the absolute value of the residual is less than the threshold $T(x_k, y_k)$, then the current estimate is not changed; if it is greater than $T(x_k, y_k)$ or less than $-T(x_k, y_k)$, then the current estimate is updated such that the updated residual is equal to the bounds $T(x_k, y_k)$ or $-T(x_k, y_k)$, respectively. The projection onto the data fidelity constraint set is illustrated in Figure 2.8.

Another constraint set is the pixel range constraint set that ensures that the resulting image has pixel values within a certain range: $C_A = \{f : 0 \le f(x, y) \le 255\}$. The projection onto this constraint set is implemented simply by setting the pixel values that are outside the specified range to the closest boundary value. There are also frequency domain constraint sets when a certain frequency range of the true signal is known. Suppose, for instance, that we are sure about the frequency content of the measured signal for a frequency range, specified by the ideal low-pass filter $P(u, v)$; then this constraint set can be defined as $C_F = \{f : |P(u, v)F(u, v) - P(u, v)G(u, v)| \le 0\}$, and implemented by setting the Fourier transform of the current estimate to $G(u, v)$ in that range.

2.3.9   Learning-Based Image Restoration

Learning-based image restoration utilizes class-specific priors, which may turn out to be more powerful than generic image priors [49–51]. This type of prior has been used with both high-level classes, such as face images [52] and text [53], and low-level classes, such as 3×3 local image patches [54, 55]. There are three main training-based approaches:

• *Constraining Solution to a Subspace:* Suppose that the image to be restored belongs to a class, and there is a set of training images $\mathbf{f}_1, ..., \mathbf{f}_N$ from that class. Using the Principal Component Analysis (PCA) technique, a low-dimensional representation of the face space can be obtained. Specifically, an image $\mathbf{f}$ can be represented in terms of the average image $\mathbf{a}$ and K basis vectors $\boldsymbol{\Psi} = [\psi_1, ..., \psi_K]$:

$$\widetilde{\mathbf{f}} = \mathbf{a} + \boldsymbol{\Psi}\mathbf{e} \tag{2.39}$$

where $\mathbf{e}$ keeps the contributions of the basis vectors and is calculated by taking the inner product with the basis vectors:

$$\mathbf{e} = \boldsymbol{\Psi}^T (\mathbf{f} - \mathbf{a}) . \tag{2.40}$$

The size of the vector $\mathbf{e}$ is typically much smaller than the total number of pixels in $\mathbf{f}$. The difference between the image $\mathbf{f}$ and its subspace representation $\widetilde{\mathbf{f}}$ can be made as small as needed by increasing the number of basis vectors. With the subspace representation, the restoration problem can be formulated in terms of the representation vector $\mathbf{e}$. For example, the regularized least squares approach finds the optimal representation vector:

$$\hat{\mathbf{e}} = \arg \min_{\mathbf{e}} \| \mathbf{g} - \mathbf{H} (\mathbf{a} + \boldsymbol{\Psi}\mathbf{e}) \|^2 + \lambda\phi(\mathbf{e}). \tag{2.41}$$

With this approach, the solution $\hat{\mathbf{f}} = \mathbf{a} + \boldsymbol{\Psi}\hat{\mathbf{e}}$ is constrained to the subspace spanned by the average image $\mathbf{a}$ and the basis vectors $[\psi_1, ..., \psi_K]$; therefore, noise that is orthogonal to the subspace is automatically eliminated. This may turn out to be very helpful in some applications, for example, face recognition from low-resolution surveillance video [50].

• *Subspace-Based Regularization:* Unlike the previous method, this approach restores the image in image space without constraining it to a subspace. However, it forces the difference between the solution and its representation in the subspace to be small. That is, the training-based prior is still used, but the solution can be outside the subspace. For instance, the least squares problem can be stated as [56]

$$\hat{\mathbf{f}} = \arg\min_{\mathbf{f}} \| \mathbf{g} - \mathbf{H}\mathbf{f} \|^2 + \lambda \| \mathbf{f} - \widetilde{\mathbf{f}} \|^2, \qquad (2.42)$$

where $\widetilde{\mathbf{f}}$ is the subspace representation as in Equation (2.39). It is possible to include additional regularization terms on $\mathbf{f}$ and/or $\mathbf{e}$. For instance, in [52], a set of features is obtained by a multiscale image decomposition, and the solution is forced to have feature vectors close to ones learned during training.

• *Class-Specific Restoration Filter:* The third approach is to classify the observed local patches into a set of classes and learn the inverse filter or the corresponding high-resolution (restored) patch for each class from training data. An example of such a method is [55], where image patches are classified based on adaptive dynamic range coding (ADRC), and for each class, the relation between the observed pixels and the corresponding true-image pixels is learned. Suppose we have a 3×3 observation patch p_1 with observations $g_{p_1}(x_1, y_1), \cdots, g_{p_1}(x_9, y_9)$ and a corresponding true-image pixel $f_{p_1}(x, y)$ at the center of the patch. The observation patch is binarized with ADRC as follows:

$$ADRC\left(g_{p_1}(x, y)\right) = \begin{cases} 1, & if \ g_{p_1}(x, y) \geq \bar{g}_{p_1} \\ 0, & otherwise \end{cases}, \qquad (2.43)$$

where $\bar{g}_{p_1}$ is the average of the patch p_1. Applying ADRC to the entire patch, a binary codeword is obtained. This codeword determines the class of that patch. That is, a 3×3 patch is coded with a 3×3 matrix of ones and zeros. There are, therefore, a total of $2^9 = 512$ classes. During training, each observation patch is classified; and for each class, linear regression is applied to learn the relation between the observation patch and the corresponding true-image pixel. Suppose, for a particular class c, there are M observation patches and corresponding true-image pixels. The regression parameters, $[\xi_{c,1} \ \xi_{c,2} \ \cdots \ \xi_{c,9}]$, are then found by solving

$$\begin{bmatrix} f_{p_1}(x, y) \\ f_{p_2}(x, y) \\ \vdots \\ f_{p_M}(x, y) \end{bmatrix} = \begin{bmatrix} g_{p_1}(x_1, y_1) & g_{p_1}(x_2, y_2) & \cdots & g_{p_1}(x_9, y_9) \\ g_{p_2}(x_1, y_1) & g_{p_1}(x_2, y_2) & \cdots & g_{p_1}(x_9, y_9) \\ \vdots & & & \\ g_{p_M}(x_1, y_1) & g_{p_M}(x_2, y_2) & \cdots & g_{p_M}(x_9, y_9) \end{bmatrix} \begin{bmatrix} \xi_{c,1} \\ \xi_{c,2} \\ \vdots \\ \xi_{c,9} \end{bmatrix}$$

$$\tag{2.44}$$

During testing, for each pixel, the local patch around that pixel is taken and classified. According to its class, the regression parameters are taken from a look-up table and applied to obtain the true-image pixel. Another classification-based algorithm is presented in [54], where a feature vector from a patch is extracted through a nonlinear transformation, which is designed to form more edge classes. And, as in the case of [55], a weighted sum of the pixels in the patch is taken to get a high-resolution pixel; the weights are learned during training. [57] uses a Markov network to find the relation between low- and high-resolution patches. Given a test patch, the best matching low-resolution patch is determined and the corresponding high-resolution patch learned from training is obtained.

2.4   Blind Image Restoration

So far in this chapter, it is assumed that the PSF of the imaging system is known. This assumption is valid in applications where the PSF can be obtained accurately from the technical design or through experiments with known imagery in case the system PSF is time invariant. Unfortunately, in most real-life problems, both the PSF and the true image must be estimated. The estimation process could be separate or joint; in the former case, the PSF is first estimated and then used to estimate the original image. In the latter case, the PSF and the true image are estimated jointly. Constraints, such as nonnegativity, symmetry, and smoothness, about the PSF and the true image need to be utilized during the estimation process.

Many blind image restoration algorithms can be viewed from the Bayesian perspective. Unlike nonblind image restoration, the PSF is also estimated. In other words, we are looking to find both the PSF $\mathbf{h}$ and the true image $\mathbf{f}$ that maximize the posterior probability

$$p\left(\mathbf{f}, \mathbf{h}|\mathbf{g}\right) = \frac{p\left(\mathbf{g}|\mathbf{f}, \mathbf{h}\right) p\left(\mathbf{f}\right) p\left(\mathbf{h}\right)}{p\left(\mathbf{g}\right)}, \tag{2.45}$$

where $p\left(\mathbf{g}|\mathbf{f}, \mathbf{h}\right)$ is the likelihood of observations, $p\left(\mathbf{f}\right)$ is the image prior, and $p\left(\mathbf{h}\right)$ is the PSF prior. The observation probability $p\left(\mathbf{g}\right)$ is dropped during optimization as it is constant with respect to $\mathbf{f}$ and $\mathbf{h}$. It is common to parameterize the prior distributions and convert the problem to estimating these parameters. Instead of deterministic (but unknown) parameterization, it is possible to introduce parameters that are also random variables and have PDFs themselves. The prior distributions depend on these so-called hyperparameters, and the posterior probability becomes

$$p\left(\mathbf{f}, \mathbf{h}, \mathbf{\Omega}|\mathbf{g}\right) = \frac{p\left(\mathbf{g}|\mathbf{f}, \mathbf{h}, \mathbf{\Omega}\right) p\left(\mathbf{f}|\mathbf{\Omega}\right) p\left(\mathbf{h}|\mathbf{\Omega}\right) p\left(\mathbf{\Omega}\right)}{p\left(\mathbf{g}\right)}. \tag{2.46}$$

Many blind deconvolution methods assume an uninformative prior model for the hyperparameters, that is, $p\left(\mathbf{\Omega}\right) = constant$. Sometimes, degenerate distribution is assumed, where $p\left(\mathbf{\Omega}\right) = \delta\left(\mathbf{\Omega} - \mathbf{\Omega_0}\right)$ with deterministic but unknown parameters $\mathbf{\Omega_0}$.

From the posterior (2.46), we can marginalize an unknown and determine its optimal value by maximizing the conditional PDF. For example, the unknown image could be obtained as

$$\hat{\mathbf{f}} = \arg \max_{\mathbf{f}} \int_{\mathbf{h}, \mathbf{\Omega}} p\left(\mathbf{f}, \mathbf{h}, \mathbf{\Omega}|\mathbf{g}\right) d\mathbf{h} d\mathbf{\Omega}. \tag{2.47}$$

An alternative to this *empirical* analysis is to first calculate the optimal PSF and hyperparameters by marginalizing over $\mathbf{f}$:

$$\hat{\mathbf{h}}, \hat{\mathbf{\Omega}} = \arg \max_{\mathbf{h}, \mathbf{\Omega}} \int_{\mathbf{f}} p\left(\mathbf{f}, \mathbf{h}, \mathbf{\Omega}|\mathbf{g}\right) d\mathbf{f}, \tag{2.48}$$

and then obtain the true image estimate

$$\hat{\mathbf{f}} = \arg \max_{\mathbf{f}} p\left(\mathbf{f}|\hat{\mathbf{\Omega}}\right) p\left(\mathbf{g}|\mathbf{f}, \hat{\mathbf{h}}, \hat{\mathbf{\Omega}}\right), \tag{2.49}$$

which is known as *evidence* analysis [58]. These two approaches involve marginalization of so-called *hidden* variables; and the classical technique to solve such problems is the expectation-maximization (EM) algorithm [59], which has been used successfully in blind image restoration [60, 61]. While the EM algorithm puts the problem in a well-defined framework that is guaranteed to converge to a local optimum, there is still the problem of calculating the marginal probabilities, which do not have closed forms in general. There are methods to overcome this issue, including the Laplace approximation technique [62, 63] and Monte Carlo methods [64]. Another approach is the variational approximation method, which approximates the posterior $p(\mathbf{f}, \mathbf{h}, \Omega|\mathbf{g})$ with a simpler tractable distribution $q(\mathbf{f}, \mathbf{h}, \Omega)$, which is obtained through the minimization of Kullback-Leibner (KL) divergence between the approximate and real distributions:

$$KL(q(\mathbf{f}, \mathbf{h}, \Omega)\,||\,p(\mathbf{f}, \mathbf{h}, \Omega|\mathbf{g})) = \int q(\mathbf{f}, \mathbf{h}, \Omega)\, log\left(\frac{q(\mathbf{f}, \mathbf{h}, \Omega)}{p(\mathbf{f}, \mathbf{h}, \Omega|\mathbf{g})}\right) d\mathbf{f} \cdot d\mathbf{h} \cdot d\Omega,$$
(2.50)

which can be interpreted as the (weighted) average or expected value of the logarithmic difference $log(q) - log(p)$ between the distributions q and p. To end up with an analytical solution, the variational distribution is approximated in factorized form as $q(\mathbf{f}, \mathbf{h}, \Omega) = q(\mathbf{f})q(\mathbf{h})q(\Omega)$. This leads to algorithms where the distribution q is updated to minimize Equation (2.50) iteratively, where a subset of variables is updated at each iteration while keeping the others fixed, as in the case of alternating minimization [65, 66].

In general, any distribution can be estimated using sampling methods, such as the Monte Carlo methods. Although these methods may provide solutions better than the alternating minimization or the variational Bayesian approaches in theory, they have high computational complexity and are not preferred in practice.

2.4.1 Alternating Minimization

It is in general difficult to determine the optimal $\mathbf{f}$ and $\mathbf{h}$ simultaneously; a commonly used technique is alternating minimization, which alternatively updates one variable while keeping the other fixed:

$$\mathbf{f}^{(i+1)} = \arg\min_{\mathbf{f}^{(i)}} C(\mathbf{f}^{(i)}, \mathbf{h}^{(i)})$$
(2.51)

$$\mathbf{h}^{(i+1)} = \arg\min_{\mathbf{h}^{(i)}} C(\mathbf{f}^{(i+1)}, \mathbf{h}^{(i)}).$$
(2.52)

For example, the least squares estimation with the Tikhonov regularization minimizes the following cost function in Fourier domain:

$$C(F(u, v), H(u, v)) = |G(u, v) - H(u, v)F(u, v)|^2 + \lambda_1|L_1(u, v)F(u, v)|^2 + \lambda_2|L_2(u, v)H(u, v)|^2,$$
(2.53)

and the corresponding alternating minimization iterations are

$$F^{(i+1)}(u, v) = \frac{H^{(i)*}(u, v)G(u, v)}{|H^{(i)}(u, v)|^2 + \lambda_1|L_1(u, v)|^2}$$
(2.54)

$$H^{(i+1)}(u, v) = \frac{F^{(i+1)*}(u, v)G(u, v)}{|F^{(i+1)}(u, v)|^2 + \lambda_2|L_2(u, v)|^2}.$$
(2.55)

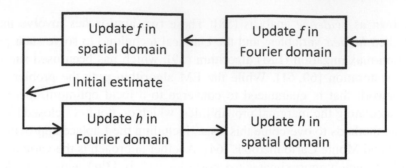

Figure 2.9 Iterative blind deconvolution with spatial and Fourier domain constraints.

In recent years, highly successful blind deconvolution algorithms based on the alternating minimization scheme have been developed [67–69]. These methods use a variety of regularization terms, including both ℓ_1- and ℓ_2- based ones, and they do not restrict the PSF to parametric models. The key factor in the success of these recent algorithms seems to be the initial true image estimation, which is obtained by applying a shock filter [70] or a bilateral filter [65], which result in sharp edges, an expected feature in true images.

2.4.2 Iterative Blind Deconvolution

A number of algorithms impose spatial and Fourier domain constraints on the PSF and true-image estimates, and update initial estimates iteratively as illustrated in Figure 2.9. In [72], nonnegativity on both the PSF and true image is imposed by setting negative values to zero in spatial domain. In Fourier domain, the current estimate $F^{(i)}$ is updated as

$$F^{(i+1)}(u, v) = (1 - \gamma)F^{(i)}(u, v) + \gamma G(u, v)/H^{(i)}(u, v), \qquad (2.56)$$

which is essentially a weighted average of $F^{(i)}(u, v)$ and $G(u, v)/H^{(i)}(u, v)$ with a weight parameter γ. The first part $F^{(i)}$ is obtained by taking the Fourier transform $\mathbf{f}^{(i)}$, and therefore imposes the constraint of spatial-domain nonnegativity. The second part $G(u, v)/H^{(i)}(u, v)$ is a result of the Fourier domain constraint $G(u, v) = H^{(i)}(u, v)F^{(i)}(u, v)$. $H^{(i)}(u, v)$ is updated in the same way as in Equation (2.56) by exchanging $H^{(i)}(u, v)$ and $F^{(i)}(u, v)$. The algorithm has heuristics: for example, there is no theoretical optimal value for γ; also the original algorithm in [72] proposes to update $1/F^{(i)}(u, v)$ instead of $F^{(i)}(u, v)$ when $H^{(i)}(u, v)$ is small to avoid $G(u, v)/H^{(i)}(u, v)$ dominating the weighted sum in Equation (2.56). In [73], the Fourier domain update (2.56) is replaced by a Wiener-like update:

$$F^{(i+1)}(u, v) = \frac{H^{(i)*}(u, v)G(u, v)}{|H^{(i)}(u, v)|^2 + \eta/|F^{(i)}(u, v)|^2}, \qquad (2.57)$$

where η is a representative of the noise energy as in the Wiener filter solution (2.30).

2.5   Other Methods of Image Restoration

We would like to mention a few other methods that are used in image restoration.

• *PDE Approach:* The PDE approach utilizes the well-established theories on partial differential equations (PDEs) to develop new models and solutions to the image restoration problem. According to this approach, an image $f(x, y, t)$ is a function of space and time, and starting with the original image at $t = 0$ it evolves in time according to a spatial functional. For example,

$$\frac{\partial f(x, y, t)}{\partial t} + \mathcal{F}\left(x, y, f(x, y, t), \nabla f(x, y, t), \nabla^2 f(x, y, t)\right) \tag{2.58}$$

is the partial differential equation governing the evolution of the image in time, where the functional $\mathcal{F}$ takes spatial position, the image, and its first- and second-order spatial derivatives as the input. With the choice of the functional, it is possible to impose smoothness or sharpness [59,70,74] on the restored image. A detailed treatment of the topic can be found in [76].

• *Total Least Squares:* The Total Least Squares (TLS) approach allows errors in both the measured (or estimated) PSF and the observation, and models the imaging process as $\mathbf{g} + \delta\mathbf{g} = (\mathbf{H} + \delta\mathbf{H})\mathbf{f}$, where $\delta\mathbf{H}$ and $\delta\mathbf{g}$ are the error components in the PSF and observation [77, 78]. The TLS problem is then formulated as finding $\mathbf{f}$, $\delta\mathbf{g}$ and $\delta\mathbf{H}$ minimizing the Frobenius norm $\|\delta\mathbf{H}, \delta\mathbf{g}\|_F$ subject to the imaging constraint $\mathbf{g} + \delta\mathbf{g} = (\mathbf{H} + \delta\mathbf{H})\mathbf{f}$. The TLS approach can be considered an extension of the least squares method, which only allows perturbations in the observation $\mathbf{g}$. While the classical TLS problem [77] has a closed-form solution based on the SVD, the weighted and structured TLS problems do not necessarily lead to closed-form solutions and are solved using numerical techniques. In the case of blind deconvolution, this is also the situation because the PSF matrix $\mathbf{H}$ and its perturbation $\delta\mathbf{H}$ have a specific (BTTB) matrix structure [79].

• *NAS-RIF:* The nonnegativity and support constraints recursive inverse filtering (NAS-RIF) technique [80] involves estimation of a finite-impulse response filter $u(x, y)$, which produces an estimate of the true image when convolved with the observed data $g(x, y)$. Constraints, such as nonnegativity and finite image support, are also incorporated into the process. The filter $u(x, y)$ is updated to minimize the squared error between the estimate $\hat{f}(x, y) = u(x, y) * g(x, y)$ and the $\hat{f}_{NL}(x, y)$, which is the result of the nonlinear constraints. More specifically, the cost function optimized for $u(x, y)$ is $\sum\left(u(x, y) * g(x, y) - \hat{f}_{NL}(x, y)\right)^2 + \sum\left(u(x, y) - \delta(x, y)\right)^2$, where the second term acts as a regularization term to penalize deviation from a delta function. A gradient descent technique is then used to achieve optimization of $u(x, y)$ at each iteration.

• *Bussgang Deconvolution:* The Bussgang deconvolution approach is an iterative filter consisting of the following steps: (1) linear (Wiener filter) estimation of the true image, (2) nonlinear minimum MSE (MMSE) update of the estimate, and (3) Wiener filter update of the inverse filter coefficients [81–83]. The approach is named after the Bussgang process, stating that the cross-correlation between the input and the output of a zero-memory nonlinearity is proportional to the autocorrelation of the input, which is the underlying idea behind the approach.

• *Spectral Zero-Crossings:* When the frequency response of the PSF has a known parametric form, which is completely characterized by its frequency domain zero-crossings the PSF can be determined from the zero crossings of the observed image $G(u, v) =$

$H(u, v)F(u, v)$. Because $G(u, v)$ has the zeros of both $H(u, v)$ and $F(u, v)$, it is critical to determine the ones that belong to $H(u, v)$. With the parametric model of the PSF, the zeros of $H(u, v)$ would display a certain pattern, which could be utilized to determine the parameter values [84, 85].

• *Homomorphic Deconvolution:* The starting point of homomorphic deconvolution is to convert the product $G(u, v) = H(u, v)F(u, v)$ into a sum by applying a logarithmic function:

$$log[G(u, v)] = log[H(u, v)] + log[F(u, v)]. \tag{2.59}$$

In certain applications, such as ultrasound imaging, the blur term $log[H(u, v)]$ is much smoother than the true image term $log[F(u, v)]$; therefore, $log[G(u, v)]$ can be considered a noisy version of $log[H(u, v)]$, and as a result, the blur can be recovered from $log[G(u, v)]$ through a denoising operation [86]. As a form of denoising, the observed image is divided into blocks $g_i(x, y)$, and homomorphic transformation is applied to each block [85]: $log[G_i(u, v)] = log[H(u, v)] + log[F_i(u, v)]$. Because the PSF is common to all blocks, when the average of $log[G_i(u, v)]$ is taken, the average is dominated by the blur term $log[H(u, v)]$.

• *ARMA Parametric Approach:* According to the autoregressive moving average (ARMA) model for blind deconvolution, the true image is a two-dimensional autoregressive (AR) process, and the PSF is a two-dimensional moving average (MA) process. The blind deconvolution problem then becomes equivalent to an ARMA parameter estimation problem, which has been studied extensively [87]. There are a few algorithms based on this approach, including [60, 88]. One critical issue with this approach is that global AR parameters may not accurately represent the entire image. This problem is overcome by [88], which divides the image into small blocks and uses a separate AR model for each block.

2.6   Super Resolution Image Restoration

Super resolution image restoration is the process of producing a high-resolution image (or a sequence of high-resolution images) from a set of low-resolution images. The process requires an image acquisition model that relates a high-resolution image to multiple low-resolution images and involves solving the resulting inverse problem. In addition to the degradation processes in single image restoration, super resolution image restoration incorporates motion and downsampling operations into the imaging process:

$$\mathbf{g}_k = \mathbf{H}_k \mathbf{f} + \mathbf{n}_k, \tag{2.60}$$

where $\mathbf{f}$ is the vectorized version of the high-resolution image, $\mathbf{g}_k$ is the kth observation, $\mathbf{n}_k$ is the kth noise, and $\mathbf{H}_k$ is the matrix that includes the following operations: geometric warping, convolution with the point spread function, and downsampling. These operations can be written in matrix notation as $\mathbf{H}_k = \mathbf{DBM}_k$, where $\mathbf{D}$ corresponds to the downsampling operation, $\mathbf{B}$ corresponds to convolution with the PSF, and $\mathbf{M}_k$ corresponds to geometric warping from $\mathbf{f}$ to the kth observation.

Sometimes, all N observations are stacked to form a simplified representation of the problem:

$$
\begin{bmatrix} \mathbf{g}_1 \\ \mathbf{g}_2 \\ \vdots \\ \mathbf{g}_N \end{bmatrix} = \begin{bmatrix} \mathbf{H}_1 \\ \mathbf{H}_2 \\ \vdots \\ \mathbf{H}_N \end{bmatrix} \mathbf{f} + \begin{bmatrix} \mathbf{n}_1 \\ \mathbf{n}_2 \\ \vdots \\ \mathbf{n}_N \end{bmatrix} \implies \bar{\mathbf{g}} = \bar{\mathbf{H}}\mathbf{f} + \bar{\mathbf{n}} \tag{2.61}
$$

It is obvious that many single image restoration techniques could be applied to a multi-frame super resolution image restoration problem. As we have already seen, some of these techniques involve the transpose of the blur matrix. In the case of super resolution, the transpose of $\mathbf{H}_k^T$ will be needed: $\mathbf{H}_k^T = \mathbf{M}_k^T \mathbf{B}^T \mathbf{D}^T$. Similar to the single-image case, these transpose operations can be implemented in spatial domain:

$\mathbf{D}^T$ is implemented as upsampling by zero-insertion,

$\mathbf{B}^T$ is implemented as convolution with transposed PSF, and

$\mathbf{M}_k^T$ is implemented as warping from the kth observation to $\mathbf{f}$.

Super resolution image restoration has been an active research field for more than two decades. There are several articles with comprehensive reviews of super resolution imaging [89, 90]. The *IEEE Signal Processing Magazine* has a special issue dedicated to super resolution imaging [91]. There are also a few books on the topic, including [92–94].

2.7   Regularization Parameter Estimation

The regularization parameter controls the trade-off between data fidelity and prior information fidelity. Choosing an appropriate regularization parameter is a critical part of the overall restoration process. When the regularization parameter is not optimal, the solution might be over-smooth or degraded with excessive noise amplification. The most commonly used methods for choosing a regularization parameter are as follows.

• *Visual Inspection:* When the viewer has considerable prior knowledge of the scene, it is reasonable to choose the regularization parameter through visual inspection of results with different parameter values. Obviously, this approach is not appropriate for all applications.

• *L-Curve Method:* Because the regularization parameter controls the trade-off between data fidelity and prior information fidelity, it makes sense to determine the parameter by examining the behavior of these fidelity terms. Assume that $\mathbf{f}(\lambda)$ is the solution with a particular regularization parameter λ. The restoration is repeated for different values of λ, and the data fidelity term $\|\mathbf{g} - \mathbf{H}\mathbf{f}(\lambda)\|^2$ is plotted against the prior term $\phi\left(\mathbf{f}(\lambda)\right)$. The plot forms an "L"-shaped curve. For some values of λ, the data fidelity term changes rapidly while the prior term does not change much; this is the over-regularized region. For some other values of λ, the data fidelity term changes very little while the prior term changes significantly; this is the under-regularized region. Intuitively, the optimal λ value is the one that corresponds to the corner of the L-curve, the transition point between the over-regularized and under-regularized regions [95, 96]. The corner point may be defined in a number of ways, including the point of maximum curvature, the point with the tangent line slope of -1, and the point closest to the origin.

• *Discrepancy Principle:* If the noise power is known, then it provides δ a bound on the residual norm $\|\mathbf{g} - \mathbf{Hf}(\lambda)\|$. As λ moves from over-regularization to under-regularization, there will be a noise amplification; therefore, one can choose the regularization parameter so that the residual norm is large but not larger than the bound [97]. In particular, the optimal λ could be chosen such that $\|\mathbf{g} - \mathbf{Hf}(\lambda)\| = \delta$.

• *Generalized Cross-Validation (GCV) Method:* GCV is an estimator that minimizes the predictive risk. The underlying idea is that the solution that is obtained using all but one observation should predict that left-out observation well if the regularization parameter is a good choice. The total error for a particular choice of the parameter is calculated by summing up the prediction errors over all observations. The optimal parameter value is the one that minimizes the total error. A search technique or an optimization method could be used to determine the optimal value [98–100].

• *Statistical Approach:* As we have seen discussed, the statistical methods may incorporate additional parameters into the estimation process. For instance, maximum a posteriori estimator maximizing $p(\mathbf{f}|\mathbf{g})$ can be modified to maximize $p(\mathbf{f}, \lambda|\mathbf{g})$. The resulting optimization problems can be solved in a number of ways, including the expectation-maximization (EM) technique [59]. The EM technique essentially involves iterative application of two steps: In the expectation step, the image is restored given an estimate of the parameter. And, in the maximization step, a new estimate of the parameter is calculated given the restored image [60].

2.8 Beyond Linear Shift-Invariant Imaging Model

So far in this chapter, it has been assumed that the degradation is linear and shift invariant. While the inverse problems with LSI systems have been studied extensively, there are indeed a nonnegligible number of scenarios where nonlinear and/or space-varying degradation happens. For instance, it is well known that the sensor response and gamma mapping result in a nonlinear camera response function, meaning that the amount of light does not linearly translate to pixel intensities. This has been considered in high-dynamic range (HDR) imaging applications, where the nonlinear camera response function must be estimated to merge differently exposed images. In [101], super resolution image restoration from differently exposed images is demonstrated and this idea requires explicit modeling and estimation of sensor nonlinearity as in the case of HDR imaging. Modeling of sensor nonlinearity may turn out to be critical in other image restoration problems as well, especially when the PSF is unknown and must be estimated. While the HDR imaging literature has focused on estimation of sensor nonlinearity from multiple images, there are also methods to estimate it from a single image [102].

Space-varying degradation is likely encountered within a variety of applications. For example, objects in a scene may move separately to create space-varying motion blur. This is a very challenging problem as it inherently involves segmentation of individual objects. One way to approach this problem is to divide the image into blocks, estimate the PSF of each block, and deconvolve each block separately [103]. The approach is apparently problematic as there could be multiple objects within a block. The next chapter in this book is entirely dedicated to space-varying restoration and will discuss the topic in detail.

The image restoration techniques that have been developed for single-channel images could also be applied to the restoration of multichannel images, including hyper-spectral and color images, by treating each channel separately. However, this would neglect the correlation among the channels and might not produce the best possible results. This has been discussed in [104], where the single-channel Wiener filter is extended to multiple channels. For color images, the human visual system should also be taken into account [105].

2.9   Summary

In this chapter, we aimed to provide an overview of image restoration. We presented mathematical modeling of linear shift-invariant image formation process, mentioned possible sources of degradation, reviewed basic deconvolution methods, and explained how to extend these ideas to super resolution image restoration. We emphasized certain ideas more than others because we personally found them essential, interesting, or to have the potential for further development. The coverage of the chapter is brief and we hope that it served as a starting point in image restoration, which is going to maintain its importance as new devices, applications, and technologies continue to emerge and bring new challenges.

Bibliography

[1] J. W. Woods, J. Biemond, and A. M. Tekalp, "Boundary value problem in image restoration," in *Proceedings of the IEEE International Conference on Acoustics, Speech, and Signal Processing*, vol. 10, pp. 692–695, April 1985.

[2] M. K. Ng, R. H. Chan, and W. C. Tang, "A fast algorithm for deblurring models with Neumann boundary conditions," *SIAM Journal of Scientific Computing*, vol. 21, no. 3, pp. 851–866, December 1999.

[3] J. Koo and N. K. Bose, "Spatial restoration with reduced boundary error," in *Proceedings of the Mathematical Theory of Networks and Systems*, August 2002.

[4] M. K. Ng and N. K. Bose, "Mathematical analysis of super resolution methodology," *IEEE Signal Processing Magazine*, vol. 20, no. 3, pp. 62–74, May 2003.

[5] G. D. Boreman, *Modulation Transfer Function in Optical and Electro-Optical Systems*. Bellingham, WA: SPIE Press, 2001.

[6] D. Sadot and N. S. Kopeika, "Forecasting optical turbulence strength on the basis of macroscale meteorology and aerosols: Models and validation," *Optical Engineering*, vol. 31, no. 2, pp. 200–212, February 1992.

[7] ——, "Imaging through the atmosphere: Practical instrumentation-based theory and verification of aerosol modulation transfer function," *Journal of the Optical Society of America - A*, vol. 10, no. 1, pp. 172–179, January 1993.

[8] Y. Yitzhaky, I. Dror, and N. S. Kopeika, "Restoration of atmospherically blurred images according to weather-predicted atmospheric modulation transfer functions," *Optical Engineering*, vol. 36, no. 11, pp. 3064–3072, November 1997.

[9] H. W. Engl, M. Hanke, and A. Neubauer, *Regularization of Inverse Problems*. Dordrecht: Kluwer Academic Publishers, 2000.

[10] C. W. Groetsch, *The Theory of Tikhonov Regularization for Fredholm Equations of the First Kind*. London: Pitman, 1984.

[11] G. H. Golub and C. F. Van Loan, *Matrix Computations*. Baltimore, MD: Johns Hopkins University Press, 1996.

[12] L. Landweber, "An iteration formula for Fredholm integral equations of the first kind," *American Journal of Mathematics*, vol. 73, no. 3, pp. 615–624, July 1951.

[13] O. N. Strand, "Theory and methods related to the singular function expansion and Landweber's iteration for integral equations of the first kind," *SIAM Journal on Numerical Analysis*, vol. 11, no. 4, pp. 798–825, September 1974.

[14] B. K. Gunturk, Y. Altunbasak, and R. M. Mersereau, "Super resolution reconstruction of compressed video using transform-domain statistics," *IEEE Transactions on Image Processing*, vol. 13, no. 1, pp. 33–43, January 2004.

[15] M. Irani and S. Peleg, "Improving resolution by image registration," *CVGIP: Graphical Models and Image Processing*, vol. 53, no. 3, pp. 231–239, May 1991.

[16] P. H. Van Cittert, "Zum einfluss der spaltbreite auf die intensitatsverteilung in spektrallinien ii," *Zeitschrift fur Physik*, vol. 69, pp. 298–308, 1931.

[17] D. G. Luenberger and Y. Ye, *Linear and Nonlinear Programming*. Berlin: Springer, 2008.

[18] R. Fletcher, *Practical Methods of Optimization*. New York: Wiley, 2000.

[19] M. K. Ng, H. Shen, E. Y. Lam, and L. Zhang, "A total variation regularization based super resolution reconstruction algorithm for digital video," *EURASIP Journal on Advances in Signal Processing*, Article ID 74585, pp. 1–16, 2007.

[20] S. D. Babacan, R. Molina, and A. K. Katsaggelos, "Variational Bayesian super resolution," *IEEE Transactions on Image Processing*, vol. 20, no. 4, pp. 984–999, April 2011.

[21] L. I. Rudin, S. Osher, and E. Fatemi, "Nonlinear total variation based noise removal algorithms," *Physica D: Nonlinear Phenomena*, vol. 60, pp. 259–268, November 1992.

[22] M. Elad, *Sparse and Redundant Representations: From Theory to Applications in Signal and Image Processing*. Berlin: Springer, 2010.

[23] J. Myrheim and H. Rue, "New algorithms for maximum entropy image restoration," *CVGIP: Graphical Models and Image Processing*, vol. 54, no. 3, pp. 223–238, May 1992.

[24] M. J. Black, G. Sapiro, D. H. Marimont, and D. Heeger, "Robust anisotropic diffusion," *IEEE Transactions on Image Processing*, vol. 7, no. 3, pp. 421–432, March 1998.

[25] S. Mallat and Z. Zhang, "Matching pursuits with time-frequency dictionaries," *IEEE Transactions on Signal Processing*, vol. 41, no. 12, pp. 3397–3415, December 1993.

[26] Y. C. Pati, R. Rezaiifar, and P. S. Krishnaprasad, "Orthogonal matching pursuit: Recursive function approximation with applications to wavelet decomposition," *Proceedings of the Asilomar Conference on Signals, Systems, and Computers*, vol. 1, pp. 40–44, November 1993.

[27] S. Chen, S. A. Billings, and W. Luo, "Orthogonal least squares methods and their application to non-linear system identification," *International Journal of Control*, vol. 50, no. 5, pp. 1873–1896, 1989.

[28] I. F. Gorodnitsky and B. D. Rao, "Sparse signal reconstruction from limited data using FOCUSS: A re-weighted minimum norm algorithm," *IEEE Transactions on Image Processing*, vol. 45, no. 3, pp. 600–616, March 1997.

[29] E. J. Candes, M. B.Wakin, and S. P. Boyd, "Enhancing sparsity by reweighted l1 minimization," *Journal of Fourier Analysis and Applications*, vol. 14, no. 5–6, pp. 877–905, 2007.

[30] J. M. Bioucas-Dias, M. A. T. Figueiredo, and J. P. Oliveira, "Total variation-based image deconvolution: A majorization-minimization approach," *Proceedings of the IEEE International Conference on Acoustics, Speech and Signal Processing*, vol. 2, pp. 861–864, May 2006.

[31] M. A. T. Figueiredo, R. D. Nowak, and S. J. Wright, "Gradient projection for sparse reconstruction: Application to compressed sensing and other inverse problems," *IEEE Journal of Selected Topics in Signal Processing*, vol. 1, no. 4, pp. 586–597, December 2007.

[32] S.-J. Kim, K. Koh, M. Lustig, S. Boyd, and D. Gorinevsky, "An interior-point method for large-scale l1-regularized least squares," *IEEE Journal of Selected Topics in Signal Processing*, vol. 1, no. 4, pp. 606–617, December 2007.

[33] J. A. Tropp and S. J. Wright, "Computational methods for sparse solution of linear inverse problems," *Proceedings of the IEEE*, vol. 98, no. 6, pp. 948–958, June 2010.

[34] I. Daubechies, M. Defriese, and C. De Mol, "An iterative thresholding algorithm for linear inverse problems with a sparsity constraint," *Communications on Pure and Applied Mathematics*, vol. 57, no. 11, pp. 1413–1457, November 2004.

[35] R. D. Nowak and M. A. T. Figueiredo, "Fast wavelet-based image deconvolution using the EM algorithm," in *Proceedings of the Asilomar Conference on Signals, Systems, and Computers*, vol. 1, pp. 371–375, November 2001.

[36] P. L. Combettes and V. R. Wajs, "Signal recovery by proximal forward-backward splitting," *Multiscale Modeling and Simulation*, vol. 4, no. 4, pp. 1168–1200, November 2005.

[37] J. M. Bioucas-Dias and M. A. T. Figueiredo, "A new twIST: Two-step iterative shrinkage/thresholding algorithms for image restoration," *IEEE Transactions on Image Processing*, vol. 16, no. 12, pp. 2992–3004, December 2007.

[38] M. A. T. Figueiredo and R. D. Nowak, "An EM algorithm for wavelet-based image restoration," *IEEE Transactions on Image Processing*, vol. 12, no. 8, pp. 906–916, August 2003.

[39] ——, "A bound optimization approach to wavelet-based image deconvolution," in *Proceedings of the IEEE International Conference on Image Processing*, vol. 2, pp. 782–785, September 2005.

[40] A. Papoulis, *Probability Random Variables and Stochastic Processes*. New York: McGraw-Hill, 1991.

[41] D. L. Donoho and J. M. Johnstone, "Ideal spatial adaptation by wavelet shrinkage," *Biometrika*, vol. 81, no. 3, pp. 425–455, September 1994.

[42] R. S. Prendergast and T. Q. Nguyen, "A novel parametric power spectral density model for images," in *Proceedings of the Asilomar Conference on Signals, Systems, and Computers*, November 2005, pp. 1671–1675.

[43] ——, "A non-isotropic parametric model for image spectra," in *Proceedings of the IEEE International Conference on Acoustics, Speech, and Signal Processing*, vol. 2, pp. 761–764, May 2006.

[44] W. H. Richardson, "Bayesian-based iterative method of image restoration," *Journal of the Optical Society of America*, vol. 62, no. 1, pp. 55–59, January 1972.

[45] L. B. Lucy, "An iteration technique for the rectification of the obscured distributions," *The Astronomical Journal*, vol. 79, no. 6, pp. 745–754, June 1974.

[46] H. Stark and P. Oskoui, "High-resolution image recovery from image-plane arrays, using convex projections," *Journal of the Optical Society of America A*, vol. 6, no. 11, pp. 1715–1726, November 1989.

[47] A. J. Patti, M. I. Sezan, and A. M. Tekalp, "Superresolution video reconstruction with arbitrary sampling lattices and nonzero aperture time," *IEEE Transactions on Image Processing*, vol. 6, no. 8, pp. 1064–1076, August 1997.

[48] A. M. Tekalp, M. K. Ozkan, and M. I. Sezan, "High-resolution image reconstruction from lower-resolution image sequences and space-varying image restoration," in *Proceedings of the IEEE International Conference on Acoustics, Speech, and Signal Processing*, vol. 3, March 1992, pp. 169–172.

[49] S. Baker and T. Kanade, "Limits on super resolution and how to break them," *IEEE Transactions on Pattern Analysis and Machine Intelligence*, vol. 24, no. 9, pp. 1167–1183, September 2002.

[50] B. K. Gunturk, A. U. Batur, Y. Altunbasak, M. H. Hayes, and R. M. Mersereau, "Eigenface-domain super resolution for face recognition," *IEEE Transactions on Image Processing*, vol. 12, no. 5, pp. 597–606, May 2003.

[51] D. Capel and A. Zisserman, "Computer vision applied to super resolution," *IEEE Signal Processing Magazine*, vol. 20, no. 3, pp. 75–86, May 2003.

[52] S. Baker and T. Kanade, "Hallucinating faces," in *Proceedings of the IEEE International Conference on Automatic Face and Gesture Recognition*, March 2000, pp. 83–88.

[53] D. Capel and A. Zisserman, "Super resolution enhancement of text image sequences," in *Proceedings of the IEEE International Conference on Pattern Recognition*, vol. 1, pp. 600–605, September 2000.

[54] C. B. Atkins, C. A. Bouman, and J. P. Allebach, "Optimal image scaling using pixel classification," *Proceedings of the IEEE International Conference on Image Processing*, vol. 3, pp. 864–867, October 2001.

[55] T. Kondo, Y. Node, T. Fujiwara, and Y. Okumura, "Picture conversion apparatus, picture conversion method, learning apparatus and learning method," US Patent No: 6,323,905, November 2001.

[56] D. Capel and A. Zisserman, "Super resolution from multiple views using learnt image models," in *Proceedings of the IEEE International Conference on Computer Vision and Pattern Recognition*, vol. 2, pp. 627–634, December 2001.

[57] W. T. Freeman, T. R. Jones, and E. C. Pasztor, "Example-based super resolution," *IEEE Computer Graphics and Applications*, vol. 22, no. 2, pp. 56–65, March/April 2002.

[58] R. Molina, "On the hierarchical bayesian approach to image restoration. Applications to astronomical images," *IEEE Transactions on Pattern Analysis and Machine Intelligence*, vol. 16, no. 11, pp. 1122–1128, November 1994.

[59] A. P. Dempster, N. M. Laird, and D. B. Rubin, "Maximum likelihood from incomplete data via the EM algorithm," *Journal of the Royal Statistical Society: Series B*, vol. 39, no. 1, pp. 1–38, 1977.

[60] R. L. Lagendijk, J. Biemond, and D. E. Boekee, "Identification and restoration of noisy blurred images using the expectation-maximization algorithm," *IEEE Transactions on Acoustics, Speech, and Signal Processing*, vol. 38, no. 7, pp. 1180–1191, July 1990.

[61] A. K. Katsaggelos and K. T. Lay, "Maximum likelihood blur identification and image restoration using the EM algorithm," *IEEE Transactions on Image Processing*, vol. 39, no. 3, pp. 729–733, March 1991.

[62] N. P. Galatsanos, V. Z. Mesarovic, R. Molina, and A. K. Katsaggelos, "Hierarchical Bayesian image restoration for partially known blur," *IEEE Transactions on Image Processing*, vol. 9, no. 10, pp. 1784–1797, October 2000.

[63] N. P. Galatsanos, V. Z. Mesarovic, R. Molina, A. K. Katsaggelos, and J. Mateos, "Hyperparameter estimation in image restoration problems with partially known blurs," *Optical Engineering*, vol. 41, no. 8, pp. 1845–1854, 2002.

[64] J. J. K. O'Ruanaidh and W. J. Fitzgerald, *Numerical Bayesian Methods Applied to Signal Processing*. Berlin: Springer, 1996.

[65] J. W. Miskin and D. J. C. MacKay, "Ensemble learning for blind image separation and deconvolution," *Advances in Independent Component Analysis* (Ed. M. Girolami), pp. 123–141, July 2000.

[66] R. Fergus, B. Singh, A. Hertzmann, S. T. Roweis, and W. T. Freeman, "Removing camera shake from a single photograph," *ACM Transactions on Graphics*, vol. 25, no. 3, pp. 787–794, July 2006.

[67] Q. Shan, J. Jia, and A. Agarwala, "High-quality motion deblurring from a single image," *ACM Transactions on Graphics*, vol. 27, no. 3, pp. 73:1–73:10, August 2008.

[68] S. Cho and S. Lee, "Fast motion deblurring," *ACM Transactions on Graphics*, vol. 28, no. 5, pp. 145:1–145:8, December 2009.

[69] L. Xu and J. Jia, "Two-phase kernel estimation for robust motion deblurring," in *Proceedings of the European Conference on Computer Vision*, vol. 1, 2010, pp. 157–170.

[70] S. Osher and L. I. Rudin, "Feature-oriented image enhancement using shock filters." *SIAM Journal on Numerical Analysis*, vol. 27, no. 4, pp. 919–940, August 1990.

[71] C. Tomasi and R. Manduchi, "Bilateral filtering for gray and color images," in *Proceedings of the IEEE International Conference on Computer Vision*, pp. 839–846, January 1998.

[72] G. R. Ayers and J. C. Dainty, "Iterative blind deconvolution method and its applications," *Optics Letters*, vol. 13, no. 7, pp. 547–549, July 1988.

[73] B. L. K. Davey, R. G. Lane, and R. H. T. Bates, "Blind deconvolution of noisy complex-valued images," *Optics Communications*, vol. 69, no. 5–6, pp. 353–356, January 1989.

[74] L. Alvarez and L. Mazorra, "Signal and image restoration using shock filters and anisotropic diffusion," *SIAM Journal on Numerical Analysis*, vol. 31, no. 2, pp. 590–605, April 1994.

[75] P. Perona and J. Malik, "Scale-space and edge detection using anisotropic diffusion," *IEEE Transactions on Pattern Analysis and Machine Intelligence*, vol. 12, no. 7, pp. 629–639, July 1990.

[76] G. Aubert and P. Kornprobst, *Mathematical Problems in Image Processing: Partial Differential Equations and the Calculus of Variations*. Berlin: Springer, 2006.

[77] G. H. Golub and C. F. Van Loan, "An analysis of the total least squares problem," *SIAM Journal on Numerical Analysis*, vol. 17, no. 6, pp. 883–893, 1980.

[78] I. Markovsky and S. Van Huffel, "Overview of total least-squares methods," *Signal Processing*, vol. 87, no. 10, pp. 2283–2302, October 2007.

[79] N. Mastronardi, P. Lemmerling, A. Kalsi, D. P. OLeary, and S. Van Huffel, "Implementation of the regularized structured total least squares algorithms for blind image deblurring," *Linear Algebra and its Applications*, vol. 391, pp. 203–221, November 2004.

[80] D. Kundur and D. Hatzinakos, "Blind image deconvolution," *IEEE Signal Processing Magazine*, vol. 13, no. 3, pp. 43–64, May 1996.

[81] R. Godfrey and F. Rocca, "Zero memory nonlinear deconvolution," *Geophysical Prospecting*, vol. 29, no. 2, pp. 189–228, 1981.

[82] S. Bellini, *Bussgang Techniques for Blind Deconvolution and Restoration*. Englewood Cliffs, NJ: Prentice-Hall, 1994.

[83] G. Panci, P. Campisi, S. Colonnese, and G. Scarano, "Multichannel blind deconvolution using the Bussgang algorithm: Spatial and multiresolution approaches," *IEEE Transactions on Image Processing*, vol. 12, no. 11, pp. 1324–1337, November 2003.

[84] T. G. Stockham, T. M. Cannon, and R. B. Ingebretsen, "Blind deconvolution through digital signal processing," *Proceedings of the IEEE*, vol. 63, no. 4, pp. 678–692, 1975.

[85] T. M. Cannon, "Blind deconvolution of spatially invariant image blurs with phase," *IEEE Transactions on Acoustics, Speech, and Signal Processing*, vol. 24, no. 1, pp. 58–63, 1976.

[86] O. V. Michailovich, and D. Adam, "A novel approach to the 2-D blind deconvolution problem in medical ultrasound," *IEEE Transactions on Medical Imaging*, vol. 24, no. 1, pp. 86–104, January 2005.

[87] D. S. G. Pollock, *A Handbook of Time-Series Analysis, Signal Processing and Dynamics*. New York: Academic Press, 1999.

[88] T. E. Bishop and J. R. Hopgood, "Blind image restoration using a block-stationary signal model," in *Proceedings of the IEEE International Conference on Acoustics, Speech, and Signal Processing*, vol. 2, pp. 853–856, May 2006.

[89] S. Borman and R. L. Stevenson, "Super resolution from image sequences–a review," in *Proceedings of the Midwest Symposium on Circuits and Systems*, vol. 5, pp. 374–378, August 1998.

[90] S. Farsiu, D. Robinson, M. Elad, and P. Milanfar, "Advances and challenges in super resolution," in *International Journal of Imaging Systems and Technology*, vol. 14, no. 2, pp. 47–57, 2004.

[91] M. G. Kang and S. Chaudhuri (Eds.), "Super resolution image reconstruction," *IEEE Signal Processing Magazine*, vol. 20, no. 3, pp. 19–86, May 2003.

[92] S. Chaudhuri (ed.), *Super Resolution Imaging*. Dordrecht: Kluwer Academic Publishers, 2001.

[93] D. Capel, *Image Mosaicing and Super Resolution*. Berlin: Springer, 2004.

[94] A. K. Katsaggelos, R. Molina, J. Mateos, and A. C. Bovik, *Super Resolution of Images and Video*. San Rafael, CA: Morgan and Claypool Publishers, 2006.

[95] P. C. Hansen, "Analysis of discrete ill-posed problems by means of the L-curve," *SIAM Review*, vol. 34, no. 4, pp. 561–580, December 1992.

[96] N. K. Bose, S. Lertrattanapanich, and J. Koo, "Advances in superresolution using l-curve," in *Proceedings of the IEEE International Symposium on Circuits and Systems*, vol. 2, pp. 433–436, May 2001.

[97] V. A. Morozov, "On the solution of functional equations by the method of regularization," *Soviet Mathematics Doklady*, vol. 7, pp. 414–417, 1966.

[98] G. Golub, M. Heath, and G. Wahba, "Generalized cross-validation as a method for choosing a good ridge parameter," *Technometrics*, vol. 21, no. 2, pp. 215–223, 1979.

[99] N. Nguyen, G. Golub, and P. Milanfar, "Blind restoration/superresolution with generalized cross-validation using Gauss-type quadrature rules," in *Proceedings of the Asilomar Conference on Signals, Systems, and Computers*, vol. 2, pp. 1257–1261, October 1999.

[100] N. Nguyen, P. Milanfar, and G. Golub, "Efficient generalized cross-validation with applications to parametric image restoration and resolution enhancement," *IEEE Transactions on Image Processing*, vol. 10, no. 9, pp. 1299–1308, September 2001.

[101] M. Gevrekci and B. K. Gunturk, "Superresolution under photometric diversity of images," *EURASIP Journal on Advances in Signal Processing*, no. 36076, pp. 1–12, 2007.

[102] H. Farid, "Blind inverse gamma correction," *IEEE Transactions on Image Processing*, vol. 10, no. 10, pp. 1428–1433, October 2001.

[103] Y. P. Guo, H. P. Lee, and C. L. Teo, "Blind restoration of images degraded by space-variant blurs using iterative algorithms for both blur identification and image restoration," *Image and Vision Computing*, vol. 15, no. 5, pp. 399–410, May 1997.

[104] B. R. Hunt, and O. Kubler, "Karhunen-Loeve multispectral image restoration. Part 1. Theory," *IEEE Transactions on Acoustics, Speech, and Signal Processing*, vol. 32, no. 3, pp. 592–600, June 1984.

[105] H. Altunbasak, and H. J. Trussell, "Colorimetric restoration of digital images," *IEEE Transactions on Image Processing*, vol. 10, no. 3, pp. 393–402, March 2001.

Chapter 3

Restoration in the Presence of Unknown Spatially Varying Blur

MICHAL ŠOREL
Academy of Sciences of the Czech Republic

FILIP ŠROUBEK
Academy of Sciences of the Czech Republic

3.1 Introduction

The past two decades brought significant progress in the development of efficient methods for classical deconvolution and super resolution problems in both single and multi-image scenarios. Most of these methods work with blurs modeled by convolution, which assumes that the properties of blur are the same in the whole image (space-invariant blur). Unfortunately, in practice, the blur is typically spatially variant. The most common types of space-variant blur are defocus, optical aberrations, and motion blur caused by either camera motion or motion of objects.

Extension of deconvolution methods to spatially varying blur is not straightforward. What makes such restoration problems more challenging than in the case of the space-invariant blur is a much higher number of unknowns that must be estimated. Consequently, the solution is ill-posed and requires additional constraints that must be chosen depending on the type of blur we wish to suppress. The requirement to remove only certain types of blur while keeping others is surprisingly common. A typical example is the removal of motion blur from portrait pictures taken in low-light conditions while keeping a small depth of focus. Similarly, it is usually desirable to remove optical aberrations but we may wish to preserve motion blur conveying the sensation of speed.

In the past few years, despite the complexity of space-variant deblurring, we can observe an increasing effort in this direction of research, including difficult problems such as the blur dependent on the depth of scene or several independently moving objects. In this

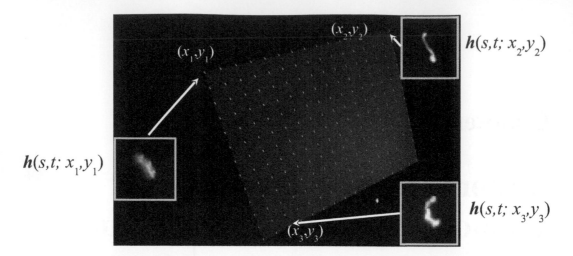

Figure 3.1 Spatially varying PSF for motion blur caused by camera shake. The image was acquired in a dark room by taking a picture of an LCD displaying regularly spaced white dots.

chapter, we give an overview of the current state-of-the-art in the space-variant restoration, addressing the latest results in both deblurring and super resolution.

The chapter is divided into two main parts. In Section 3.2, we describe mathematical models used for spatially varying blur and basic restoration approaches connected with these models. Our purpose is not to give a complete survey of all known algorithms; instead, we just briefly outline the models and point out interesting papers that have appeared in the past several years to indicate the current trends in the research of space-variant restoration. Among others, we introduce models describing the blur caused by camera motion by three-dimensional kernels analogous to those used in standard deconvolution algorithms and models used for complex scenes consisting of several independently moving objects.

The second part of this chapter (Section 3.3) details a new approach to space-variant super resolution for images blurred by camera motion or any other blur changing slowly enough so that it can be locally modeled by convolution. It is one of the first attempts to apply true super resolution to data with space-variant blur.

3.2 Blur models

The previous chapter addressed general problems of image restoration, including deconvolution that assumes space-invariant blurring, working with model

$$g = u * h + n, \tag{3.1}$$

where g and u are the blurred and sharp (original) images, respectively; h is a convolution kernel; and n is a white Gaussian noise $N(0, \sigma^2)$.

Spatially varying blur can be described by a general linear operator

$$g = Hu + n. \tag{3.2}$$

Figure 3.2 Defocus PSF acquired by deliberately defocusing an LCD covered by regularly spaced white dots. Vignetting clips the polygonal shape of the PSF, especially on the periphery of the field of view.

The operator H can be written in a form naturally generalizing standard convolution as

$$[Hu](x,y) = \int u(x-s, y-t)h(s,t,x-s,y-t)dsdt \tag{3.3}$$

with the point-spread function (PSF) h now dependent on the position (third and fourth variable) in the image. Vice versa, convolution with a kernel $\tilde{h}(s,t)$ is a special case of Equation (3.3) with $h(s,t,x,y) = \tilde{h}(s,t)$ for an arbitrary position (x,y). Note that the convolution kernel h in Equation (3.1) is sometimes also called a PSF. In this chapter, we reserve the expression PSF for h in (3.3), which is a function of four variables. It is consistent, however, to use the term *kernel* for the function $h(s,t,x_0,y_0)$ taken as a function of two variables with fixed x_0 and y_0, when the operator H can be locally approximated by convolution with $h(s,t,x_0,y_0)$ in a neighborhood of the point (x_0,y_0).

In practice, we work with a discrete representation, where the same notation can be used with the following differences: PSF h is defined on a discrete set of coordinates, the integral sign in Equation (3.3) becomes a sum, and operator H corresponds to a sparse matrix and u to a vector obtained by stacking columns of the image into one long vector. For convolution, H is a block-Toeplitz matrix with Toeplitz blocks, and each column of H corresponds to the same kernel. In the space-variant case, as each column corresponds to a different position (x,y), it may contain a different kernel $h(s,t,x,y)$.

Figures 3.1 and 3.2 show two examples of space-variant PSFs. Both images were acquired by taking a picture of an LCD displaying regularly spaced white dots. Figure 3.1 was taken by hand with a long shutter time. The smears one can see correspond directly to locally valid convolution kernels; that is, the PSF h on fixed positions. Three close-ups show the PSF in the top left, top right and bottom left corners of the LCD for fixed coordinates (x_1,y_1), (x_2,y_2), and (x_3,y_3), respectively. Figure 3.2 is an out-of-focus image of the same LCD, which gives again the corresponding PSF. Notice the irregular shape of

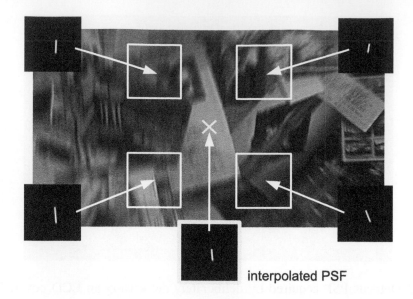

interpolated PSF

Figure 3.3 If the PSF varies slowly, we can estimate convolution kernels on a grid of positions and approximate the PSF in the rest of the image by interpolation of four adjacent kernels.

the PSF, caused by vignetting. While the darkening of corners due to vignetting is a well-known problem of wide-angle lenses at wide apertures, it is less known that vignetting affects also the PSF, cutting off part of its circular (for full aperture) or polygonal shape, giving significantly asymmetrical PSFs.

If we know the PSF, we are able to restore the image. As described in the previous chapter, the most common Bayesian approach achieves this by picking the most probable image u, which is equivalent to minimization

$$\arg\min_u \frac{1}{2\sigma^2} \|g - Hu\|^2 - \log p(u), \tag{3.4}$$

where $p(u)$ is an estimate of the prior probability of u. For super resolution, which requires multiple input images g_i, the solution is of the form

$$\arg\min_x \frac{1}{2\sigma^2} \sum_i \|g_i - DH_i u\|^2 - \log p(x), \tag{3.5}$$

where D is an operator modeling resolution loss and H_i is the operator of blurring corresponding to input g_i. Note that all efficient numerical algorithms minimizing (3.4) and (3.5) require the knowledge of the operator adjoint to H

$$[H^* u](x, y) = \int u(x - s, y - t) h(-s, -t, x, y) ds dt. \tag{3.6}$$

For details, see [1] describing a more general form of (3.4) and (3.5) that includes the process of image registration, image distortion, etc.

In practice, we usually do not know the PSF and it must be estimated. For the space-invariant blur, we can use one of many elaborated blind deconvolution methods [2–4].

Figure 3.4 Original sharp image (left) and the same image blurred by simulation of camera rotation (right).

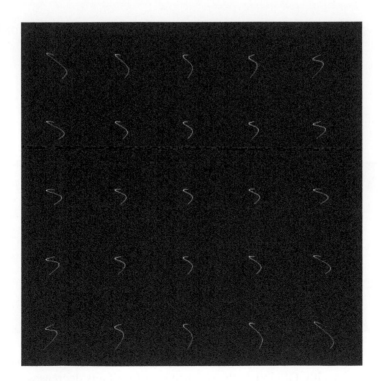

Figure 3.5 PSF $h(s, t, x_i, y_i)$ rendered at 25 positions (x_i, y_i).

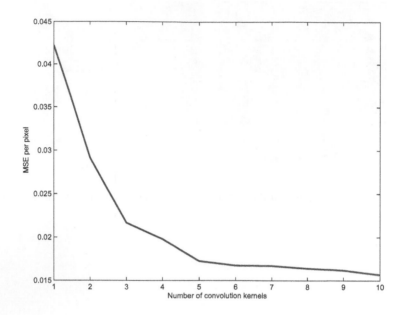

Figure 3.6 Graph of the MSE as a function of the number of kernels used for PSF approximation.

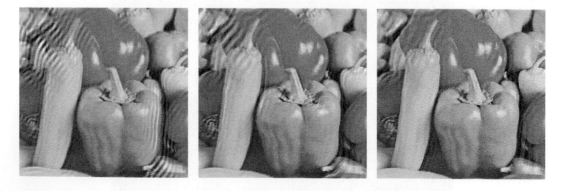

Figure 3.7 Deblurring of the right image from Figure 3.4 using (from left to right) 1, 2, and 3 kernels per image width. Obviously the occurrence of artifacts decreases with the number of kernels.

Estimation of the space-variant PSF in its general form of Equation (3.3) is too complex and ambiguous. As a rule, it cannot be expressed by an explicit formula but in many cases it has a special structure that can be exploited. For example, the blur caused by camera rotation is limited by three degrees of freedom of rigid body rotation. If we have an estimate of the camera rotation from inertial sensors, we are able to reconstruct the PSF and deblur the image using Equation (3.4) as described by Joshi et al. in [5]. Nevertheless, the authors also report that motion tracking using inertial sensors is prone to significant errors accumulating over the time of capture and resulting PSF is not precise enough to get artifact-free images. However, it is not clear if these problems are due to principal limitations, camera hardware, or the algorithm used for integration. Another possibility is to attach an auxiliary high-speed camera of lower resolution to estimate the PSF using, for example, optical flow techniques [6–8].

Unfortunately, in practice, we mostly do not have external sensors or devices providing information about the PSF, and the PSF must be estimated directly from the input images. In the rest of this section, we describe the properties of the PSF for the most frequent types of blur and corresponding strategies to estimate the PSF.

There are several types of blur that can be assumed to change slowly with position, which allows to approximate the blurring locally by convolution. Under this assumption we can estimate locally valid convolution kernels that give us a local estimate of the PSF. This usually holds for motion blurs caused by camera shake and optical aberrations. For defocus it holds for approximately planar scenes.

The kernels are usually estimated at a set of regularly spaced positions (see Figure 3.12), using one of already mentioned blind deconvolution methods, working either with a single image [2, 3] or more precise methods working with multiple images [4]. If it is possible to change camera settings, we can also fuse information from one blurred and one noisy/underexposed image [9–11].

Once the local convolution kernels are computed, they can be used to estimate the PSF for an arbitrary position (x, y). The simplest possibility is to divide the image to a set of regularly spaced patches, each with an assigned convolution kernel. The main problem with this approach are blocking artifacts appearing at patch boundaries.

One way to deal with these artifacts is to assign the estimated kernels just to patch centers (instead of the whole patch) and approximate the PSF h in intermediate positions by bilinear interpolation, as indicated in Figure 3.3. Accuracy can be improved by first bringing the estimated kernels into normalized positions (which can be advantageously accomplished by means of regular or complex moments of the PSFs, see [12]), performing bilinear interpolation, and then returning the kernels to their original position.

An advantage of this solution is that the PSF changes smoothly, thus avoiding the blocking artifacts. Moreover, the corresponding operator H can be computed in time comparable with the time of standard convolution using the fast Fourier transform (FFT) [13, 14]. The main reason is simple formulas for blur operators created as a linear combination of a finite set of convolution kernels. Indeed, if the PSF h is defined as a combination of kernels $\sum_i w_i h_i$, then

$$Hu = \sum_i (w_i u) * h_i \tag{3.7}$$

$$H^* u = \sum_i w_i (u * h_i^c), \tag{3.8}$$

where the symbol h_i^c denotes the convolution kernel h_i rotated by 180 degrees. For linear interpolation, the weight functions w_i satisfy the constraint $\sum_i w_i(x, y) = 1$ for an arbitrary position (x, y), and $w_i(x, y) = 1$ in the center (x, y) of the window where the kernel h_i was estimated. In our paper [11], we show how to use this model to deblur an image if another underexposed image taken with sufficiently short shutter time is available. The same model in a more versatile setup working with only one input image and using a recent blind deconvolution method [15] is shown in [16].

Hirsch et al. [17] show that this model also allows us to compute the operator adjoint to an operator U, acting on a vector h consisting of all vectorized kernels h_i and satisfying $Uh = Hu$. In theory, this could be useful in blind algorithms estimating simultaneously all kernels h_i. The practical value of this relation was not shown, however.

Here we demonstrate a simple deblurring experiment that indicates the feasibility of the interpolation of kernels. The left image in Figure 3.4 was blurred by simulating an irregular camera rotation about the x and y axes, giving the right blurred image. Figure 3.5 depicts the PSF $h(s, t, x, y)$ by rendering the function $h(s, t, x_i, y_i)$ as a function of the first two parameters at 25 positions (x_i, y_i).

Next, we deblurred the image in the standard Bayesian manner (3.4) with total variation regularization [18]. The PSF was approximated by linear interpolation of k convolution kernels, where the number k was varied from 1 to 10, using relations (3.7) and (3.8). For $k = 1$, it corresponds to standard deconvolution using the kernel valid in the image center. For $k = 5$, the algorithm works with the kernels shown in Figure 3.5. Figure 3.6 shows the mean square error of the result decreasing as k increases. Figure 3.7 shows the resulting images for $k = 1, \ldots, 3$. In Section 3.3 we demonstrate that the same approach can be used for super resolution as well.

This approach is relatively fast and can be used for most types of blur. A certain limitation is that the local estimation of the convolution kernels can fail because of weak texture, sensor saturation, or, for example, misregistration. In the following section, we show the special properties of the blur caused by camera motion that can be used to estimate the PSF in a more robust way. However, this is achieved at the expense of higher time complexity.

3.2.1 Camera Motion Blur

The blur caused by camera motion is limited by six degrees of freedom of a rigid body motion, most commonly decomposed to three rotations and three translations. The main obstacle when dealing with this type of blur is that for translations, the blur depends on scene distance (depth). As a consequence, under general camera motion, we need to estimate the depth map, which makes the algorithm complicated and very time consuming. Nevertheless, there are algorithms that work satisfactorily, assuming certain additional constraints on the camera motion. For example, [1] shows results of deblurring for the blur caused by camera translatation along an arbirtrary curve in one plane and for the out-of-focus blur. We refer interested readers to [1, 19, 20] and references therein.

In the following paragraphs we describe more practical approaches that use the fact that certain types of camera motion can be neglected in practice. The most common assumption is that all camera translations can be neglected, thus making the blur independent of scene depth. For example, for pictures taken by hand, Joshi et al. [5] tested exposures up to half a second and found that the typical translation in the direction of view (z-axis) was just a few millimeters in depth and had only a very small effect for lenses of common focal length.

The assumption of negligible translations was used to compute the PSF, for example in the above mentioned paper [6], getting the information about camera motion from an auxiliary high-speed camera and in [5] using inertial sensors for the same purpose. The same holds for papers [11] and [16] mentioned in the previous section.

Assuming only rotations, the simplest way is to transform the image so that the blur becomes a convolution and apply common deconvolution techniques. An immediate example is rotation about the optical axis (z-axis) that, expressed in polar coordinates, corresponds to one-dimensional translation. Therefore, any blur caused by such a rotation can be described by one-dimensional convolution. Similar transforms can be written for rotations about an arbitrary fixed axis. In practice, however, camera motion is rarely limited to one axis. Moreover, the interpolation necessary to transform the image is an additional source of error introduced into the process.

A more versatile approach applied only recently is to express the operator of blurring in a specially chosen set of basis images u_i,

$$Hu = \sum_i u_i k_i, \qquad (3.9)$$

which allows us to work with spatially varying blur in a manner similar to common convolution. The images u_i correspond to all possible transforms (rotations in our case) within a specified range of motions. Note, however, that unlike common convolution, such operators do not commute. The functions k_i are referred to as kernels or motion density functions.

Whyte et al. [21] consider rotations about three axes up to several degrees and describe blurring by the corresponding three-dimensional kernel. For blind deconvolution, the algorithm uses a straightforward analogy of the well-known blind deconvolution algorithm [2] based on marginalization over the latent sharp image. The only difference is the use of Equation (3.9) instead of convolution. For deblurring, following the kernel estimation phase, it uses the corresponding modification of the Richardson-Lucy algorithm.

Gupta et al. [22] adopt a similar approach but instead of rotations about x- and y-axes, they consider translations in these directions. Because of the dependence of translation on depth, they require that the scene is approximately planar and perpendicular to the optical axis. Interestingly, in this case it is not necessary to know the real distance because the corresponding kernel works in pixel units. They first estimate locally valid convolution kernels by the original blind deconvolution algorithm [2] and compute the corresponding sharp image patches. In the second step, they estimate the kernels k_i from Equation (3.9) using knowledge of both the observed image and an estimate of the sharp image made up of the uniformly distributed patches from the previous step. They do not use all patches but choose iteratively a subset of patches and check consistency with the rest by a RANSAC-like algorithm. The image is regularized by standard smoothness priors applied separately on derivatives in the x and y directions. The kernel is regularized to be sparse by a $\|.\|_p$ norm applied on kernel values and to be continuous using a quadratic penalty on kernel gradient.

An obvious advantage of the kernel model in Equation (3.9) is that it is very robust with respect to local non-existence of texture as well as local inaccuracies of the used model — sensor saturation, for example. On the other hand, it may be considerably more time consuming than algorithms based on local approximations by convolutions described in the previous section and used also in the super resolution algorithm we propose in Section 3.3. The bottleneck is the computation of all possible image transforms that must be repeated many times — basically in each step of the algorithm. Another disadvantage is that the actual motion may be more complicated and it is difficult to combine Equation (3.9) with other models.

3.2.2   Scene Motion Blur

An especially difficult situation is that of the blur caused by object motion, as objects usually move independently of each other, often in different directions. In order to achieve good-quality deblurring, the object must be precisely segmented, taking into account partial occlusion close to the object outline. Moreover, object blur may appear simultaneously with the blur due to camera motion, causing another, possibly spatially varying, blurring in the background.

Similar to other types of blur, the algorithms differ according to the available input. Single-image approaches are attractive because of its ability to work, for example, with pictures already taken. The quality of restoration is limited, however, because of insufficient information and principle ambiguity. True super resolution is impossible. Multi-image approaches, working usually with a video stream, can achieve a higher quality of restoration. On the other hand, the involved registration is an additional source of possible errors.

A frequent simplifying assumption is that the background is sharp, which holds when the camera is fixed (surveillance cameras) or mounted on a tripod stand. In addition, the blur of each object is often modeled by a standard convolution. An attempt in this direction was [23], using level-sets to segment the objects and ignoring the occlusion. Super resolution was considered in [24], limited to known and negligible PSF.

The blur caused by one moving object on a sharp background, including occlusion, can be described by a relatively simple formula [25, 26]:

$$g = f * h + (1 - w * h)b, \tag{3.10}$$

where f and b are the foreground and background images, h the convolution kernel of the foreground, and w the support of f corresponding to the area of the foreground object. The values of w are 0 for the background and 1 for the foreground.

Deblurring using even such a simple model is not easy. Raskar et al. [25] give a detailed analysis of possible cases, including the most ambiguous when the extent of blur is larger than the size of the moving object. Of course, for several moving objects or a blurred background, the situation becomes even more complicated.

In recent years, several papers appeared describing deblurring methods that work with only one image. Of course, the model is even more simplified by ignoring the oclussion effect. The objects are segmented based on various intensity, color, or blur cues.

Most of the methods follow the pioneering paper of Levin [27]. She assumed that objects move with a constant velocity and are therefore blurred by a one-dimensional rectangular kernel. The distribution of derivatives then changes as a function of the width of the blur kernel. The algorithm first estimates the direction of motion as the direction with minimal variation of derivatives [28]. The image is then segmented based on the statistics of derivatives expected in an average natural image.

Chakrabarti et al. [29] extend the ideas from [27]. Again, all objects are assumed to be blurred by the same motion kernel, being chosen from a discrete set of possible candidates corresponding to horizontal or vertical box filters of a certain length. To describe the likelihood of a given window to be blurred by a candidate kernel, they propose a more elaborate likelihood measure based on Gaussian scale mixtures. This likelihood is combined with statistical distributions for object and background colors described by a mixture of Gaussians into an MRF model with smoothness constraints.

Dai and Wu [30] came with an interesting constraint, analogous to the optical flow constraint equation, that links the gradient of the blurred image with a difference of the sharp image in the direction of blur. The formula again holds for motion blurs that can be locally described as a convolution with a rectangular impulse. Derivation is straightforward, based on the fact that the derivative of a rectangular impulse is zero everywhere except the beginning and end of the impulse, where it is equal to plus and minus Dirac delta functions. The relation can be written as

$$(\nabla g) \cdot b = u(x + b/2) - u(x - b/2), \qquad (3.11)$$

where b is the locally valid vector of motion blur — it has the same direction as the motion kernel and its size is equal to the length of the kernel. The dot in Equation (3.11) denotes the dot product of vectors. The authors show that almost any optical flow algorithm that uses the optical flow constraint can be modified to work with Equation (3.11) as well. They detail variants considering global motion models (affine and rotational) as well as a nonparametric model.

Liu et al. [31] locally detect and classify motion blur using a set of cues, including the local slope of the power spectrum, gradient histogram, maximum saturation, and variance of autocorrelation function in different directions. The image is classified into blurred and non-blurred regions based on the decision of a Bayes classifier, previously trained using a set of training images. The authors combine the results with a segmentation method based on graph cuts [32] and show examples of results superior to [27].

3.2.3 Defocus and Aberrations

The last important type of blur is defocus and related optical aberrations. Readers are probably familiar with the basic properties of out-of-focus blur, in particular that the ideal PSF has a circular shape (often called pillbox in the literature) and the inverse of its radius grows linearly with the distance from the plane of focus. In practice, the PSF is given by convolution of a diaphragm shape with a diffraction pattern. Basic equations related to these PSFs

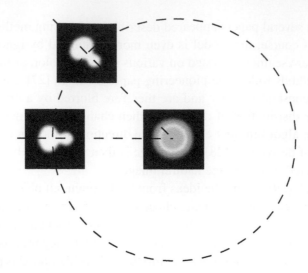

Figure 3.8 Three types of PSF symmetry for defocus and optical aberrations for a symmetrical optical system.

are summarized in [1]. Note that here we do not consider more complicated optical systems such as confocal microscopes, where the PSF is three-dimensional [33].

Compared to motion blur, there are significantly fewer papers addressing the restoration tasks for out-of-focus blur. There are several probable reasons. The first one is that the PSF depends on object distance. Consequently, to deblur the image, we need to estimate also the depth map, which makes the problem complex and time consuming. The respective algorithms are described in [1,19,20] and references therein. Second, the single-image blind deconvolution algorithms needed to estimate the PSF are less reliable than for the motion blur and using more than one image is less practical. Finally, the results of deblurring are not so impressive, because defocus destroys most information contained in high frequencies.

Interestingly, there are special cases where the problem of defocus is formally the same as the problem with moving objects mentioned in the previous section. Gu et al. [34] remove the blur caused by dirty lenses or thin occluders such as window shutters or grilles. Both cases can be modeled by the image formation model in Equation (3.10), originally formulated for an object moving on a sharp background. In this case, the dirt or an occluder corresponds to the foreground object f, this time blurred by a defocus PSF. For thin occluders in a known distance, they need two images with different apertures. For the dirty lenses, they estimate the occluder f by calibration using projected patterns or by autocalibration from a long sequence of video images.

The PSF of optical aberrations is much more complicated than that of pure defocus and it is mostly impossible to describe it by an explicit formula. Quite often, all we can say about the PSF are three symmetries caused by the rotational symmetry of common optical systems — see Figure 3.8. First, the PSF in the image center is radially symmetric. Second, the PSF is axially symmetrical about the axis connecting the image center and the position where the PSF is measured. Finally, the PSF at the same distance from the image center differ only by orientation. In practice, the PSF is usually measured experimentally.

In spite of the complexity of optical aberrations, paradoxically, removing the optical aberrations is an easier problem than pure defocus because their PSF can be measured for a particular lens in advance and does not depend so much on depth. Moreover, we usually need to remove aberrations only where the image is in focus, and this distance is provided by many modern cameras in the EXIF data. Indeed, there are several commercial applications for this task.

In theory, the PSF can be measured by observing very small light sources placed regularly across the field of view. Oddly enough, our experience is that such measurement is relatively difficult to do in practice, and it is simpler and more precise to compute the PSF from a picture of a known calibration pattern. The same observation is mentioned in [35]. There are several papers using a set of slanted edges to get a line-spread function that corresponds to a 1-D projection of the 2-D PSF. However, for the purpose of deblurring or super resolution, it is usually preferable to directly compute the two-dimensional PSF [35–37], which is possible even with sub-pixel resolution. The basic principle is to look locally for a convolution kernel h minimizing

$$\|g - D(u * h)\|^2, \tag{3.12}$$

where g is a patch of the observed image, u the known sharp calibration pattern, and D a downsampling operator. Minimization of relation (3.12) is a solution of a system of linear equations in the least squares sense. Further references can be found in [35].

From the recent literature, we would like to point out the paper of Kee et al. [38], showing that aberrations of high-quality lenses (Canon $24 - 105$ mm $f/4$, Canon $17 - 40$ mm $f/4$) can be locally described as a convolution with an oriented Gaussian specified by a covariance matrix. An unpleasant property of aberrations is that they change with spatial location and both the focal length and aperture. Fortunately, [38] also shows that the covariance matrix changes smoothly and can be described by a low-order polynomial of four variables (x, y, focal length, and aperture).

3.3 Space-Variant Super Resolution

In Section 3.2 we went through the basic types of space-variant blur and models used to describe it, and indicated approaches to how these models could be used for deblurring. To the best of our knowledge, there are basically no true super resolution algorithms working with spatially varying blur.

In the rest of this chapter we propose such a space-variant super resolution algorithm, based on the local detection of convolution kernels and approximating the PSF by linear interpolation (3.7) between the positions where the kernels were detected. The use of this approach for deblurring was demonstrated in Figures 3.4 through 3.7.

The proposed algorithm works for an arbitrary blur that changes slowly enough that it can be locally approximated by convolution. It requires the input images (it needs at least five input images for the SR factor of 2) to be at least approximately registered.

3.3.1 Algorithm

We have K input frames g_k ($k = 1, \ldots, K$), which are noisy, blurred, and downsampled representations of some "ideal" image u. Let us divide each image g_k into overlapping patches denoted as g_k^p, where p is the patch index, $p = 1, \ldots, P$. Blurring H_k is space variant, but we assume that the convolution (space invariant) model holds in each patch. The formation model thus writes locally as

$$g_k^p = DH_k^p u^p + n_k. \tag{3.13}$$

Both the degradation D and noise n_k are space invariant and therefore the index p is omitted. The blurring operator acting on the original patch u^p is space invariant and thus H_k^p denotes convolution with some PSF h_k^p. Smaller patches achieve a better approximation of the true space-variant case. However the patch size is limited from below by the PSF size.

Solving the above equation for each patch becomes a multichannel blind deconvolution and super resolution problem. A flowchart of the proposed algorithm is summarized in Figure 3.9. The algorithm consists of four steps:

1. **Splitting**. We split the input frames into patches. Restrictions on the patch size implied by the PSF size are discussed in Section 3.3.2.

2. **PSF estimation**. We estimate the convolution kernels (h_k^p) in each patch p and frame k. For this purpose we apply patch-wise the blind super resolution algorithm [39] and use the estimated kernels. The reconstructed patches are ignored. This is described in Section 3.3.3.

3. **PSF refinement**. We refine the estimated kernels by replacing those with erroneous masks with interpolated ones. Estimating PSFs in each patch, especially if the overlapping scheme is considered, would be very time consuming. Therefore, we divide the frames in a nonoverlapping fashion and use the interpolation method to generate PSFs in every location; see Figure 3.3. The interpolation scheme is covered in Section 3.3.4.

4. **Space-variant deblurring and super resolution**. The final step takes the estimated PSFs and computes the high-resolution sharp image using the Bayesian approach in Equation (3.5) as described in Section 3.3.5.

Estimating PSFs locally has several advantages. We can apply robust PSF estimation algorithm, such as [39], which works also in the presence of the decimation operator and which compensates for misregistration in the form of translation by shifting PSFs. If we assume that locally any misregistration manifests itself as translation, then by estimating shifted PSFs locally we can compensate for a more complex geometrical transformation than just translation.

The decimation operator simulates the behavior of digital sensors by performing convolution with a sensor PSF followed by downsampling with some step ε. We also refer to the sampling step ε as an *SR factor*. It is important to underline that ε is a user-defined parameter. The sensor PSF is modeled as a Gaussian function, which is experimentally justified in [40]. A physical interpretation of the sensor blur is that the sensor is of finite size and it integrates impinging light over its surface. The sensitivity of the sensor is highest in the middle and decreases toward its borders with a Gaussian-like decay. We can insert the sensor PSF inside the blurring H_k and regard D solely as a downsampling operator.

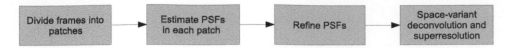

Figure 3.9 Flowchart of the algorithm.

3.3.2 Splitting

From a theoretical point of view, small patches better approximate local space-invariant nature of degradation H_k. However, small patches may lack the information necessary to estimate PSFs (we have more unknowns than equations). It is therefore important to derive limitations on the patch size. Intuitively we see that the minimum patch size must depend on the PSF size and the number of input frames K. However, to derive exact constraints, we need to analyze the subspace method proposed in blind super resolution [39] for PSF estimation in the presence of decimation D.

First we rewrite the local observation model of Equation (3.13) in terms of convolution operators and use a vector-matrix notation. To simplify the notation, we assume all image and PSF supports square, and the downsampling factor ε integer ($\varepsilon = 2, 3, \ldots$) are the same in both directions. An extension to rectangular supports is straightforward. Rational SR factors ε can be considered as well if polyphase decomposition is used; see [41]. We omit the patch index p and the reader should keep in mind that in the following discussion we refer to patches and not to whole images. The size of all patches in all frames g_k are assumed to be equal and denoted as G. The size of the corresponding original patch u is U. The maximum PSF size of h_k is denoted as H. Next, we define a convolution matrix. Let u be an arbitrary discrete image of size U; then $\mathbf{u}$ denotes an image column vector of size U^2 and $\mathbf{C}_A\{u\}$ denotes a matrix that performs convolution of u with an image of size A. The convolution matrix can have a different output size. Adopting the MATLAB naming convention, we distinguish two cases: "full" convolution $\mathbf{C}_A\{u\}$ of size $(U + A - 1)^2 \times A^2$ and "valid" convolution $\mathbf{C}_A^v\{u\}$ of size $(U - A + 1)^2 \times A^2$. In both cases, the convolution matrix is a Toeplitz-block-Toeplitz matrix. Let $\mathcal{G} := [\mathbf{G}_1, \ldots, \mathbf{G}_K]$, where $\mathbf{G}_k = \mathbf{C}_A^v\{g_k\}$ is the "valid" convolution matrix of g_k acting on an $A \times A$ support. Assuming no noise, we can express $\mathcal{G}$ using the observation model in Equation (3.13) as

$$\mathcal{G} = \mathbf{DU}\mathcal{H}, \qquad (3.14)$$

where $\mathbf{D}$ is the downsampling matrix with the sampling factor ε and the appropriate size to match the term on the right, $\mathbf{U} := \mathbf{C}_{\varepsilon A + H - 1}^v\{u\}$ and

$$\mathcal{H} := [\mathbf{C}_{\varepsilon A}\{h_1\}\mathbf{D}^T, \ldots, \mathbf{C}_{\varepsilon A}\{h_K\}\mathbf{D}^T]. \qquad (3.15)$$

Note that the transposed matrix $\mathbf{D}^T$ behaves as an upsampling operator that interlaces the original samples with $(\varepsilon - 1)$ zeros. Estimating the PSFs h_k from the observed patches g_k as proposed in blind super resolution relies on the following observation. If $\mathbf{DU}$ is of full column rank, then the null space of $\mathcal{G}$ is equal to the null space of $\mathcal{H}$ and because we can calculate the null space of $\mathcal{G}$, we get information about PSFs h_k. If the original patch is not degenerated (e.g., constant image), and contains some details, then $\mathbf{DU}$ is generally of full column rank if it has at least as many rows as columns. The difference between the

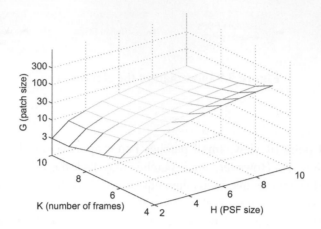

Figure 3.10 The minimum patch size as a function of number of input frames and PSF size. The SR factor is 2. Note that the patch size axis is on the logarithmic scale.

number of columns and rows of $\mathcal{H}$, which is $KA^2 - (\varepsilon A + H - 1)^2$, bounds from below the null-space dimension of $\mathcal{H}$ and thus the null-space size of $\mathcal{G}$. From the bound follows that $A \geq \frac{H-1}{\sqrt{K}-\varepsilon}$, where A is the window size, for which we construct the convolution matrices $\mathbf{G}_k$. The necessary condition for keeping the null space of $\mathcal{G}$ equal to the null space of $\mathcal{H}$ is to have at least as many rows as columns in $\mathcal{G}$. Thus, from the size of $\mathcal{G}$ it follows that $(G - A + 1)^2 \geq A^2 K$ and after substituting for A we obtain

$$G \geq \left\lceil \frac{\sqrt{K}(H - 2) + H + \varepsilon - 1}{\sqrt{K} - \varepsilon} \right\rceil . \tag{3.16}$$

As one would expect, the necessary minimum patch size G decreases with the decreasing PSF size H and/or increasing number of input frames K. For a better interpretation of the size constraint, Figure 3.10 shows G as a function of H and K. For example, in the case of $\varepsilon = 2$ and blurs of maximum size 10×10, the minimum patch size is over 100×100 if only five input frames are used, but much smaller patch sizes roughly 30×30 are sufficient if ten input frames are used.

3.3.3   PSF Estimation

As in the previous section, the following discussion applies to individual patches and the index p is omitted. Because we can estimate the null space of $\mathcal{H}$ from the null space of $\mathcal{G}$ in Equation (3.14), we have means to determine the original PSFs directly from the observed input patches. In the blind super resolution algorithm [39], it was shown that the PSFs must satisfy

$$\mathcal{N}\mathbf{h} = \mathbf{0} , \tag{3.17}$$

where the matrix $\mathcal{N}$ contains convolution matrices with images lying in the null space of $\mathcal{G}$ and $\mathbf{h}$ are K PSFs stacked in one vector. PSFs are not uniquely determined by the above equation, as $\mathcal{N}$ is rank deficient by at least ε^2 due to the presence of the downsampling matrix $\mathbf{D}$ in Equation (3.14). In addition, the matrix rank further decreases if the PSF support is larger than the maximum support of the true PSFs. In practice, we rarely know the correct support size exactly and therefore we work with overestimated sizes. Last but not least, the null space approach neglects noise in its derivation. These facts prevent using Equation (3.17) for PSF estimation directly. Instead, we use the MAP approach to estimate both the latent patch u and the PSFs h_k, and combine two observation models. The first observation model is in a standard form $M_1(u, \{h_k\}) \equiv \frac{\mu_1}{2} \sum_{k=1}^{K} \|D(u * h_k) - g_k\|^2$, where μ_1 is inversely proportional to the variance of noise n_k and $\|\cdot\|$ denotes the ℓ_2 norm. For simplicity, we assume the same noise variance in all frames and patches, and therefore the single parameter μ_1 suffices. The second observation model is defined by Equation (3.17) as $M_2(h_1, \ldots, h_K) \equiv \frac{\mu_2}{2} \|\mathcal{N}\mathbf{h}\|^2$, where μ_2 is inversely proportional to the variance of noise. Then the MAP approach is equivalent to solving the optimization problem:

$$\min_{u, \{h_k\}} M_1(u, \{h_k\}) + M_2(\{h_k\}) + R_u(u) + R_h(\{h_k\}), \tag{3.18}$$

where R_u, R_h are image and PSF regularizers.

A popular recent approach to image regularization is to assume that the unknown image u is represented as a linear combination of few elements of some frame (usually an overcomplete dictionary) and force this sparse representation by using the ℓ_1 norm (or ℓ_0). Arguably, the best-known and most commonly used image regularizer, which belongs to the category of sparse priors, is the total variation (TV) norm [18]. The isotropic TV model is the ℓ_1 norm of image gradient magnitudes and takes the form

$$R_u(u) = \phi(\nabla u) = \sum_i \sqrt{(\nabla_x u(i))^2 + (\nabla_y u(i))^2}, \tag{3.19}$$

where $\phi(x) = \|x\|$. The TV regularizer thus forces the solution to have sparse image gradient. Depending on the type of data, one can have sparsity in different domains. This modification is, however, easy to achieve. All we have to do is replace derivatives with a transformation (e.g., wavelet-like multi-scale transform), which gives sparse representation of our data.

For the PSF regularization R_h, we found it sufficient to enforce in this term only positivity. One can add TV regularization as in the case of images (or sparsity of intensity values) if the expected blurs are, for example, motion blurs.

The standard approach to solve the optimization problem (3.18) is called *alternating minimization* (AM) and will be adopted here as well. We split the problem into two subproblems:

$$\text{"}u\text{-step":} \quad \min_u M_1(u, \{h_k\}) + R_u(u) \tag{3.20}$$

and

$$\text{"}h\text{-step":} \quad \min_{\{h_k\}} M_1(u, \{h_k\}) + M_2(\{h_k\}) + R_h(\{h_k\}), \tag{3.21}$$

and alternate between them. Convergence to the global minimum is theoretically not guaranteed because the unknown variables are coupled in the data term M_1. However, in our formulation all the terms are convex and thus each subproblem separately converges to its global minimum and it can be solved efficiently, for example, by the augmented Lagrangian method [42].

We must underline that the primary output of this step are estimated PSFs in each patch and frame. We are not interested in the reconstructed patches u, which are in this stage merely by-products of AM. We observe that AM can be greatly accelerated (reducing the number of alternations necessary for convergence) if the weight μ_1 is set smaller than appropriate for the current noise level. This means that we give more weight to the TV regularization term, and small details in the estimated patch u are wiped out. Only main features (strong edges) inside the estimated patch u are reconstructed. The advantage is that the h-step becomes more stable with such u, and PSFs can be estimated with the same accuracy. This idea is similar to recent improvements proposed in single-channel blind deconvolution [3]. In order for the single-channel case to work, we must perform a few tricks. One trick is to slightly enhance the edges of the estimated image after each iteration, which avoids the trivial solution, that is, the estimated image being equal to the input blurry image and PSF being a delta function. Another trick is to remove edges that correspond to small objects, which would otherwise favor smaller PSFs than the true ones. Using TV regularization with a larger weight than necessary simulates in a way the above tricks and boosts the PSF estimation step.

3.3.4   PSF Refinement

Ideally, we want to have estimated PSFs in every position (pixel) of the input frames. This is computationally very demanding as the number of patches is equal to the number of pixels, and patches heavily overlap. However, we assume that PSFs vary slowly so that locally (inside each patch) they can be considered constant. Calculating thus the PSFs in every position is unnecessary labor. We estimate PSFs on a coarser mesh and find the remaining ones by the interpolation method described in Section 3.2. Another reason for PSF interpolation is if the PSF estimation fails in some patches. Then we must use adjacent PSFs to interpolate and replace the erroneous one. There are basically two reasons why kernel estimation fails [11]. The first reason is due to patches with no details, which corresponds to the situation that **DU** in Equation (3.14) is column rank deficient. To identify such cases, we compute the entropy of the estimated kernels and take those with the entropy above some threshold. The other case of failure is pixel saturation caused by light levels above the sensor range. This situation can be identified by computing the kernel energy, that is, the sum of kernel values. For valid kernels, the energy should be unity. Therefore, we simply remove kernels whose sum is too different from unity, again above some threshold. These two thresholds are user parameters.

3.3.5   Deconvolution and Super Resolution

Once we have the convolution kernel estimated (or interpolated from neighboring patches) in each patch of every input frame, we can use the kernels to approximate the PSF in an arbitrary position using linear interpolation. The respective blurring operators can be easily implemented using relations (3.7) and (3.8). As mentioned in Section 3.2, this approach was used for deconvolution in [11] and [16].

We propose to extend this approach to super resolution. The idea is that we can perform space-variant deblurring and super resolution in one step and directly obtain the high-

(a) 1st input image   (b) 2nd input image is slighty rotated

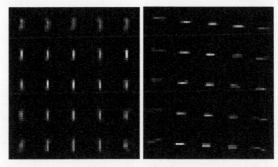

(c) PSF of the 1st image   (d) PSF of the 2nd image

(e) result of deconvolution   (f) result of space-variant deblurring

Figure 3.11 Experiment illustrating an important property of the proposed approach — ability to compensate for registration inaccuracies. Notice how the estimated PSF of the second image compensates for rotation.

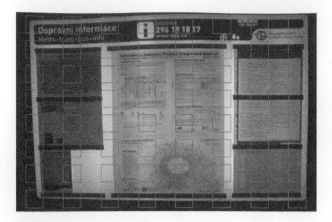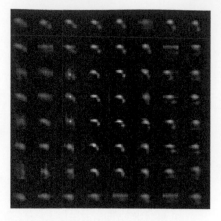

Figure 3.12 First of six input images (1700 × 1130 pixels) used for super resolution (left) and 8 × 8 local convolution kernels computed by the SR algorithm [39] (right). Spatial variance of the PSF is obvious. Squares in the image depict patches, in which the kernels were estimated.

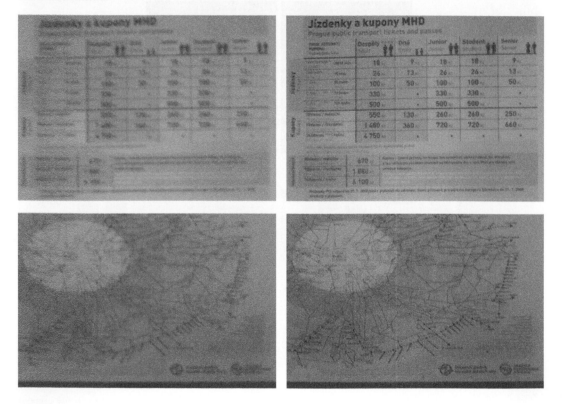

Figure 3.13 Two details of the blurred low-resolution image from Figure 3.12 (left column) and the resulting high-resolution image of 3400 × 2260 pixels (right column).

resolution image using relation (3.5). Indeed, the experiments show that the PSF approximated in this manner gives satisfactory results, locally comparable with the result of space-invariant super resolution methods. Moreover, because the PSF changes smoothly between the patches, there are almost no visible artifacts.

This final step is analogous to the u-step in relation (3.20) of the kernel estimation stage except that this time we work with whole frames and not just patches. We solve

$$\min_u \frac{\mu_1}{2} \sum_{k=1}^{K} \|DH_k u - g_k\|^2 + \phi(\nabla u), \qquad (3.22)$$

where H_k is the space-variant convolution operator, which is created from estimated PSFs $h_k^p, p = 1, \ldots P$. In theory, each column of H_k contains a PSF corresponding to that position and estimated in the second step or interpolated in the third step. In our implementation, we used regularization by total variation (3.19) as in the PSF estimation step.

3.3.6 Experiments

This section describes two experiments. The first one in Figure 3.11 illustrates an important ability of the proposed method to work with input images that are slightly misregistered. The second experiment (Figures 3.12 and 3.13) is an example of real results.

Our super resolution method requires multiple input images, which are properly registered. Super resolution by itself is very sensitive to registration and sub-pixel accuracy is generally needed. However, registration methods rarely achieve such accuracy. It was shown in [39] that blind deconvolution and super resolution can automatically register shifted images (misregistration in the form of translation) by shifting the centers of estimated convolution kernels.

In practice, we often have more serious misregistration, such as rotation or projective distortion. It turns out that our local approach can handle such complex geometric transforms. The reason is that with rotations up to several degrees, any similarity transform can be locally approximated by translation. The same is true even with more complicated transforms we might meet in practice.

This important feature can be demonstrated by the following experiment. An image was degraded by two different space-variant blurs simulating camera rotation around the x- and y-axes, respectively. The second image was first rotated by 1 degree. The resulting degraded and misregistered images are in Figure 3.11(a) and (b). We divided the images into 5×5 patches and applied the blind super resolution algorithm [39]. The estimated convolution kernels in the first and second images are in Figure 3.11(c) and Figure 3.11(d), respectively. Notice that the estimated blurs in Figure 3.11(d) are shifted to compensate for rotation of the second image. Figure 3.11(f) shows the estimated sharp image. For comparison, Figure 3.11(e) shows also the result of the space-invariant method applied to the whole image, which is equivalent to the case of one large patch. Artifacts are noticeable compared to the space-variant case. However, even for space-variant deblurring, because the rotation violates the convolution model, some artifacts appear in areas farther away from the center of rotation.

In the second experiment, we took a sequence of six images. One of them is shown in Figure 3.12, together with $8 \times 8 = 64$ kernels estimated in the first phase of the algorithm. We can see that the PSF is significantly space-variant and the image becomes aberrated along boundaries. The result of the space-variant super resolution is shown in two close-ups in Figure 3.13. We do not show the whole image because its size (3400×2260) would be too large to tell the difference compared to the original low-resolution image in Figure 3.12. The improvement in resolution is clearly visible and the image contains almost no artifacts.

3.4 Summary

In this chapter we outlined current approaches to deblurring in the presence of spatially varying blur. Until now, these algorithm had not been extended to super resolution.

Approximately half of this chapter was devoted to a new super resolution algorithm working for slowly varying blurs. Its main advantage is the possibility to combine several types of blur, such as camera motion blur, defocus, and aberrations, and even compensate for small registration errors. In addition, it is much faster than approaches working with more restrictive models, and it is not much slower than standard deconvolution methods. On the other hand, it does not cope well with the blur caused by object motion.

The most interesting direction of future research is probably the extension of deblurring methods designed to work with moving objects to super resolution, including the rigorous treatment of occlusion effects along object boundaries.

Acknowledgment

This work was supported in part by the Grant Agency of the Czech Republic project P103/11/1552.

Bibliography

[1] M. Sorel, F. Sroubek, and J. Flusser, "Towards super resolution in the presence of spatially varying blur," in *Super Resolution Imaging* (P. Milanfar, Ed.), pp. 187–218, Boca Raton: CRC Press, 2010.

[2] R. Fergus, B. Singh, A. Hertzmann, S. T. Roweis, and W. Freeman, "Removing camera shake from a single photograph," *ACM Transactions on Graphics*, vol. 25, pp. 787–794, 2006.

[3] L. Xu and J. Jia, "Two-phase kernel estimation for robust motion deblurring," in *Proceedings of the 11th European Conference on Computer Vision: Part I*, ECCV'10, pp. 157–170, Berlin, Springer-Verlag, 2010.

[4] F. Sroubek and J. Flusser, "Multichannel blind deconvolution of spatially misaligned images," *IEEE Transactions on Image Processing*, vol. 14, pp. 874–883, July 2005.

[5] N. Joshi, S. B. Kang, C. L. Zitnick, and R. Szeliski, "Image deblurring using inertial measurement sensors," *ACM Transactions on Graphics*, vol. 29, pp. 30:1–30:9, July 2010.

[6] M. Ben-Ezra and S. Nayar, "Motion deblurring using hybrid imaging," in *IEEE Computer Society Conference on Computer Vision and Pattern Recognition*, vol. 1, pp. 657–664 , June 2003.

[7] M. Ben-Ezra and S. K. Nayar, "Motion-based motion deblurring," *IEEE Transactions Pattern Analysis and Machine Intelligence*, vol. 26, pp. 689–698, June 2004.

[8] Y.-W. Tai, H. Du, M. S. Brown, and S. Lin, "Image/video deblurring using a hybrid camera," in *IEEE Computer Society Conference on Computer Vision and Pattern Recognition*, 2008.

[9] M. Tico, M. Trimeche, and M. Vehvilainen, "Motion blur identification based on differently exposed images," in *Proc. IEEE International Conference Image Processing*, pp. 2021–2024, 2006.

[10] L. Yuan, J. Sun, L. Quan, and H.-Y. Shum, "Image deblurring with blurred/noisy image pairs," in *SIGGRAPH '07*, (New York, NY, USA), article no. 1, ACM, 2007.

[11] M. Sorel and F. Sroubek, "Space-variant deblurring using one blurred and one underexposed image," in *Proceedings of the 16th IEEE International Conference on Image Processing*, pp. 157–160, 2009.

[12] J. Flusser, T. Suk, and B. Zitová, *Moments and Moment Invariants in Pattern Recognition*. New York: J. Wiley, 2009.

[13] T. G. Stockham, Jr., "High-speed convolution and correlation," in *Proceedings of the April 26-28, 1966, Spring Joint Computer Conference*, AFIPS '66 (Spring), (New York, NY, USA), pp. 229–233, ACM, 1966.

[14] J. G. Nagy and D. P. O'Leary, "Restoring images degraded by spatially variant blur," *SIAM Journal on Scientific Computing*, vol. 19, no. 4, pp. 1063–1082, 1998.

[15] S. Cho and S. Lee, "Fast motion deblurring," *ACM Transactions on Graphics (SIGGRAPH ASIA 2009)*, vol. 28, no. 5, article no. 145, 2009.

[16] S. Harmeling, H. Michael, and B. Scholkopf, "Space-variant single-image blind deconvolution for removing camera shake," in *Advances in Neural Information Processing Systems 23* (J. Lafferty, C. K. I. Williams, J. Shawe-Taylor, R. Zemel, and A. Culotta, Eds.), pp. 829–837, 2010.

[17] M. Hirsch, S. Sra, B. Scholkopf, and S. Harmeling, "Efficient filter flow for space-variant multiframe blind deconvolution," in *IEEE Conference on Computer Vision and Pattern Recognition (CVPR)*, pp. 607 –614, June 2010.

[18] L. I. Rudin, S. Osher, and E. Fatemi, "Nonlinear total variation based noise removal algorithms," *Physica D*, vol. 60, pp. 259–268, 1992.

[19] M. Sorel and J. Flusser, "Space-variant restoration of images degraded by camera motion blur," *IEEE Transactions on Image Processing*, vol. 17, pp. 105–116, Feb. 2008.

[20] P. Favaro, M. Burger, and S. Soatto, "Scene and motion reconstruction from defocus and motion-blurred images via anisothropic diffusion," in *ECCV 2004, LNCS 3021*, Berlin: Springer Verlag, (T. Pajdla and J. Matas, Eds.), pp. 257–269, 2004.

[21] O. Whyte, J. Sivic, A. Zisserman, and J. Ponce, "Non-uniform deblurring for shaken images," in *IEEE Conference on Computer Vision and Pattern Recognition (CVPR)*, pp. 491 –498, June 2010.

[22] A. Gupta, N. Joshi, C. L. Zitnick, M. Cohen, and B. Curless, "Single image deblurring using motion density functions," in *Proceedings of the 11th European Conference on Computer Vision (ECCV)*, (Berlin, Heidelberg), pp. 171–184, Springer-Verlag, 2010.

[23] L. Bar, N. A. Sochen, and N. Kiryati, "Restoration of images with piecewise space-variant blur," in *Scale Space and Variational Methods in Computer Vision*, pp. 533–544, 2007.

[24] H. Shen, L. Zhang, B. Huang, and P. Li, "A MAP approach for joint motion estimation, segmentation, and super resolution," *IEEE Transactions on Image Processing*, vol. 16, pp. 479–490, Feb. 2007.

[25] R. Raskar, A. Agrawal, and J. Tumblin, "Coded exposure photography: Motion deblurring using fluttered shutter," in *ACM SIGGRAPH 2006 Papers*, (New York, NY, USA), pp. 795–804, ACM, 2006.

[26] A. Agrawal, Y. Xu, and R. Raskar, "Invertible motion blur in video," *ACM Transactions on Graphics*, vol. 28, no. 3, pp. 1–8, 2009.

[27] A. Levin, "Blind motion deblurring using image statistics," in *Advances in Neural Information Processing Systems (NIPS)*, pp. 841–848, 2006.

[28] Y. Yitzhaky, I. Mor, A. Lantzman, and N. S. Kopeika, "Direct method for restoration of motion-blurred images," *Journal of the Optical Society of America A: Optics Image Science, and Vision*, vol. 15, no. 6, 1512–1519, 1998.

[29] A. Chakrabarti, T. Zickler, and W. T. Freeman, "Analyzing spatially-varying blur," in *IEEE Conference on Computer Vision and Pattern Recognition (CVPR)*, (San Francisco, CA), pp. 2512–2519, June 2010.

[30] S. Dai and Y. Wu, "Motion from blur," in *IEEE Conference on Computer Vision and Pattern Recognition (CVPR)*, June 2008.

[31] R. Liu, Z. Li, and J. Jia, "Image partial blur detection and classification," in *IEEE Conference on Computer Vision and Pattern Recognition (CVPR)*, pp. 1 –8, June 2008.

[32] V. Kolmogorov and R. Zabih, "What energy functions can be minimized via graph cuts?" *IEEE Transactions on Pattern Analysis and Machine Intelligence*, vol. 26, pp. 147 –159, Feb. 2004.

[33] M. J. Nasse and J. C. Woehl, "Realistic modeling of the illumination point spread function in confocal scanning optical microscopy," *Journal of the Optical Society of America A*, vol. 27, pp. 295–302, Feb. 2010.

[34] J. Gu, R. Ramamoorthi, P. Belhumeur, and S. Nayar, "Removing image artifacts due to dirty camera lenses and thin occluders," *ACM Transactions on Graphics*, vol. 28, pp. 144:1–144:10, Dec. 2009.

[35] R. Szeliski, *Computer Vision: Algorithms and Applications (Texts in Computer Science)*. Berlin: Springer, 2010.

[36] T. L. Williams, *The Optical Transfer Function of Imaging Systems*. London: Institute of Physics Publishing, 1999.

[37] N. Joshi, R. Szeliski, and D. J. Kriegman, "PSF estimation using sharp edge prediction," *IEEE Computer Society Conference on Computer Vision and Pattern Recognition (CVPR)*, 2008.

[38] E. Kee, S. Paris, S. Chen, and J. Wang, "Modeling and removing spatially-varying optical blur," in *Proceedings of IEEE International Conference on Computational Photography (ICCP)*, 2011.

[39] F. Sroubek, G. Cristobal, and J. Flusser, "A unified approach to superresolution and multichannel blind deconvolution," *IEEE Transactions on Image Processing*, vol. 16, pp. 2322–2332, Sept. 2007.

[40] D. Capel, *Image Mosaicing and Super Resolution (Cphc/Bcs Distinguished Dissertations)*. Berlin: SpringerVerlag, 2004.

[41] F. Sroubek, J. Flusser, and G. Cristobal, "Super resolution and blind deconvolution for rational factors with an application to color images," *The Computer Journal*, vol. 52, pp. 142–152, 2009.

[42] M. Afonso, J. Bioucas-Dias, and M. Figueiredo, "Fast image recovery using variable splitting and constrained optimization," *IEEE Transactions on Image Processing*, vol. 19, pp. 2345–2356, Sept. 2010.

Chapter 4

Image Denoising and Restoration Based on Nonlocal Means

PETER VAN BEEK
Sharp Laboratories of America

YEPING SU
Apple Inc.

JUNLAN YANG
Marseille Inc.

4.1 Introduction

In this chapter we review recent work in image denoising, deblurring, and super resolution reconstruction based on the *nonlocal means* approach. The nonlocal means (NLM) filter was originally proposed for denoising by Buades, Coll, and Morel [1, 2]. The basic NLM filter is a form of weighted averaging over a large window, where the weights are determined by the similarity between image patches or neighborhoods. The NLM algorithm has a very simple form, has few parameters, and is easy to implement. Its denoising performance is very good and was considered state-of-the-art at the time of its introduction. The NLM filter is able to remove additive white noise while preserving sharp edges and fine texture details. The approach attracted significant attention in the image processing community and was quickly improved upon using refinements and iterative extensions [3–5]. The state-of-the-art in denoising is currently formed by a group of methods that are related to NLM [6–9] and are discussed later in the chapter. Furthermore, the NLM approach has been applied in other signal processing areas such as video denoising [10], image segmentation [4], super resolution [11], and deblurring [12]. Due to the high computational cost required by the NLM filter, several variations of the algorithm have been proposed that aim to reduce the time complexity [13–15].

In this chapter we aim to introduce the NLM approach to image restoration and describe the connections between this approach, existing restoration methods, and recently proposed methods. We focus mainly on the problems of image denoising and deblurring, while also reviewing work on super resolution. We now briefly discuss basic models of the observed image and of the "ideal" image (to be recovered), which form the basis of common image restoration methods as well as many of the nonlocal methods we discuss in this chapter.

We consider images that are degraded by blur and noise, inherent in the image capturing process. Blurring involves loss of high-frequency detail in the image due to, for example, the camera optics and imaging sensor. Random noise may be caused by the sensing process (e.g., photon noise in CCD and CMOS sensors), by the sensor or camera electronics, by analog-to-digital conversion, and by other factors. For image deblurring and denoising purposes, we assume the following discrete observation model:

$$g(\mathbf{x}_i) = \sum_{j \in \Omega} h(\mathbf{x}_i - \mathbf{x}_j) f(\mathbf{x}_j) + v(\mathbf{x}_i), \tag{4.1}$$

where $\mathbf{x}_i = (x_i, y_i)$ is the location of the ith pixel, $f(\mathbf{x})$ is the "ideal" image, and $g(\mathbf{x})$ is the degraded image data. The various blurring effects are combined into the point spread function (p.s.f.) $h(\mathbf{x})$, which has spatial support Ω, and the effects of different pointwise noise sources are modeled by an additive noise term $v(\mathbf{x})$. For the image denoising problem (without deblurring), the p.s.f. $h(\mathbf{x})$ can effectively be removed from the model. The observation model is often written in matrix-vector form as $\mathbf{g} = \mathbf{H}\,\mathbf{f} + \mathbf{v}$. The goal is to estimate $\mathbf{f}$, given $\mathbf{g}$ and assuming $\mathbf{H}$ is (approximately) known. Hence, one aims to reduce the blur and noise in $\mathbf{g}$, while avoiding the introduction of artifacts or further loss of detail.

A common image restoration framework [16, 17] is to estimate $\mathbf{f}$ by minimizing a cost function:

$$\hat{\mathbf{f}} = \arg \min_{\mathbf{f}} \Phi(\mathbf{f}), \tag{4.2}$$

where the cost function $\Phi(\mathbf{f})$ captures the relevant elements of the model of the observed and the desired image. A common form of the cost function is as follows:

$$\Phi(\mathbf{f}) = \|\mathbf{g} - \mathbf{H}\,\mathbf{f}\|_2^2 + \lambda\, C(\mathbf{f}). \tag{4.3}$$

The first term in $\Phi(\mathbf{f})$ ensures fidelity of the estimate to the data $\mathbf{g}$. The second term is a *regularizing* or *stabilizing constraint* $C(\mathbf{f})$. A regularization parameter λ controls the trade-off between the two terms. The regularization functional stabilizes the inverse problem of estimating $\mathbf{f}$ from $\mathbf{g}$, which otherwise may be ill-posed or ill-conditioned due to the properties of $\mathbf{H}$. In practice, one of the main goals of using a regularizing constraint is to avoid noise amplification during deblurring. In fact, the regularization functional can be used for noise removal, outside of the deblurring context. Furthermore, $C(\mathbf{f})$ may be formulated to exploit prior knowledge of the underlying image $\mathbf{f}$. In this sense, such a regularization functional implicitly embodies an image model, and the choice of this model will be reflected in the quality of the reconstructed image. Over many years, several types of such prior image models have been developed in the image processing community, resulting in improved reconstruction results in applications such as denoising, deblurring, inpainting, super resolution, and others.

A common regularization constraint takes the form $C(\mathbf{f}) = \|\mathbf{L}\mathbf{f}\|_2^2$, combining a linear high-pass filter $\mathbf{L}$ with the ℓ_2 norm. This is known as the Tikhonov-Miller regularization approach. This approach enforces smoothness of the solution and suppresses noise by penalizing high-frequency components. More recently, regularization based on the Total Variation (TV) norm has been introduced. Total Variation minimization was originally introduced for noise reduction [18, 19] and has also been used for image deblurring [20] and super resolution image reconstruction [21]. The TV constraint is computed as the ℓ_1 norm of the gradient magnitude: $C(\mathbf{f}) = \|\nabla \mathbf{f}\|_1$. TV minimization is better able to preserve sharp edges and fine detail in the image. However, both approaches have the potential for oversmoothing and re-introducing some blur, as both models penalize local differences between pixel values.

The NLM filter emphasized two innovations over earlier approaches. The first idea is the use of image *patches* to estimate the structural relations between the data, as opposed to individual pixel value differences. Specifically, a patch-based distance metric is used in NLM, and is described in the next section. This has influenced the development of patch-based image models (extending pixel-based models) and patch-based restoration methods. The second idea is often referred to as *self-similarity*; that is, the notion that for any small patch of pixels, one can often find several patches at other locations in the data that have very similar structure. Hence, pixel data from all over an image may be beneficial in predicting pixels of interest, leading to nonlocal image processing. In recent work, these nonlocal means principles have been used in a cost function minimization framework [4, 5, 11, 12, 22, 23]. In [4], NLM weights are included in regularization functionals for image denoising and segmentation. A general framework that includes local and nonlocal regularization as well as the NLM filter as special cases is proposed in [22]. NLM-based super resolution is proposed in [11]. In [12], NLM-based regularization is used for image deblurring. An NLM-based regularization constraint has the potential to suppress noise, while preserving edges and fine texture detail, more effectively than conventional regularization constraints.

In related work, image models and regularization methods have been developed based on sparse and redundant representations [24]. These methods are based on the notion of *sparsity* of image decompositions, for example, in the wavelet domain or in terms of a redundant dictionary of signal atoms — typically image patches. In recent work, such patch dictionaries may be *learned* from the image data, thus promising improved performance over standard decompositions [7, 25].

The remainder of this chapter is organized as follows. In Section 4.2 we describe the NLM filter and its application to image denoising in more detail. In Section 4.3 we briefly review existing iterative deblurring methods and describe the use of nonlocal means regularization for image deblurring and denoising. In Section 4.4 we review related recent patch-based methods based on the nonlocal approach or sparse modeling approach, as well as their application to denoising and single-image super resolution reconstruction. In Section 4.5 we review relevant work to reduce the high computational cost of the nonlocal means approach. We conclude in Section 4.6 with a brief discussion of applications of the nonlocal means approach beyond 2-D image processing.

4.2   Image Denoising Based on the Nonlocal Means

In this section, we first describe the NLM filter itself and provide some visual examples of its denoising performance. Then we describe several iterative denoising methods based on the NLM approach and show how these methods fit within the cost function minimization framework mentioned in the previous section. Such iterative methods extend the basic NLM filter and have shown improved denoising performance.

4.2.1   NLM Filter

The discrete NLM filter [1, 2] is a form of weighted averaging over a set of pixels. The family of weighted averaging filters also includes the basic linear Gaussian filter, the Sigma filter [26], the SUSAN filter [27], and the bilateral filter [28]. The weighted averaging operation can be defined as follows. Let $g_i \equiv g(\mathbf{x}_i)$ be the value of the ith pixel. The estimate of the denoised pixel value $\hat{f}_i \equiv \hat{f}(\mathbf{x}_i)$ is given by

$$\hat{f}_i = \frac{\sum_{j \in N_i} w_{i,j}\, g_j}{\sum_{j \in N_i} w_{i,j}}, \tag{4.4}$$

where N_i is the set of pixels involved in the averaging process and $w_{i,j}$ is the weight for the jth pixel in N_i. Gaussian linear filter weights are defined using a spatial kernel $G_s(\mathbf{x}) = exp(-\|\mathbf{x}\|^2/2s^2)$:

$$w_{i,j} = G_s(\mathbf{x}_i - \mathbf{x}_j), \tag{4.5}$$

where s controls the spatial weighting. The bilateral filter uses a Gaussian kernel applied to the pixel values $G_r(g)$ in addition to the spatial Gaussian kernel:

$$w_{i,j} = G_s(\mathbf{x}_i - \mathbf{x}_j)G_r(g_i - g_j). \tag{4.6}$$

The so-called *range kernel* restricts the averaging process to pixels with value g_j similar to g_i; this data-adaptive weighting (range weighting) is controlled by r. The important result is that the bilateral filter is able to preserve image edges.

In both cases, N_i contains pixels in a small local spatial window around $\mathbf{x}_i$. The NLM filter, on the other hand, allows N_i to include pixels from a very large window or even the entire image. Furthermore, in the NLM filter, the weights are based on similarity between image *patches*, rather than individual pixel values. Let $\mathbf{g}_i$ denote a vector of pixel values in a small (e.g., 7×7 or 9×9) image patch centered on $\mathbf{x}_i$. The NLM weights are defined based on the (weighted) Euclidean distance between image patches:

$$w_{i,j} = G_h(\|\mathbf{g}_i - \mathbf{g}_j\|_{2,a}) = exp\{-\frac{1}{h^2}\|\mathbf{g}_i - \mathbf{g}_j\|_{2,a}^2\}, \tag{4.7}$$

where h controls the denoising strength. The larger the value of h, the higher the strength of denoising. For example, Buades et al. [2] suggest values between $10 \times \sigma$ and $15 \times \sigma$, where σ is the standard deviation of the noise in the image data. The (Gaussian weighted) patch distance is given by

$$\|\mathbf{g}_i - \mathbf{g}_j\|_{2,a}^2 = \frac{1}{\sum_t G_a(\mathbf{t})} \sum_t G_a(\mathbf{t})\, [g(\mathbf{x}_i + \mathbf{t}) - g(\mathbf{x}_j + \mathbf{t})]^2, \tag{4.8}$$

where t indicates the offset of a pixel in an image patch from the center pixel and a is a weighting parameter. The patch distance in Equation (4.8) is sometimes defined without weights $G_a(\cdot)$.

The use of image patches to compute the filter weights makes the NLM filter more robust to noise, compared to the bilateral filter in Equation (4.6). Furthermore, the filter is expected to better preserve important image structures, including edges and texture details.

It is also important to note that the definition of NLM weights in Equation (4.7) does not include a spatial weighting kernel, such as included in Equations (4.5) and (4.6). The set of pixels N_i involved in denoising the ith pixel is nonlocal and can potentially include samples from all over the image. Earlier work in the area of texture synthesis by Efros and Leung [29] exploited predictions from anywhere in an image in a similar manner. In practice, a large "search window" may be used to limit computational cost; for example, Buades et al. [1] suggest a window of 21×21 pixels (centered on x_i).

In Figure 4.1 we illustrate how the NLM weights act to implicitly identify pixels with

(a)

(b)

(c)

(d)

Figure 4.1 Comparison of weight distribution of different methods for image *Houston*: (a) original image; (b) enlarged portion corresponding to the dotted rectangle in (a); where small solid rectangles show the center patch and examples of similar patches; (c) weight distribution corresponding to the bilateral filter; (d) weight distribution corresponding to the NLM filter.

(a) Original 512×512 image and two enlarged 128×128 portions

(b) Image corrupted by additive white Gaussian noise, PSNR = 24.6 dB

Figure 4.2 Original and degraded image *Barbara*, with two enlarged portions each containing both texture-less regions and highly textured regions.

similar surrounding image patches. Figure 4.1(a) is a Houston skyline image containing a large number of repeated patterns. Figure 4.1(b) shows an enlarged portion, where the center contains the reference pixel and reference patch, based on which the weights are calculated. Figure 4.1(c) illustrates the distribution of weights across the image calculated by the bilateral filter without spatial penalty; that is, we show only $G_r(g_i - g_j)$ as in Equation (4.6), for fair comparison. The distribution of NLM weights across the image are shown in Figure 4.1(d). It can be seen clearly that the NLM identifies fewer but highly reliable pixels that are similar to the central pixel in terms of both intensity and local structure, due to the use of image patches.

In Figures 4.2 and 4.3, we compare the denoised image quality for three weighted-averaging filter, techniques: the Gaussian filter, bilateral filter and NLM filter. Figure 4.2(a) shows the original noise-free grayscale image *Barbara* and two enlarged portions for detailed viewing. The image was corrupted with additive white Gaussian noise with zero mean and standard deviation 15, resulting in the noisy image shown in Figure 4.2(b), with PSNR of 24.6 dB. The results of applying the above three filters are shown Figure 4.3(a), (b), and (c). The parameters were optimized individually for each method for highest PSNR. It can be seen that the Gaussian filter removes noise but at the same time blurs the textures. The bilateral filter preserves edges and some textures, but relatively more noise remains. The NLM filter provides the best quality among these three, achieving a good balance between

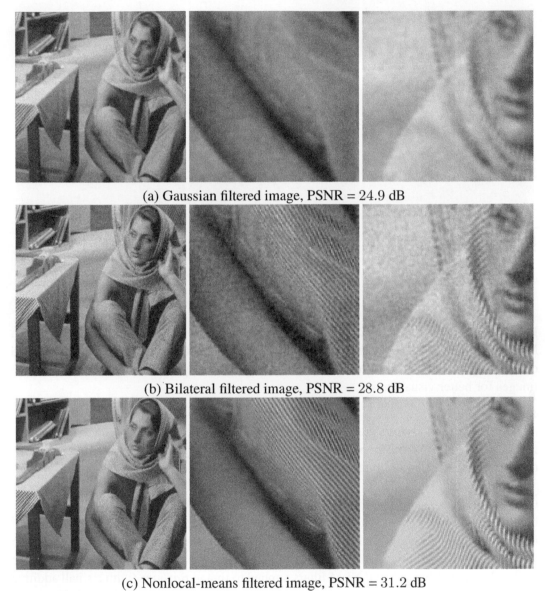

(a) Gaussian filtered image, PSNR = 24.9 dB

(b) Bilateral filtered image, PSNR = 28.8 dB

(c) Nonlocal-means filtered image, PSNR = 31.2 dB

Figure 4.3 Comparison of denoising performance of Gaussian, bilateral, and NLM filters for image *Barbara* shown in Figure 4.2.

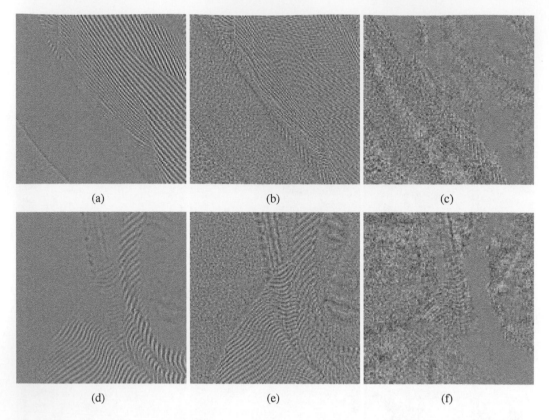

(a)   (b)   (c)

(d)   (e)   (f)

Figure 4.4 Comparison of method noise for two rectangular areas in image *Barbara*: (a)(d) Method noise of Gaussian filter; (b)(e) method noise of bilateral filter; (c)(f) method noise of NLM filter. Pixel noise values were multiplied by a factor of 10 and offset by 128 in all images for better visualization.

noise removal and texture preservation. The NLM denoised image also has the highest PSNR, showing a 5.6-dB improvement over the noisy image and a 2.4-dB improvement over the bilateral denoised image.

Buades et al. suggested to visualize the "method noise" as an intuitive way to further understand the properties of the NLM filter [2]. Method noise is defined as the difference between a noise-free (or slightly noisy) image and its filtered output. An ideal filter should not remove any of the original image content and produce a method noise that is close to white noise, that is, structureless and content independent. To illustrate, the method noise of the Gaussian, bilateral, and NLM filters applied to image *Barbara* (with a small additive noise of standard deviation 2.5) is shown in Figure 4.4. The parameters for each filter were chosen so that the resulting method noise has approximately a standard deviation of 2.5. We present enlarged portions similar to Figure 4.3 for better visualization of the details. It can be seen that the method noise for both the Gaussian filter and bilateral filter contain image edges and corners. The method noise for the NLM filter contains the least amount of image structure among the three. One can also see that the NLM filter removes less noise in some textured areas, where few similar patches can be found.

4.2.2 Iterative NLM Denoising

The basic NLM filter can be extended by applying it in an iterative manner [4, 5, 22, 23]. Such iterative algorithms can in many cases be seen to minimize a cost function. As an example, we can derive an iterative approach based on the following cost function:

$$C(\mathbf{f}) = \frac{1}{2} \sum_i \frac{1}{\sum_{j \in N_i} w_{i,j}} \sum_{j \in N_i} w_{i,j} (f_i - g_j)^2. \tag{4.9}$$

In this equation, the weights $w_{i,j}$ are NLM weights computed from the (noisy) data $\mathbf{g}$, as in Equation (4.7). To explicitly indicate the data being used when computing the weights, we use the notation $w_{i,j} = w_{i,j}^{\mathbf{g}}$. Hence, the above cost function penalizes weighted differences between pixel value f_i in the denoised image and pixels g_j in the input image for which the corresponding patch $\mathbf{g}_j$ is similar to patch $\mathbf{g}_i$. The weights themselves are not subject to optimization in this case, as they are not dependent on $\mathbf{f}$.

The above cost function can be minimized in an iterative manner, for example using a gradient descent approach. The estimate of pixel value f_i at the kth iteration is denoted by $\hat{f}_i^k$, while the estimate for the entire image is denoted by $\hat{\mathbf{f}}^k$. The initial estimate can be chosen to be the noisy input image $\hat{\mathbf{f}}^0 = \mathbf{g}$. At each iteration, the estimate is updated in the opposite direction of the gradient of the cost function:

$$\hat{f}_i^{k+1} = \hat{f}_i^k - \beta \left. \frac{\partial C(\mathbf{f})}{\partial f_i} \right|_{\hat{f}_i^k} \quad \text{for all } i,$$

where β is a step-size parameter. In this basic example, this results in

$$\hat{f}_i^{k+1} = \hat{f}_i^k - \beta \left[\frac{1}{\sum_j w_{i,j}} \sum_{j \in N_i} w_{i,j} (\hat{f}_i^k - g_j) \right] = \hat{f}_i^k - \beta \left[\hat{f}_i^k - \frac{1}{\sum_j w_{i,j}} \sum_{j \in N_i} w_{i,j} \, g_j \right]. \tag{4.10}$$

The expression on the right clearly shows the application of the NLM filter to $\mathbf{g}$ and the update of $\mathbf{f}$ by moving it a step in that direction.

In the above trivial example, the NLM filter provides the direct solution to minimizing Equation (4.9), and the iterative algorithm actually does not provide any gain in performance. We now turn to discuss more useful iterative approaches, as proposed by Gilboa and Osher [4] and Elmoataz et al. [22]. A cost function that fits in the frameworks of References. [4, 22] is as follows:

$$C(\mathbf{f}) = \frac{1}{4} \sum_i \sum_{j \in N_i} w_{i,j} (f_i - f_j)^2. \tag{4.11}$$

This cost function penalizes differences between pixel values f_i and f_j, *both* in the denoised image, for which the corresponding patch $\mathbf{g}_j$ is similar to patch $\mathbf{g}_i$. This is in contrast to Equation (4.9). Minimization based on a gradient descent approach leads to the following type of iteration:

$$\hat{f}_i^{k+1} = \hat{f}_i^k - \beta \left[\sum_{j \in N_i} w_{i,j} (\hat{f}_i^k - \hat{f}_j^k) \right]. \tag{4.12}$$

Again, weights are given by $w_{i,j} = w_{i,j}^{\mathbf{g}}$ as in (4.7). Note that the weight values stay constant for every iteration. This iterative smoothing of $\hat{\mathbf{f}}$ is a type of weighted diffusion, and eventually will lead to smearing out all image details. To prevent this, the number of iterations needs to be limited. Another solution is to add a data fidelity term to the cost function:

$$\Phi(\mathbf{f}) = \frac{\lambda}{2} \sum_i (f_i - g_i)^2 + \frac{1}{4} \sum_i \sum_{j \in N_i} w_{i,j}(f_i - f_j)^2. \qquad (4.13)$$

This can again be minimized with a gradient descent approach. Gilboa et al. [4] proposed algorithms based on (continuous domain) functionals similar to Equation (4.11) and (4.13), and reported improved denoising performance over the standard NLM filter. A very general framework that includes the above examples as special cases is proposed by Elmoataz et al. [22]. This framework also includes the bilateral filter and a discrete form of Total Variation regularization as special cases, and extends further to nonquadratic norms. They propose to use a different iterative method for minimization and apply the resulting smoothing algorithms to various types of data, including color images, (2-D) polygonal curves, and (3-D) surface mesh data.

A different type of extension is proposed by Brox et al. [5] and Peyré et al. [23]. Both propose to define the weights $w_{i,j}$ based on the image to be reconstructed $\mathbf{f}$ (instead of the observed image $\mathbf{g}$):

$$w_{i,j} = w_{i,j}^{\mathbf{f}} = G_h(\|\mathbf{f}_i - \mathbf{f}_j\|_{2,a}). \qquad (4.14)$$

A cost function defined on this basis is:

$$C(\mathbf{f}) = \sum_i \left[f_i - \frac{1}{\sum_{j \in N_i} w_{i,j}^{\mathbf{f}}} \sum_{j \in N_i} w_{i,j}^{\mathbf{f}} g_j \right]^2. \qquad (4.15)$$

In contrast to Equation (4.9), this penalizes differences between f_i and g_j at locations where the corresponding patch $\mathbf{f}_j$ is similar to patch $\mathbf{f}_i$. The idea is that the weights should be adapted based on the ideal image $\mathbf{f}$ rather than the degraded image $\mathbf{g}$.

Note that the cost function in Equation (4.15) is not quadratic anymore, as it now depends in a nonlinear manner on $\mathbf{f}$ through Equation (4.14). An approximate solution is proposed in [5] using a fixed-point iteration scheme. In this solution, weights $w_{i,j}$ are computed using the previous solution at iteration k, that is, $w_{i,j} = w_{i,j}^{\hat{\mathbf{f}}^k}$. Then, an update on $\hat{\mathbf{f}}$ is obtained in a manner similar to Equation (4.10), keeping $w_{i,j}$ fixed:

$$\hat{f}_i^{k+1} = \hat{f}_i^k - \beta \left[\hat{f}_i^k - \frac{1}{\sum_j w_{i,j}^{\hat{\mathbf{f}}^k}} \sum_j w_{i,j}^{\hat{\mathbf{f}}^k} g_j \right]. \qquad (4.16)$$

Because the denoising weights $w_{i,j}^{\mathbf{f}^k}$ are now updated on the basis of $\hat{\mathbf{f}}^k$, the weights are continuously improved as the iterative scheme progresses. At the same time, the noisy input image $\mathbf{g}$ is still used as the data for the actual denoising step in each iteration. Denoising results reported in [5] with this algorithm slightly outperformed the algorithms proposed in [4].

In a similar iterative algorithm proposed by Kervrann and Boulanger [3], weights are also computed on a (partially) denoised image. The basic method is further extended by adapting the size of the search window N_i and a criterion to stop the iterative process. As reported by Brox et al., this algorithm outperforms the basic NLM filter as well as the method in [5].

4.3 Image Deblurring Using Nonlocal Means Regularization

In this section we consider deblurring and denoising; hence, we return to the full observation model of Equation (4.1), including the blur operator $h(\mathbf{x})$. We consider cost functions such as Equation (4.3), reviewing classical Total Variation regularization in Subsection 4.3.1, and then describe NLM regularization in Subsection 4.3.2.

4.3.1 Iterative Deblurring

The so-called unconstrained least-squares (LS) estimate of $\mathbf{f}$ is obtained when the regularization constraint $C(\mathbf{f})$ is omitted. In this case, iterative minimization based on the gradient descent approach is as follows:

$$\hat{\mathbf{f}}^{k+1} = \hat{\mathbf{f}}^k - \beta \left[\mathbf{H}^T (\mathbf{H}\, \hat{\mathbf{f}}^k - \mathbf{g}) \right], \tag{4.17}$$

where β is the step size and $\hat{\mathbf{f}}^0 = \mathbf{g}$. We can write this in more explicit form as

$$\hat{f}_i^{k+1} = \hat{f}_i^k - \beta \left[\sum_{m \in \Omega} h(\mathbf{x}_m - \mathbf{x}_i)(\sum_{n \in \Omega} h(\mathbf{x}_m - \mathbf{x}_n)\hat{f}^k(\mathbf{x}_n) - g(\mathbf{x}_m)) \right]. \tag{4.18}$$

In the basic minimization scheme, β is held constant. At the cost of additional computation, an optimal value of β can be computed at each iteration. The method of conjugate gradients and other, more advanced minimization methods can be used alternatively [16].

Using Total Variation regularization is now a classical and popular technique to stabilize the estimation of $\mathbf{f}$ and suppress noise during deblurring. A discrete version of the Total Variation regularization constraint $C(\mathbf{f}) = \|\nabla \mathbf{f}\|_1$ is defined in [19] by

$$C(\mathbf{f}) = \sum_i \gamma_i(\mathbf{f}), \tag{4.19}$$

where $\gamma_i(\mathbf{f})$ is the *local variation* of $\mathbf{f}$ at pixel i. The local variation at $\mathbf{x}_i$ is defined as

$$\gamma_i(\mathbf{f}) \equiv \sqrt{\sum_{m \in S_i} [f(\mathbf{x}_i) - f(\mathbf{x}_m)]^2 + \epsilon}, \tag{4.20}$$

with $S_i = \{(x_i + 1, y_i), (x_i, y_i + 1), (x_i - 1, y_i), (x_i, y_i - 1)\}$ a local neighborhood around $\mathbf{x}_i = (x_i, y_i)$ and ϵ a small positive constant. An iterative scheme for minimizing the resulting cost function can be defined as follows:

$$\hat{f}_i^{k+1} = \hat{f}_i^k - \beta \left[\sum_{m \in \Omega} h(\mathbf{x}_m - \mathbf{x}_i) (\sum_{n \in \Omega} h(\mathbf{x}_m - \mathbf{x}_n) \hat{f}^k(\mathbf{x}_n) - g(\mathbf{x}_m)) \right]$$

$$- 0.5 \, \beta \, \lambda \, \left[\sum_{j \in S_i} \omega_{i,j} \, (\hat{f}^k(\mathbf{x}_i) - \hat{f}^k(\mathbf{x}_j)) \right], \tag{4.21}$$

where

$$\omega_{i,j} = \frac{1}{\gamma_i(\hat{\mathbf{f}}^k)} + \frac{1}{\gamma_j(\hat{\mathbf{f}}^k)}. \tag{4.22}$$

The $\omega_{i,j}$ values follow directly from the discrete total variation framework, but could also be seen as data-adaptive weights for the local neighbors in S. The above iteration is a so-called fixed-point iteration scheme, where computation of the TV term lags one iteration behind; that is, the weights $\omega_{i,j}$ are computed on the basis of the previous estimate $\hat{\mathbf{f}}^k$ and are updated after each iteration. Again, other iterative minimization schemes, with faster convergence, can be utilized [21].

4.3.2   Iterative Deblurring with Nonlocal Means Regularization

In this section we present regularization constraints based on the NLM principle and describe a deblurring and denoising scheme using NLM-based regularization. We start by formulating a first regularization constraint as follows:

$$C(\mathbf{f}) = \sum_i \frac{1}{\sum_{j \in N_i} w_{i,j}^{\mathbf{f}}} \sum_{j \in N_i} w_{i,j}^{\mathbf{f}} (f_i - f_j)^2. \tag{4.23}$$

This function penalizes differences between pixel values of f_i and its neighbors f_j in the reconstructed image. Furthermore, these differences are weighted by nonlocal weights $w_{i,j}^{\mathbf{f}}$ computed on the reconstructed image as well, as in Equation (4.14). This cost function is different from Equation (4.11), where the weights are computed on the noisy image $\mathbf{g}$, and different from Equation (4.15), which penalizes differences between pixel values f_i and neighbors g_j in the noisy image. In Equation (4.15), this penalization provided a coupling with the observation data, to prevent smoothing out all details of the image. In the formulation in this section, coupling with the observation data is enforced by a data fidelity term. A similar cost function was proposed by Peyré et al. [23], based on a model of the image as a weighted graph (as in [19] and [22]). A slightly different cost function was considered in [12]:

$$C(\mathbf{f}) = \sum_i \left[f_i - \frac{1}{\sum_{j \in N_i} w_{i,j}^{\mathbf{f}}} \sum_{j \in N_i} w_{i,j}^{\mathbf{f}} \, f_j \right]^2. \tag{4.24}$$

This penalizes differences between f_i and its NLM-denoised version, with weights defined using the reconstructed image $\mathbf{f}$.

In both cases, the regularization function depends on the image to be reconstructed in a nonlinear manner (due to the weights). Hence, the full cost function $\Phi(\mathbf{f}) = \|\mathbf{g} - \mathbf{H}\mathbf{f}\|^2 + \lambda\, C(\mathbf{f})$ is difficult to minimize exactly. An approximate gradient descent approach is proposed in [12], using the following iteration:

$$\hat{f}_i^{k+1} = \hat{f}_i^k - \beta \left[\sum_{m \in \Omega} h(\mathbf{x}_m - \mathbf{x}_i) \Big(\sum_{n \in \Omega} h(\mathbf{x}_m - \mathbf{x}_n) \hat{f}^k(\mathbf{x}_n) - g(\mathbf{x}_m) \Big) \right]$$

$$- \beta\,\lambda \left[\hat{f}_i^k - \frac{1}{\sum_{j \in N_i} w_{i,j}^{\hat{\mathbf{f}}^k}} \sum_{j \in N_i} w_{i,j}^{\hat{\mathbf{f}}^k} \hat{f}_j^k \right]. \tag{4.25}$$

Here, the denoising weights are computed on the (partially) denoised and deblurred image at the previous iteration $\hat{\mathbf{f}}^k$. Hence, the weights are continuously improved as iterative deblurring progresses. Also, $\hat{\mathbf{f}}^k$ is now used as the data for the denoising step in each iteration, unlike Equation (4.16). This is because we must compute an estimate of the denoised *and* deblurred image. The data fidelity term of the cost function and proposed iterative algorithm will ensure consistency with the input image data $\mathbf{g}$, as mentioned above. Derivation of this iterative scheme from Equation (4.24) requires an approximation in which specific terms are neglected. As reported in [12], this scheme outperforms Total Variation regularized deblurring. A different optimization algorithm is proposed in [23] to solve Equation (4.23). In that work, the method is applied to several other inverse problems, such as inpainting, super resolution reconstruction, and compressive sampling.

Example results of some of the above deblurring algorithms are shown in Figure 4.5. In this example, the original "Lena" image is degraded by Gaussian blur with $\sigma = 1.5$ and white Gaussian noise with std. dev. = 15. Subsequently, we applied deblurring using several different methods. We included a result of unconstrained least-squares deblurring (Equation (4.17), 10 iterations) in Figure 4.5 (c). We also included a result of a two-stage method, using the NLM filter ($h = 150$, patch size 11×11) followed by unconstrained least-squares deblurring (10 iterations) in Figure 4.5(d). Finally, we included the result of TV-regularized deblurring (Equation (4.21), $\lambda = 3.0$, 50 iterations) in (e), and that of NLM-regularized deblurring (Equation (4.25), $\lambda = 0.5$, $h = 120$, patch size 9×9, 50 iterations) in (f).

4.4   Recent Nonlocal and Sparse Modeling Methods

In this section we look at related patch-based methods that have been developed more recently and have been reported to outperform the basic NLM filter, at least for denoising. These methods have been applied to several other image restoration problems, such as demosaicking, inpainting, and super resolution reconstruction, and currently are part of the state-of-the-art in several areas. Several of these methods are based on sparse and redundant representations, reviewed in [24].

One of the state-of-the-art methods, proposed by Dabov et al. [6, 30], is called BM3D. This approach is patch based and relies on self-similarities within an image — similar to the NLM filter. Similar image patches are found in the image using the block-matching (BM) technique. Unlike NLM filtering, similar patches are grouped together and organized

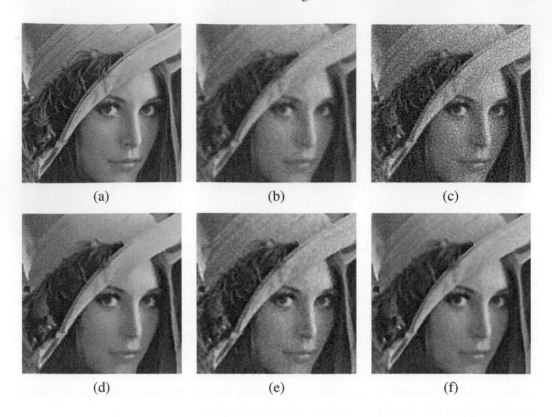

(a)	(b)	(c)
(d)	(e)	(f)

Figure 4.5 Example image deblurring results (each image shown is a 256×256 crop out of a 512×512 image). (a) original "Lena" image; (b) image degraded by Gaussian blur with $\sigma = 1.5$ and white Gaussian noise with std. dev. = 15 (PSNR = 23.27 dB); (c) unconstrained LS deblurring (PSNR = 21.88 dB); (d) NLM filter followed by unconstrained LS deblurring (PSNR = 28.96 dB); (e) TV-regularized deblurring (PSNR = 27.58 dB); (f) NLM-regularized deblurring (PSNR = 29.86 dB).

into a 3-D signal block by stacking them. Denoising can be achieved by applying a 3-D orthogonal transform on each block followed by hard thresholding or Wiener filtering. Final results are obtained by applying the inverse transform and aggregating overlapping block estimates. In BM3D, this procedure is applied two times, where the second stage benefits from improved estimates from the first stage. The grouping and 3-D transform operations result in a sparse representation of the data, which can be denoised effectively while retaining relevant structures [6]. BM3D provides some of the best reported PSNR results for image denoising, and is still considered state-of-the-art. The authors extended the approach to image restoration [30], by regularized inversion of the p.s.f. in the Fourier transform domain, followed by an extended BM3D filter to suppress the remaining, colored, noise.

Chatterjee and Milanfar [8] recently proposed a patch-based locally optimal Wiener filter (PLOW) based on an earlier study of denoising performance bounds [31]. PLOW is based on a patch-based linear minimum mean square error (LMMSE) estimator, using clustering of geometrically similar patches to estimate means and covariances. The filter also includes patch weighting based on photometric similarity, akin to NLM weighting. The visual and PSNR performance was found to be comparable to that of BM3D.

Elad and Aharon [25] proposed a patch-based method relying on explicit modeling of image data as allowing a sparse representation over a redundant dictionary and using this model as a type of regularization constraint or prior for image denoising. Furthermore, a method called K-SVD is proposed to *learn* the dictionary from either the corrupted image itself or from a set of high-quality images, resulting in improved denoising results.

We briefly describe the sparse representation model in the following. The model relies on a dictionary of image patches, represented by an $M \times N$ matrix $\mathbf{D}$, where M is the number of pixels in an image patch and $N > M$. The basic assumption is that an image patch $\mathbf{f}_i$ can be represented as a linear combination of a few of the N columns (called atoms) of $\mathbf{D}$. That is, $\mathbf{f}_i \approx \mathbf{D}\alpha_i$, where α_i is a *sparse* vector (the number of its nonzero elements $\ll N$). Denoising an image patch $\mathbf{g}_i$ corrupted by additive white noise can be achieved by solving

$$\hat{\alpha}_i = \arg\min_{\alpha_i} \|\alpha_i\|_0 \text{ subject to } \|\mathbf{g}_i - \mathbf{D}\alpha_i\|_2^2 \leq T, \tag{4.26}$$

where $\|\alpha\|_0$ denotes the number of nonzero elements of the vector α. The denoised image patch is given by $\hat{\mathbf{f}}_i = \mathbf{D}\hat{\alpha}_i$. In some approaches, the ℓ_0 pseudo-norm is replaced by the ℓ_1 norm, which is sparsity inducing as well. Denoising of the entire image can proceed in a patch-by-patch manner [25]. As patches are typically designed to overlap, each pixel is associated with multiple estimated values; these estimates can be averaged to arrive at the final result. This basic denoising process can be iterated, each pass utilizing an improved estimate of the image $\mathbf{f}$, updated during the previous iteration.

The dictionary itself may be based on orthogonal transforms, for example, by using the basis vectors of the 2-D Discrete Cosine Transform or Discrete Wavelet Transform, or based on overcomplete (redundant) representations. The K-SVD algorithm [25] extends the basic algorithm by iteratively updating a dictionary based on the given data, thus training the dictionary as denoising progresses. The sparse modeling approach and K-SVD algorithm were extended further in [32] by using *multi-scale* learned dictionaries.

Mairal et al. [7] proposed to combine the nonlocal means and sparse modeling approaches by enforcing similar patches to be decomposed in similar manner, in terms of the dictionary elements. In other words, similar patches are encouraged to share the same sparse coefficient vector α_i. Patch similarity is determined based on the familiar patch distance measure $\|\mathbf{g}_i - \mathbf{g}_j\|_2^2$, as in NLM and BM3D. This method, which also utilized dictionary learning, was termed "learned simultaneous sparse coding" (LSSC). The resulting algorithm was applied to denoising as well as image demosaicking and compared to several state-of-the-art algorithms for both tasks. For suppressing additive white Gaussian noise with standard deviation ranging from 5 to 100, LSSC slightly outperformed BM3D [6] in terms of average PSNR over a set of standard images.

Based on a related idea, Dong et al. [9] proposed a method called clustering-based sparse representation (CSR). The method extends the basic sparse representation approach by clustering the sparse coefficient vectors α_i (using k-means clustering). Using a global cost function, an additional regularization term is defined that penalizes differences be-

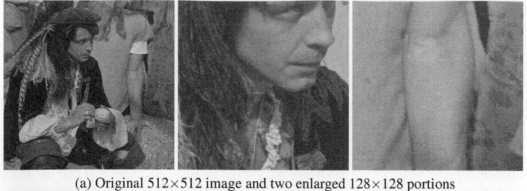

(a) Original 512×512 image and two enlarged 128×128 portions

(b) Image corrupted by additive white Gaussian noise, PSNR = 24.63 dB

Figure 4.6 Original and degraded input image *Man*, with two cropped and enlarged portions, containing a highly-textured region as well as a mostly flat region with some weak textures.

tween coefficient vectors and the centroid β_k of the kth cluster C_k:

$$\sum_{k=1}^{K} \sum_{i \in C_k} \|\alpha_i - \beta_k\|_1.$$

The resulting iterative algorithm achieved very good denoising performance, with PSNR that slightly outperformed BM3D on average and was very competitive with BM3D on individual images.

In Figures 4.6, 4.7, and 4.8, we illustrate the denoising performance of several state-of-the-art techniques and compare them to the NLM filter. Specifically, we generated example results for K-SVD [25], BM3D [6], CSR [9], and LSSC [7] using implementations provided by their authors with default parameter settings. We used our own implementation of the NLM filter and optimized the parameters for highest PSNR. Figure 4.6(a) shows the original gray-scale image *Man* and two enlarged regions for detailed viewing. Zero-mean white Gaussian noise with standard deviation of 15 was added to the original image, producing the noisy image shown in Figure 4.6(b), with a PSNR of 24.63 dB. Figures 4.7 and 4.8 show the results of applying NLM, K-SVD, BM3D, CSR, and LSSC to the noisy image, respectively. We can see from the images that all five techniques removed a good

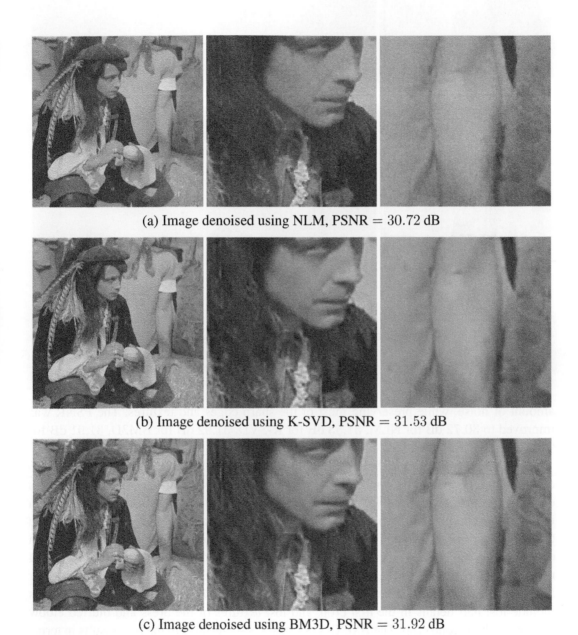

(a) Image denoised using NLM, PSNR = 30.72 dB

(b) Image denoised using K-SVD, PSNR = 31.53 dB

(c) Image denoised using BM3D, PSNR = 31.92 dB

Figure 4.7 Results of NLM, K-SVD, and BM3D on the noisy image *Man* shown in Figure 4.6.

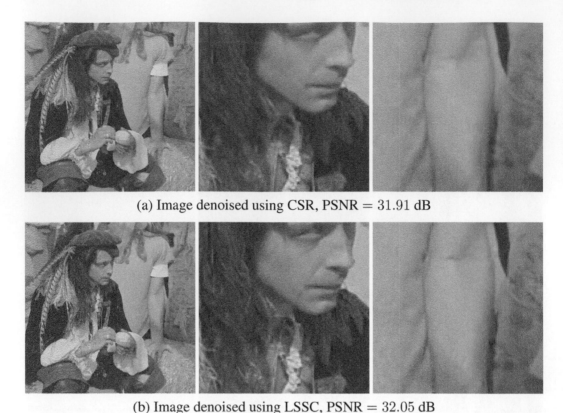

(a) Image denoised using CSR, PSNR $=$ 31.91 dB

(b) Image denoised using LSSC, PSNR $=$ 32.05 dB

Figure 4.8 Results of CSR and LSSC on the noisy image *Man* shown in Figure 4.6.

amount of noise, while preserving most of the sharp edges and textures. The PSNR was improved to 30.72 dB for NLM, 31.53 dB for K-SVD, 31.92 dB for BM3D, 31.91 dB for CSR, and 32.05 dB for LSSC.

While all the algorithms showed big improvements in terms of the PSNR (more than 6 dB), K-SVD, BM3D, CSR, and LSSC provided further PSNR increases over NLM in this example. As visually noticeable, the NLM results in general allowed more noise to remain in the image compared to the other three techniques (especially visible in flat regions). On the other hand, NLM avoided over-smoothing in regions containing weak textures. As an example, this can be seen when carefully examining the second cropped portion: the NLM result has slightly less blurring over the floral pattern in the dress. K-SVD improved over NLM, but resulted in relatively more artifacts than BM3D, CSR, and LSSC. Examples include lost hair details in the first crop, and lost textures of the dress and blurred elbow shadow in the second crop. The BM3D, CSR, and LSSC results are the best results in terms of PSNR and visual quality, illustrating the current state-of-the-art in the area of denoising.

The sparse representation model has also been applied to single-image super resolution reconstruction [33, 34]. In this case, a high-resolution image patch $\mathbf{f}_i$ is assumed to have a sparse representation in terms of a dictionary of high-resolution patches $\mathbf{D}_h$, that is. $\mathbf{f}_i \approx \mathbf{D}_h \alpha_i$. The observed low-resolution image patch $\mathbf{g}_i$ can be modeled as $\mathbf{g}_i = \mathbf{SH}\mathbf{f}_i \approx \mathbf{SHD}_h \alpha_i$, where $\mathbf{S}$ is a subsampling operator and $\mathbf{H}$ is the p.s.f. blurring operator. This means that the low-resolution patch can be sparsely represented using a dictionary of low-resolution image patches $\mathbf{D}_l = \mathbf{SHD}_h$. Hence, by solving a sparse decomposition problem similar to that of Equation (4.26) using the low-resolution image patches $\mathbf{g}_i$, it is possible to estimate the high-resolution data effectively using the calculated vector α_i. The method proposed in [33] relies on a dictionary of a large number of raw image patches randomly sampled from natural images. The resulting algorithm obtained state-of-the-art image super resolution results. The speed of this algorithm was significantly improved later by learning a compact set of dictionaries, instead of using raw patch pairs directly [34]. The dictionaries of low- and high-resolution patches are trained jointly, such that corresponding image patch pairs have the same sparse representation with respect to $\mathbf{D}_l$ and $\mathbf{D}_h$.

The earliest works utilizing dictionaries or databases of images for super resolution from a single image were by Freeman et al. [35, 36]. This approach is also known as *example-based super resolution* and generally involves learning correspondences between low- and high-resolution image patches from a database of low- and high-resolution image pairs. Example-based approaches are considered able to exceed the limits of *classical* super resolution methods. Classical super resolution methods [21, 37] often utilize multiple low-resolution images of the same scene that are first registered (at sub-pixel accuracy) and then combined into a high-resolution image. Example-based methods exploit high-resolution detail from the example images in the database to synthesize a plausible high-resolution image. Such methods (also called "hallucination" methods) have the potential to recreate fine textures that still match the low-resolution input. Recent work in this area [38] emphasizes reconstruction of texture content by searching texturally similar segments in the training image database.

A framework to combine classical and example-based super resolution has been proposed by Glasner, Bagon, and Irani [39]. Furthermore, this approach utilizes only the (single) low-resolution input image, without any dictionary or database. Similar to NLM, the approach exploits patch redundancies within the input image: patches in natural images tend to recur several times, both within the same resolution scale as well as across different scales. Similarity of patches across resolution scales within the input image provides examples of low-resolution and high-resolution patch pairs for example-based super resolution. The output images resulting from the proposed algorithm have very good visual quality. This approach was extended to space-time super resolution from a single video in [40]. This method includes temporal super resolution and is shown to resolve difficult temporal artifacts such as motion aliasing and motion blur.

4.5   Reducing Computational Cost of NLM-Based Methods

In this section we review several methods that have been proposed to reduce the high computational cost of NLM-based methods. We focus this discussion on the basic NLM denoising filter, although these ideas can be extended for application to other NLM-based algorithms, such as iterative denoising algorithms or nonlocal regularized deblurring algorithms.

Let's assume there are N pixels in the search window over which the NLM filter performs averaging, and that each patch $\mathbf{g}_i$ contains M pixels. A naive NLM implementation would result in a computational (time) complexity of $O(N \cdot M)$ for each pixel g_i that is to be denoised. This number of operations can be prohibitively high, especially considering how large N and M can be in order for the NLM filter to achieve high denoising performance. Clearly, there is a need to accelerate the algorithm in order to reduce the processing time. Several approaches have been reported in the literature and are briefly discussed here. Note that we use the per-pixel time complexity throughout this section as the measure to compare among different approaches.

One approach to reduce the computational complexity is to reduce the number of pixels involved in the calculation, hence reducing the number of patches on which the patch distance is evaluated. In [4], image patches are first binned into a two-dimensional table using their mean and standard deviation, and only patches in the neighboring bins are involved in the subsequent computation. Mahmoudi et al. [13] proposed to eliminate unrelated patches through a set of heuristic rules. Specifically, average gray values and gradient orientation information are utilized to quickly prescreen all image patches involved in the weighted averaging. Although a precise analysis is not available from the paper, the time complexity of the approach in [13] is claimed to be $O(c \cdot M)$, where c is a small constant relative to N. Note there is overhead involved in evaluating the set of conditions designed to reject a given patch. Some of the corresponding data can be precomputed and stored in look-up tables to reduce the extra complexity. A related approach in reducing complexity is to explore certain domain knowledge of the images on which the NLM filter is being applied. As an example, in restricted application domains, one may have a priori knowledge about periodic patterns in the input images [12]. In such cases, the subset of the pixels in the search window involved in the computation can be restricted significantly, resulting in a large speed-up.

Another acceleration approach is to organize image patches in a data structure that allows fast search. For example, an efficient implementation is proposed by Brox et al. [5] by arranging image patches in a cluster tree. The basic idea is to build up a binary decision tree of image patches, which enables the algorithm to quickly discard a large amount of dissimilar patches in NLM filtering. The cluster tree is built by recursively splitting a starting cluster into two smaller clusters, using the well-known k-means clustering algorithm (with $k = 2$). The cluster splitting process stops when the maximum distance of a member to the cluster center is smaller than a threshold, or when the number of patches in a cluster gets too small. The depth of the tree is therefore constrained as a consequence of the minimum cluster size. After the tree is built, one can quickly look-up the cluster containing a given patch, and subsequently this cluster is used to come up with the subset of patches involved in the NLM weighted averaging stage. The idea of being able to quickly narrow down the set of candidate patches to a small subset of "good" patches is similar to that in [13], which relies on a heuristic approach. In contrast, the scheme proposed in [5] makes use of the same distance measure for clustering and for filtering, which conceptually is a more consistent design. In terms of computational complexity, the approach based on clustering yields $O(log(N) \cdot M)$, which is mainly due to the cost of building up the tree.

In [14], a Monte Carlo kd-tree sampling algorithm is used to implement accelerated versions of the bilateral filter and the NLM filter. The general idea is to first reformulate the core problem in nonlinear filtering (such as the NLM filter) as nearest neighbor search in some high-dimensional space. A modified kd-tree is constructed in the high-dimensional space to facilitate fast queries, which is very similar to the binary tree approach in [5].

Due to the fact that all stages of the algorithm are data-parallel once the tree is built, a GPU implementation is described in [14] reporting $10\times$ speed-up over single-threaded CPU implementation. The time complexity in [14] is $O(log(N) \cdot M)$. Both approaches in [5, 14] require $O(M \cdot N)$ storage space for the tree data structure for an entire image.

Another acceleration approach is to compute the patch distance $\|\mathbf{g}_i - \mathbf{g}_j\|_2^2$ in a lower-dimensional subspace, as proposed by Tasdizen [15]. The subspace is determined by applying Principle Component Analysis (PCA) on a data covariance matrix $\mathbf{C_g}$. The process of PCA-based NLM can be summarized as follows: (1) a random subset of all image patches is used to compute a data covariance matrix; (2) analysis is performed on the eigenvalues of $\mathbf{C_g}$, with the goal of reducing the dimension of the subspace to a suitable value d; (3) all image patches are projected onto the lower-dimensional subspace and the resulting coefficients are stored; (4) NLM averaging is performed by computing the approximated patch distance using the projection coefficients. The time complexity of PCA-based NLM can be approximated as $O(N \cdot d + 0.1 \cdot M^2 + M \cdot d)$, considering the additional cost in computing the covariance matrix ($O(0.1 \cdot M^2)$) and in projecting the coefficients onto the subspace ($O(M \cdot d)$). Another advantage of this approach is that computing NLM weights in the subspace is less sensitive to noise and can lead to improved performance.

In [41], an early termination algorithm is used to speed up NLM filtering by terminating patch distance computation when the expected distance is too high. Specifically, a χ^2 distribution is used to model the partial distance of pixels in the remaining region given the partial distance of already summed pixels. When the probability is higher than a threshold, the current patch is rejected. The approach in [41] is similar in spirit to the approach in [15], as they both try to compute patch distance faster. No complexity analysis is given in [41].

Gilboa and Osher [4] note that applying an iterative NLM filter with a small search window effectively achieves interactions across a larger (nonlocal) region. This could achieve filtering over the larger region with a smaller computational cost than the direct noniterative approach, provided the search window in each iteration is small and the number of iterations is limited.

To summarize, quite a few approaches have been proposed to accelerate the NLM filter. A common design characteristic of these algorithms is to explore statistical redundancies of the data as relevant in the evaluation of the NLM filter. Note that, in this section, we have considered mainly the time complexity of the proposed algorithms. In real-world implementations, it is often the case that other dimensions of a given algorithm merit more attention. Additional considerations include space/memory complexity, code complexity in implementing a given algorithm, and different kinds of parallelism that can be exploited on modern processor architectures. Another cautionary note in choosing a reduced-complexity approach is image quality: reduced complexity may lead to quality degradation. Comparisons of the visual quality resulting from the various approaches are beyond the scope of this chapter.

4.6 Conclusions

In this chapter we have reviewed the nonlocal approach to image denoising and deblurring. We have described the NLM filter as well as closely related denoising methods directly derived from the NLM. We have also described the use of NLM-based regularization constraints, in both image denoising and deblurring. Most of these algorithms can be derived from a cost-function formulation including penalty or regularization terms with nonlocal weights. We have also reviewed several recent nonlocal and related sparse representation methods based on more sophisticated image models. Furthermore, we have reviewed techniques to reduce the computational cost of NLM-based methods.

Beyond 2-D image processing, the NLM approach has been applied to a wider variety of data. In part, this is due to the generality of the nonlocal means approach. Also, some of this work is due to the recognition that *graphs* can be used as a common representation for many types of data in signal processing. For example, the nonlocal regularization framework by Elmoataz et al. [22] is based on graphs, and applied to image data, polygonal curves, surface meshes, high-dimensional manifolds, and other data. Likewise, the NLM approach has been applied to geometric computer graphics data [14]. Furthermore, the approach was extended to 3-D volumetric data in [42] and [43].

For application to *video* sequences, a straightforward extension of the NLM image denoising filter is to expand the set of pixels involved in the averaging process N_i from a spatial window to a spatio-temporal window, spanning multiple video frames. Such a video denoising algorithm would consist of comparing a 2-D image patch $\mathbf{g}_i$ to other image patches in the same and other video frames, and average their respective center pixels based on their similarity. Other than the extension of N_i, the main steps of the process as defined by Equations (4.8), (4.7) and (4.4) would remain the same. This is, in fact, the approach proposed by Buades et al. [10].

While this is a natural approach in the context of the NLM, the resulting method perhaps is in opposition to the established notion that motion estimation is required for high-quality video denoising. Classical video denoising algorithms often incorporate explicit motion estimation and motion-compensated filtering. Further testing of the NLM-based idea is probably needed to understand whether such an approach remains effective under more difficult conditions, including local object motion, large frame-to-frame object displacements, significant occlusion, as well as other difficulties that have traditionally made motion estimation into a very difficult problem in general. Still, this somewhat unconventional approach and its potential has generated further interest in the video processing community. The reason is the ill-posed nature of motion estimation and the various difficulties encountered when attempting to use motion-compensated video processing in practical applications. Hence, other researchers have been motivated to explore the potential of NLM and related methods to avoid the explicit use of motion estimation or at least reduce its accuracy requirements [11, 44–46].

The NLM approach was extended to the video super resolution problem in the work by Protter et al. [11]. Again, for video super resolution, it had been assumed that explicit and accurate motion estimation is mandatory. However, in classical multiframe super resolution algorithms, it is difficult to handle general local object motion. While classical methods are often based on generative image models, the NLM-based method is instead based on implicit models captured by carefully constructed cost functions or a regularization term in a Bayesian estimation framework. The approach simply assumes that, to estimate a pixel in a high-resolution frame, several pixels in the low-resolution frames in a spatio-temporal neighborhood may be useful. Their respective contributions are determined by the similarity, of the patches around these pixels and the patch around the pixel to be estimated. That is, pixel weights are computed based on patch similarity as in the NLM. It can be noted that computing patch distances is similar to the well-known block-matching technique applied commonly in motion estimation. However, in motion estimation, it is common to select one final match as it is assumed to correspond to the true motion or simply because it provides the best predicting block. In the NLM-based algorithm, on the other hand, no specific patch is singled out; instead, it is recognized that several good matches may exist and can all be exploited. Hence, this type of approach is said to avoid explicit motion estimation. The results reported in [11] appear visually of high quality and artifact-free, while the quantitative PSNR gain seems mild.

As evident from the results reported above and in this chapter, significant progress has been made in image reconstruction in recent years. Several algorithms referenced here represent the current state-of-the-art and provide very high-quality results, especially in image denoising. This has prompted the authors in [31] to pose the question, "Is denoising dead?", That is, "Have we reached some limit of performance?" Through a study of performance bounds, their answer is that, for now, some room for improvement remains for a wide class of images. Likewise, in related areas such as deblurring and super resolution, it would appear that there is still room for significant improvement as well. Another challenge is that methods based on nonlocal means as well as sparse modeling have very high computational cost. Further work to reduce this computational burden appears necessary for successful application of these and related methods in practice.

Bibliography

[1] A. Buades, B. Coll, and J.-M. Morel, "A non-local algorithm for image denoising," in *Proc. IEEE International Conference on Computer Vision and Pattern Recognition (CVPR 2005)*, June 2005.

[2] A. Buades, B. Coll, and J.-M. Morel, "A review of image denoising algorithms, with a new one," *Multiscale Modeling and Simulation* vol. 4, no. 2, pp. 490–530, 2005.

[3] C. Kervrann and J. Boulanger, "Optimal spatial adaptation for patch-based image denoising," *IEEE Transactions on Image Processing*, vol. 15, pp. 2866–2878, October 2006.

[4] G. Gilboa and S. Osher, "Nonlocal linear image regularization and supervised segmentation," Tech. Rep. CAM-06-47, Dept. of Math., University of California, Los Angeles, 2006.

[5] T. Brox, O. Kleinschmid, and D. Cremers, "Efficient nonlocal means for denoising of textural patterns," *IEEE Transactions on Image Processing*, vol. 17, pp. 1083–1092, July 2008.

[6] K. Dabov, A. Foi, V. Katkovnik, and K. Egiazarian, "Image denoising by sparse 3-D transform-domain collaborative filtering," *IEEE Transactions on Image Processing*, vol. 16, pp. 2080–2095, August 2007.

[7] J. Mairal, F. Bach, J. Ponce, G. Sapiro, and A. Zisserman, "Non-local sparse models for image restoration," in *Proc. IEEE International Conference on Computer Vision 2009 (ICCV 2009)*, September 2009.

[8] P. Chatterjee and P. Milanfar, "Patch-based locally optimal denoising," in *Proc. IEEE International Conference on Image Processing 2011 (ICIP 2011)*, September 2011.

[9] W. Dong, X. Li, L. Zhang, and G. Shi, "Sparsity-based image denoising via dictionary learning and structural clustering," in *Proc. IEEE International Conference on Computer Vision and Pattern Recognition (CVPR 2011)*, June 2011.

[10] A. Buades, B. Coll, and J.-M. Morel, "Denoising image sequences does not require motion estimation," in *Proc. of IEEE Conference on Advanced Video and Signal Based Surveillance (AVSS)*, pp. 70–74, September 2005.

[11] M. Protter, M. Elad, H. Takeda, and P. Milanfar, "Generalizing the non-local-means to super resolution reconstruction," *IEEE Transaction on Image Processing*, vol. 18, pp. 36–51, January 2009.

[12] P. van Beek, J. Yang, S. Yamamoto, and Y. Ueda, "Deblurring and denoising with non-local regularization," in *Visual Information Processing and Communication 2010, Proc. SPIE*, vol. 7543 (A. Said and O. G. Guleryuz, Eds.), January 2010.

[13] M. Mahmoudi and G. Sapiro, "Fast image and video denoising via nonlocal means of similar neighborhoods," *IEEE Signal Processing Letters*, vol. 12, pp. 839–842, 2005.

[14] A. Adams, N. Gelfand, J. Dolson, and M. Levoy, "Gaussian KD-trees for fast high-dimensional filtering," *ACM Transactions on Graphics*, vol. 28, p. 21, August 2009.

[15] T. Tasdizen, "Principal neighborhood dictionaries for nonlocal means image denoising," *IEEE Transactions on Image Processing*, vol. 18, pp. 2649–2660, 2009.

[16] J. Biemond, R. L. Lagendijk, and R. M. Merserau, "Iterative methods for image deblurring," *Proceedings of the IEEE*, vol. 78, pp. 856–883, May 1990.

[17] M. Banham and A. K. Katsaggelos, "Digital image restoration," *IEEE Signal Processing Magazine*, vol. 14, pp. 24–41, May 1997.

[18] L. I. Rudin, S. Osher, and E. Fatemi, "Nonlinear total variation based noise removal algorithms," *Physica D*, vol. 60, 1992.

[19] T. Chan, S. Osher, and J. Shen, "The digital TV filter and nonlinear denoising," *IEEE Transactions on Image Processing*, vol. 10, pp. 231–241, February 2001.

[20] Y. Li and F. Santosa, "A computational algorithm for minimizing total variation in image restoration," *IEEE Transactions on Image Processing*, vol. 5, no. 6, pp. 987–995, June 1996.

[21] M. K. Ng, H. Shen, E. Y. Lam, and L. Zhang, "A total variation regularization based super resolution reconstruction algorithm for digital video," *EURASIP Journal on Advances in Signal Processing*, Article ID 74585, 2007.

[22] A. Elmoataz, O. Lezoray, and S. Bougleux, "Nonlocal discrete regularization on weighted graphs: A framework for image and manifold processing," *IEEE Transactions on Image Processing*, vol. 17, pp. 1047–1060, July 2008.

[23] G. Peyré, S. Bougleux, and L. Cohen, "Non-local regularization of inverse problems," in *Proc. European Conference on Computer Vision 2008, LNCS*, vol. 5304, Part III (D. Forsyth, P. Torr, and A. Zisserman, Eds.), pp. 57–68, 2008.

[24] M. Elad, M. A. T. Figueiredo, and Y. Ma, "On the role of sparse and redundant representation in image processing," *Proceedings of the IEEE*, vol. 98, pp. 972–982, June 2010.

[25] M. Elad and M. Aharon, "Image denoising via sparse and redundant representation over learned dictionaries," *IEEE Transactions on Image Processing*, vol. 15, pp. 3736–3745, December 2006.

[26] J. S. Lee, "Digital image smoothing and the sigma filter," *Computer Vision, Graphics and Image Processing*, vol. 24, pp. 255–269, 1983.

[27] S. Smith and J. Brady, "SUSAN — A new approach to low level image processing," *International Journal of Computer Vision*, vol. 23, no. 1, pp. 45–78, 1997.

[28] C. Tomasi and R. Manduchi, "Bilateral filtering for gray and color images," in *Proc. IEEE International Conference on Computer Vision (ICCV 1998)*, January 1998.

[29] A. A. Efros and T. K. Leung, "Texture synthesis by non-parametric sampling," in *Proc. IEEE International Conference on Computer Vision (ICCV 1999)*, September 1999.

[30] K. Dabov, A. Foi, V. Katkovnik, and K. Egiazarian, "Image restoration by sparse 3-D transform-domain collaborative filtering," in *Image Processing: Algorithms and Systems VI, Proc. SPIE*, vol. 6812-07 (J. T. Astola, K. O. Egiazarian, and E. O. Dougherty, Eds.), January 2008.

[31] P. Chatterjee and P. Milanfar, "Is denoising dead?," *IEEE Transactions on Image Processing*, vol. 19, pp. 895–911, April 2010.

[32] J. Mairal, G. Sapiro, and M. Elad, "Multiscale sparse image representation with learned dictionaries," in *Proc. IEEE International Conference on Image Processing 2007 (ICIP 2007)*, September 2007.

[33] J. Yang, J. Wright, Y. Ma, and T. Huang, "Image super resolution as sparse representation of raw image patches," in *Proc. IEEE International Conference on Computer Vision and Pattern Recognition (CVPR 2008)*, June 2008.

[34] J. Yang, J. Wright, Y. Ma, and T. Huang, "Image super resolution via sparse representation," *IEEE Transactions on Image Processing*, vol. 19, pp. 2861–2871, November 2010.

[35] W. T. Freeman, E. C. Pasztor, and O. T. Carmichael, "Learning low-level vision," *International Journal of Computer Vision*, vol. 40, no. 1, pp. 25–47, 2000.

[36] W. T. Freeman, T. R. Jones, and E. C. Pasztor, "Example-based super resolution," *IEEE Computer Graphics and Applications*, pp. 56–65, March/April 2002.

[37] S. C. Park, M. K. Park, and M. G. Kang, "Super resolution image reconstruction — A technical overview," *IEEE Signal Processing Magazine*, pp. 21–36, May 2003.

[38] J. Sun, J. Zhu, and M. F. Tappen, "Context-constrained hallucination for image super resolution," in *Proc. IEEE International Conference on Computer Vision and Pattern Recognition (CVPR 2011)*, June 2011.

[39] D. Glasner, S. Bagon, and M. Irani, "Super resolution from a single image," in *Proc. IEEE International Conference on Computer Vision 2009 (ICCV 2009)*, September 2009.

[40] O. Shahar, A. Faktor, and M. Irani, "Space-time super resolution from a single video," in *Proc. IEEE International Conference on Computer Vision and Pattern Recognition (CVPR 2011)*, June 2011.

[41] R. Vignesh, B. T. Oh, and C.-C. J. Kuo, "Fast non-local means (NLM) computation with probabilistic early termination," *IEEE Signal Processing Letters*, vol. 17, pp. 277–280, 2010.

[42] J. V. Manjón, J. Carbonell-Caballero, J. J. Lull, G. García-Martí, L. Martí-Bonmatí, and M. Robles, "MRI denoising using non-local means," *Medical Image Analysis*, vol. 12, no. 4, pp. 514–523, 2008.

[43] P. Coupé, P. Yger, and C. Barillot, "Fast non local means denoising for 3D MR images," in *Proc. Medical Image Computing and Computer-Assisted Intervention (MIC-CAI)*, pp. 33–40, October 2006.

[44] H. Takeda, P. van Beek, and P. Milanfar, "Spatio-temporal video interpolation and denoising using motion-assisted steering kernel (MASK) regression," in *Proc. IEEE International Conference on Image Processing (ICIP 2008)*, October 2008.

[45] H. Takeda, P. Milanfar, M. Protter, and M. Elad, "Superresolution without explicit subpixel motion estimation," *IEEE Transactions on Image Processing*, vol. 18, pp. 1958–1975, September 2009.

[46] H. Takeda, P. van Beek, and P. Milanfar, "Spatiotemporal video upscaling using motion-assisted steering kernel (MASK) regression," in *High-Quality Visual Experience: Creation, Processing and Interactivity of High-Resolution and High-Dimensional Video Signals* (M. Mrak, M. Grgic, and M. Kunt, Eds.), Springer Verlag, 2010.

Chapter 5

Sparsity-Regularized Image Restoration: Locality and Convexity Revisited

WEISHENG DONG
Xidian University

XIN LI
West Virginia University

5.1 Introduction

Many image-related inverse problems deal with the restoration of an unknown image from its degraded observation. Depending on the model of the degradation process, we could have a variety of restoration problems, such as denoising, deblurring, demosaicking, deblocking/deringing, inverse halftoning, and so on. It has been widely recognized that the a priori knowledge, in the form of either regularization functional in a deterministic setting or prior probability distribution in a statistical setting, plays a critical role in the accuracy of image processing algorithms. The task of representing the a priori knowledge for the class of photographic images is particularly challenging due to the diversity of various structures (e.g., edges, corners, lines, and textures) in natural scenes. Extensive effort has been devoted to the pursuit of good mathematical models for photographic images in recent decades.

Early good image models are based on heuristic observations such as the local smoothness of the intensity field. Such an intuitive concept of local smoothness has been quantified by several different mathematical tools in the literature, including (1) partial differential equation (PDE) based models — the replacement of l_2 by l_1 norm in total-variation (TV) diffusion [1, 2] has shown better edge-preserving capabilities and the magic of l_1-optimization has restimulated interest in sparse representations in the context of compressed sensing since 2006 [3, 4]; (2) transform-based models — advances of wavelet theory in 1990s have shown the connection between smoothness characterization by Lipschitz regularity [5, pp. 165–171] and nonlinear approximation of images by Besov-space functions [6]. The simple idea of thresholding [7, 8] has been long influential and led many applications of sparse representations (e.g., inverse halftoning [9], recovery [10], deblurring [11], inpainting [12]).

Despite the conceptual simplicity and mathematical rigor of PDE or transform-based models, their optimality of characterizing complex structures in natural images is often questionable because self-repeating patterns of edges and textures violate the locality assumption. In recent years, exploiting the self-similarity of images by patch-based models has shown great potential in several applications, including texture synthesis [13, 14], image inpainting [15] (also called image completion [16]), image denoising [17, 18], super resolution [19, 20] and image restoration [21, 22]. All those findings have suggested that nonlocal self-similarity is supplementary to local smoothness for the task of abstracting the a priori knowledge about photographic images. Patch-based models, when viewed as the extension of transform-based models, lead to a new class of nonlocal sparse representations (NSR) useful for regularized restoration.

In this chapter we attempt to provide a gentle introduction of NSR from a manifold point of view. A manifold perspective is conceptually appealing for engineering students without advanced mathematical training (e.g., variational calculus [23]). Photographic images, when decomposed into a collection of maximally overlapped patches, have been empirically shown to behave like a low-dimensional manifold embedded in the patch space [24, 25]. Such patch-based image models [26] have received increasingly more attention from the vision community in the past decade. We discuss two contrasting strategies (*dictionary learning* versus *structural clustering*) of discovering the manifold structure underlying image signals and two competing approaches (*local PCA* versus *local embedding*) of exploiting the local subspace constraint of the image manifold. Simple toy examples in 2-D (e.g., spiral data) are used to illustrate their differences and connections.

The manifold perspective allows us to maintain a healthy skepticism toward the increasing enthusiasm about convex optimization. If the collection of photographic images do not form a manifold (a nonconvex set), we must be extra cautious when approximating it by convex tools. Loosely speaking, organizational principles underlying photographic images might give rise to some *saddle points* of an energy function, which would defeat any effort of pursuing local optima [27]. For nonconvex optimization problems, we argue that deterministic annealing (DA) [28] (a.k.a. graduated nonconvexity [29]) inspired by statistical physics is a more appropriate tool for sparsity-regularized image restoration [10, 12, 22]. We resort to a Monte Carlo method to supply an intuitive explanation of why DA-based nonconvex sparsity optimization is particularly effective for the restoration of photographic images.

The rest of this chapter is organized as follows. We first provide a historical review of sparse representations in Section 5.2, which might appear enlightening to some readers despite our own subjective bias. Then we tell the tale of two sparsity models: from local smoothness to nonlocal similarity in Section 5.3. Because NSR is the focus of our tale, we introduce a manifold-based geometric language to facilitate a deeper understanding. In Section 5.4 we briefly take a glimpse of the theoretic connection between cluster Monte

Carlo methods in statistical physics and nonlocal sparse representations in image processing, which motivates us to reinspect the role of convexity in image restoration. For readers who are less theory oriented, we use Monte Carlo simulations to help them gain a more concrete understanding of DA-based nonconvex sparsity optimization. The potential of NSR and DA is demonstrated for four sets of fully reproducible image restoration experiments in Section 5.5. We make some concluding remarks and discuss future research directions in Section 5.6.

5.2 Historical Review of Sparse Representations

Where does sparse representation come from? An accurate recount of the history seems impossible but it is still enlightening to trace back various ideas and terms connected with sparsity. The so-called Pareto principle (a.k.a. 80-20 rule or the principle of factor sparsity) was among the earliest to highlight the importance of *exceptions*. It roughly states that "80% of the effects come from 20% of the causes": for example, Pareto observed in 1906 that 80% of the land in Italy was owned by 20% of the population (or it was also said that 80% of the wealth was owned by 20% of the richest people in the world). Originally a principle in economics, similar observations were made later about physical and social sciences. For instance, in linguistics, Zipf's law says that "given a corpus of natural language utterances, the frequency of any word is inversely proportional to its rank in the frequency table."

The connection between the Pareto principle and Zipf's law can be seen more clearly by studying the class of heavy-tailed distributions in probability theory. Unlike common distributions such as Gaussian or Laplacian, heavy tails imply that tails are not exponentially bounded. Despite the fact that a rigorous definition of heavy-tail distribution is still debatable, its practical significance has been widely agreed upon. Without a heavy tail, the 80-20 rule would not be valid; and what Zipf's law exemplifies is a special class of heavy-tail distributions called the power law. Power-law distributions have interesting properties, such as scale-invariance and universality, whose significance in statistical physics has been long known [30]. If one believes that "nature uses only the longest threads to weave her patterns, so that each small piece of her fabric reveals the organization of the entire tapestry" (cited from legendary Richard Feynman), he will not be surprised by the finding of power-law distributions in chemical, economical, social, and engineering systems.

Historically it was the French mathematician Paul Levy who first formalized the concept of heavy tails into probability theory. Levy's ideas have influenced his students, including Benoit Mandelbrot who pioneered the fractal theory and advocated the importance of self-similarity in nature. As Mandelbrot said in [31], "Nature has played a joke on the mathematicians the same pathological structures that the mathematicians invented to break loose from 19th-century naturalism turn out to be inherent in familiar objects all around us." The importance of "pathological structures" seemed to be better appreciated by TV engineers in the early 1980s while Ted Adelson worked at RCA Lab (according to Eero Simoncelli) but their impact on the scientific community had to wait until the birth of wavelet theory.

In the late 1980s, three schools of researchers from applied math, electrical engineering, and computer science independently worked out wavelet theory, filter bank, and multiresolution analysis. The mathematical equivalence among those three trains of thoughts was not difficult to establish but the significance of sparsity to wavelet-based representation was only recognized later in the early 1990s, thanks to a maverick image coding algorithm EZW [32] and an orthodox nonlinear shrinkage idea [7]. Since then, our understanding of photographic images (images of natural scenes) has improved, for example, the connection with Besov-space functions in nonlinear approximation [6] and generalized Gaussian distribution in statistical modeling [33]. The good localization property of wavelet bases in both space and frequency turns out to be the key contributor underlying sparse representations in the wavelet space. Numerous wavelet-based image processing algorithms have been developed since then; wavelet has also found connections with the fields of neuroscience and vision research [34, 35]. In the past decade, disciples of Donoho, Vetterli, and Mallat have continued their effort of constructing so-called geometric-wavelets (e.g., curvelets [36], contourlets [37], ridgelets [38], bandelets [39]). New basis functions have shown convincing improvements over conventional separable wavelets for certain classes of images (e.g., fingerprints); however, their effectiveness on generic photographic images remains uncertain.

Most recently, there have been two flurries of competing thoughts related to sparse representations. On the one hand, patch-based models originated from vision/graphics communities (e.g., texture synthesis [13, 14], image inpainting [15, 40]) started to leverage into signal processing community and reshaped our thinking about the locality assumption behind transform-based image models. In particular, a flurry of work on nonlocal image denoising (e.g., nonlocal mean [17], K-SVD [41], BM3D [18, 42], K-LLD [43], LSSC [21]) has gained increasing attention even though its connection with sparse representations is not always obvious. Despite their outstanding performance, our understanding about why they work remains lacking.

On the other hand, sparsity optimization, when formulated by the l_0-norm of transform coefficients, is nonconvex and computationally intractable. The idea of approximating l_0-based nonconvex optimization by its l_1-based convex counterpart has given birth to a new field called compressed sensing (CS) [3, 4] and stimulated tremendous interest in l_1-based optimization (e.g., refer to [44] and a recent tutorial article [45]). It is beyond many people's imagination that CS has evolved so rapidly in a short period of five years (many of its applications can be found at the website of CS Resources (*http://dsp.rice.edu/cs*)). However, does CS provide fundamental insight about signal processing more than serving as a computational tool? Is there any unidentified connection between the above two lines of thoughts (representation versus optimization)? What changing role do *locality* and *convexity* play in our understanding of image modeling and processing techniques? Those are the primary motivating questions behind this tutorial article.

5.3   From Local to Nonlocal Sparse Representations

5.3.1   Local Variations: Wavelets and Beyond

Ever since Leibniz and Newton invented Calculus to analyze the variations in the world, one of the most basic premises in physical science is that nature can be described locally [46]. Such a locality principle has well served the need to analyze signals (e.g., geophysical, acoustic, image) acquired from the physical world in the past. To motivate our discussion, let's consider the task of modeling photographic images (images of natural scenes) and ask ourselves: what makes the collection of photographic images different from that of random noise? An intuitive answer would be that we can find various objects (e.g., people, buildings, flowers, etc.) in a photo. Then, what consists of those objects? One might say edges (the boundary of an object) and textures (the interior of an object). Then, how do we detect edges? Any student who has learned about edge detection in their undergraduate image processing courses would be eager to propose the use of various edge-detection operators (Sobel, Prewitt, Canny) [47]. However, all edge-detection operators are essentially high-pass filters detecting local changes; but random noises are also a type of local change. So here comes a tantalizing question: How do we tell apart transient but regular events (meaningful changes in a signal) apart from lasting but irregular ones (random changes in contaminating noise)?

The motivation behind wavelet shrinkage is twofold. First, we note that the regularity of transient events can be observed *across different scales* — such an idea at least originated from Marr's vision theory [48] and is at the foundation of scale-space theory [49]. What wavelets have offered is a principled way of decomposing signals in the scale space with certain desirable properties [50]. Second[1], although it is often difficult to do signal-noise separation for a single sample or coefficient (*microscopic* level), the ensemble (*macroscopic* level) property of a signal is often sufficient to distinguish itself from that of noise. For example, the tail of a Gaussian distribution is always light (i.e., decaying at the rate of $exp(-\frac{(x-\mu)^2}{\sigma^2})$), regardless of change-of-coordinate (as we will elaborate next); while the tail of an empirical distribution of photographic images is often heavy after the transform if the basis functions (new coordinates) have good localization properties in space and frequency. Local changes (called singularities in [51]) would only produce a small number of exceptions, often called *sparse significant coefficients*, the physical characterization of singularities in an image (e.g., edges, corners, lines etc.).

From this perspective, the high impact of nonlinear thresholding strategies [7, 52] is easy to appreciate because they opt to preserve the heavy tail only. It should be noted that the above two lines of reasoning produce *strong* results applicable to a wide range of image restoration tasks. Image denoising — removal of additive white Gaussian noise (AWGN) — represents the simplest scenario where no matter how you change the coordinate, the ensemble property of AWGN would forbid the production of a heavy tail (sufficient number of exceptions). More generally, we can claim that the strategy of nonlinear shrinkage is applicable to other restoration tasks as long as the noise under consideration (e.g., colored Gaussian noise, quantization/halftoning noise) is unlikely to produce a heavy tail as signals do. In fact, this view helps us understand the ubiquity of wavelet thresholding (e.g., deblurring [11], post-processing [53], inverse half-toning [9]).

Despite the effectiveness of the locality principle, we note that it only reflects one plausible approach to understanding signals acquired from the natural world. As Ingrid Daubechies, a pioneer of wavelet theory, once said in [54], "This fails when you talk about textures, where fine scale features can have a very large correlation length." Indeed, it is often a difficult task to characterize the correlation length of observation data - not only tex-

---

[1]This is is where a statistical view becomes more preferred over a deterministic one.

tures, but also edges could show long-range dependency beyond the reach of local models. The importance of understanding nonlocal connections was again recognized by physicists first (e.g., nonlocal interpretation of quantum mechanics [55]). As noted by Nobel Laureate P. Anderson commented in *More is Different* [56], "Each level can require a whole new conceptual structure." As the complexity of a system moves up the hierarchy (physical$\rightarrow$ chemical $\rightarrow$ biological), the locality principle becomes more and more questionable. If we think of image processing as a scientific endeavor of understanding the organizational principle of neural systems, we must be extra cautious about the locality principle because synaptic connections among neurons are often not constrained by their geometric proximity.

5.3.2 Nonlocal Similarity: From Manifold Learning to Subspace Constraint Exploitation

The *relativity* of defining the local neighborhood is the primary motivation behind the class of nonlocal sparse representations. To see this, we first introduce a simple concept called *manifold*. We opt to tell our story using a manifold-based geometric language because we believe it is conceptually simpler to follow for readers without advanced mathematical training (e.g., calculus of variation and functional analysis). Loosely speaking, a manifold is "a mathematical space that on a small enough scale resembles the Euclidean space of a specific dimension" (cited from Wikipedia). Some toy examples in our daily lives: an electric cord or a plastic string is a 1-D manifold embedded in 3-D space; a folded aluminum foil paper or any torus serves as an example of a 2-D manifold in 3-D space. It is often difficult to envision a manifold embedded in a space whose dimensionality is more than three; but with some imagination we can still conjecture that if the collection of all $B \times B$ image patches form a manifold in $R^{B \times B}$, the dimensionality of its local subspace could be dramatically less than B^2.

In fact, there are several ways of developing a manifold intuition about image processing. First, just like the idea of bilateral filtering [57], one could think of an image not as a mapping $(x, y) \rightarrow f(x, y)$ but its inverse mapping, f^{-1}. This line of thought led to an empirical study verifying the manifold constraint for 8×8 patches in photographic images [24] in 1999. In the same year, the pioneering work on texture synthesis by nonparametric sampling [13] clearly showed the advantage of searching nearest neighbor or k-nearest-neighbor for modeling textures. Just another year later, two influential works by the machine learning community — ISOMAP [58] and locally linear embedding (LLE) [59] — demonstrated the feasibility of discovering low-dimensional manifolds embedded in a high-dimensional space. Unlike the norm used by the signal processing community (local neighborhood is defined with respect to the *domain*), such a manifold perspective represents a significant departure because the local neighborhood on a signal manifold is defined with respect to the *range* of a function. In other words, two points within the same local neighborhood on a manifold could correspond to two image patches distant from each other (therefore nonlocal) in the spatial domain (please refer to Figure 5.1(a)). Note that such an observation can be made to both image structures of edges and textures (in other words, their self-similarity contributes to the local subspace constraint of an image manifold in the patch space $R^{P \times P}$).

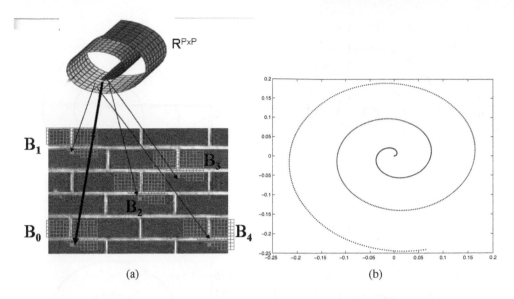

(a)   (b)

Figure 5.1 Illustration of manifold: (a) two adjacent points on a manifold could be spatially distant (e.g., B_i vs. B_j ($0 \leq i \neq j \leq 4$)); (b) spiral data (a 1-D manifold embedded in R^2).

5.3.2.1   Manifold Discovery: Global versus Local

With some necessary background on manifold, a logical follow-up question is: How do we discover the manifold structure underlying a given signal? There are two classes of approaches toward such a problem of manifold learning: *global* versus *local*. Suppose $\mathbf{x} \in \mathbf{R}^N$ is the signal of interest and $\mathbf{A}$ denotes the dictionary. A global approach attempts to represent $\mathbf{x}$ by the linear combination of a few atoms from $\mathbf{A}$, that is,

$$\mathbf{x} = \mathbf{A}_{N \times M} \mathbf{X}, \qquad (5.1)$$

where $\mathbf{X} \in \mathbf{R}^M$ is expected to be sparse. Note that in the case of complete expansion ($N = M$), linear transforms are nothing but change-of-coordinates and therefore have a limited capability of handling a low-dimensional manifold (i.e., rotating a torus is useless). A more powerful strategy is to learn an *overcomplete* dictionary $N < M$ (e.g., via K-SVD [60]) with the sparsity constraint on $\mathbf{X}$. To illustrate the idea of dictionary learning, we have designed a toy example of 2-D spiral data as shown in Figure 5.1(b) and used it to get some hands-on experience with the K-SVD algorithm. Figure 5.2 (a) displays the learned dictionaries (highlighted by "+") as the redundancy increases; we can see that the dictionary attempts to cover the spiral structure from more and more directions as its size goes from $M = 5$ to $M = 40$. It has been experimentally found that as the redundancy increases, the sparsity (as measured by the percentage of nonzero coefficients) improves — from 40% ($M = 5$) to merely 4.8% ($M = 40$). Therefore, dictionary learning does offer an approach toward improved sparsity by exploiting the redundancy. It seems to us that K-SVD might be connected with support vector machines, which admit a sparse-approximation interpretation [61].

By contrast, a local approach attempts to first segment the points on the manifold into different clusters and then characterize each cluster/neighborhood based on the local ge-

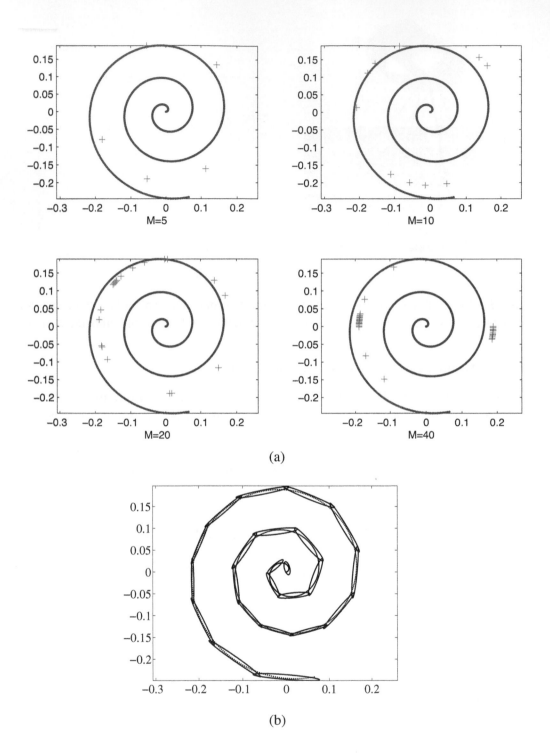

Figure 5.2 Two ways of discovering the manifold structure underlying a signal: (a) dictionary learning (atoms in dictionary $\mathbf{A}$ are highlighted in "+"); (b) structural clustering (each segment denotes a different cluster $p(\mathbf{x}|\theta_{\mathbf{m}})$).

ometry of the manifold (e.g., via parametric Gaussian models):

$$p(\mathbf{x}|\theta) = \sum_{m=1}^{k} \alpha_m p(\mathbf{x}|\theta_m), \tag{5.2}$$

where α_m denotes the mixing probabilities. Some popular clustering techniques include exemplar-based (each cluster consists of the k-Nearest-Neighbors of a given exemplar) and K-means (K clusters with representative codewords are found) [62]. Figure 5.2 (b) includes an example of segmenting the spiral data into 23 clusters (mixture of Gaussian) by the EM algorithm [63]. We can observe that the spiral structure in the original data is approximated by a finite number of constellations (local clusters). A geometrically equivalent interpretation of clustering is embedding, that is, to embed high-dimensional data into a lower-dimensional space [59, 64]. Unlike K-SVD, the sparsity constraint implicitly translates to the local subspace constraint with each cluster, namely, it can be better approximated by a lower-dimensional object after the clustering or embedding. Apparently, the issue of sparsity boils down to the question of how to exploit the subspace constraint.

5.3.2.2 Subspace Constraint Exploitation: Learning versus Embedding

There are two competing approaches of exploiting the local subspace constraint of signal manifold: learning such as local principal component analysis (PCA) [65] and embedding[2] [66] as shown in Figure 5.3. Loosely speaking, the PCA approach is the same as the classical idea of maximizing the energy compaction (therefore sparsity) by learning a collection of signal-dependent basis functions; while the embedding approach saves the effort of dictionary learning by mapping the local neighborhood of a point to a higher-dimensional space - e.g., data points $\{B_0, B_1\}$ and $\{B_0', B_1'\}$ respectively map to $\mathbf{D}$ and $\mathbf{D}'$ (refer to Figure 5.3(b)). It can be observed that $\mathbf{D}$ and $\mathbf{D}'$ can be sparsified by a signal-independent transform in a higher-dimensional space, regardless of the local geometric variations associated with B_i or B_i'. Geometrically local PCA is less sensitive to the outliers in the local neighborhood (e.g., due to segmentation errors) than local embedding and therefore has been adopted in our most recent work on CSR denoising [67], while some constraint on the size of kNN (parameter k) is often needed as a strategy of outlier rejection for local embedding (e.g., in BM3D denoising [18]).

It is our hope that the manifold perspective could offer a unified and enlightening interpretation of several recently developed image restoration techniques in the literature. If structural clustering is combined with local embedding, we obtain an approximation of BM3D denoising [18]; the combination of dictionary learning and structural clustering would lead to K-locally-learned- dictionaries (K-LLD) denoising [43], learned simultaneous sparse coding (LSSC) denoising [21] , as well as our own clustering-based sparse representation (CSR) denoising algorithm [67]. The potential of nonlocal sparse representations has been gradually recognized by the community, which is likely to stimulate further research along the line of understanding nonlocal similarity. To summarize, the key to the effectiveness of nonlocal sparse representations is their capability of *discovering and*

---

[2]Note that an embedding $f : X \to Y$ can refer to both the case of $Dim(X) < Dim(Y)$ and $Dim(X) > Dim(Y)$ (here we mean the latter).

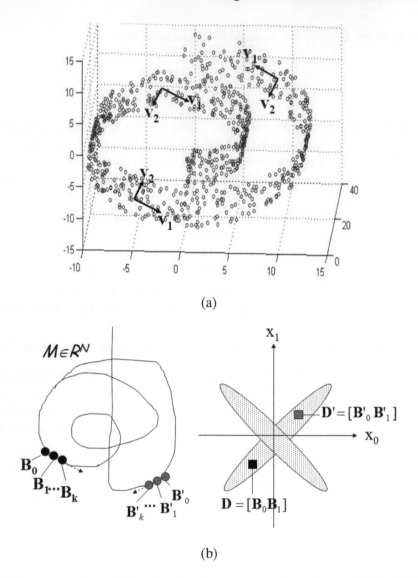

(a)

(b)

Figure 5.3 Two ways of exploiting local subspace constraint of a manifold: (a) local PCA (note that the basis functions are adaptive to local geometry and therefore signal dependent); (b) local embedding to a higher-dimensional space.

exploiting the manifold constraint underlying image signals, which originates in the self-similarity of edges and textures.

5.4 From Convex to Nonconvex Optimization Algorithms

Many image restoration tasks can be formulated as a constrained optimization problem such as the minimization of $||\mathbf{X}||_{l_0}$ such that $\mathbf{x} = \mathbf{A}\mathbf{X}$ satisfies the observation constraint $||\mathbf{y} - \mathbf{D}[\mathbf{x}]||_{l_2}^2 \leq \epsilon$ ($\mathbf{D}$ denotes an operator characterizing the degradation process). A common practice in solving the above constrained optimization problem is to convert it into an equivalent unconstrained optimization problem as follows:

$$\mathbf{X} = \arg\min_{\mathbf{X}} \frac{1}{2}||\mathbf{y} - \mathbf{D}[\mathbf{AX}]||_{l_2}^2 + \lambda||\mathbf{X}||_{l_0}. \tag{5.3}$$

Due to the computationally intractability of l_0-optimization (due to its nonconvexity), an attractive idea is to replace the l_0-norm by its l_1 counterpart (so-called l_1-magic; *http://www.acm.caltech.edu/l1magic/*). Despite the mathematical appeal of convexity, one can argue it has missed the point from an image modeling perspective, as image manifold is necessarily *nonconvex*, the tool of convex optimization might lead us to tackle the wrong problem and overlook more fundamental issues at the level of image representations. To back up the above claim, let's recall the lesson we just learned from local to nonlocal sparse representation. Because NSR often involves data clustering and objective functions associated with clustering are often nonconvex, it is often necessary to approximate a nonconvex problem by one or more convex counterparts. Therefore, logically speaking, strategies of *convex approximation* are as important as those of solving convex optimization alone.

Another line of argument comes from the observation of the evolution in the biological world. It is often said that "Nature does not have foresight"; instead, biological organisms manage to survive and prosper in a constantly changing environment through adaptation. In fact, from an evolutionary point of view, seeking a *saddle point* instead of a local minimum (cited from Ivar Ekeland [27]) is more favorable in the long run because it facilitates the organism to adapt to the changing physical environment. Under the context of image modeling, the nonstationarity of photographic images can often be interpreted as spatially varying statistics of the source. If rephrased by a manifold language, projection onto one subspace (e.g., more smoothness as one zooms out) might observe the decrease of some conceptually defined energy while projection onto another subspace (e.g., more jaggedness as one zooms in) could observe the opposite. Therefore, the saddle-point thinking at least offers us a more "scale-invariant" perspective toward understanding photographic images.

Although less well known, the problem of nonconvex minimization has been studied by mathematicians, scientists, and engineers. The most well-known optimization technique for the nonconvex problems is likely to be simulated annealing [68] inspired by the analogy between combinational optimization and statistical mechanics. It was argued in [68] that "as the size of optimization problems increases, the worst-case analysis of a problem will become increasingly irrelevant, and the average performance of algorithms will dominate the analysis of practical applications." Such observation has motivated the introduction of an artificial "temperature" parameter to simulate the annealing process in nature. Simulated annealing was often implemented by the Metropolis algorithm based on local random perturbation; more powerful nonlocal/cluster Monte Carlo methods (e.g., Swendsen-Wang algorithm [69], Wolff algorithm [70]) were developed afterward. A key insight behind cluster algorithms is that collective updating could eliminate the problem of "critical slowing down" [71].

For image restoration applications, we argue that it is also the *average* performance that matters because any restoration algorithm is supposed to work for a class of images instead of an individual one. In fact, it is often difficult to come up with the worst-case example in the first place (i.e., what kind of image is the hardest to recover? — no one knows). What seems an unknown connection between nonlocal Monte Carlo methods and

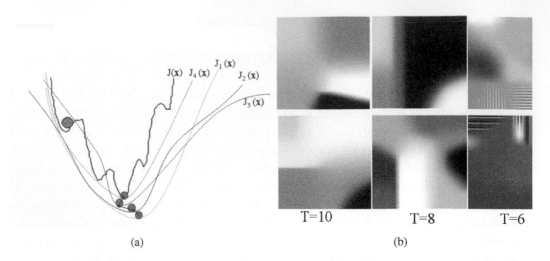

Figure 5.4 Illustration of deterministic annealing (DA): (a) the minimum of a non-convex function J may be found by a series of convex approximations $J_1, J_2, J_3, \dots$ (color-coded in the diagram); (b) sample stable states (local-minimum) corresponding to varying temperatures.

nonlocal sparse representations is the role played by *clustering*. More specifically, the observation constraint appears to suggest that we do not need to do annealing in a stochastic but deterministic fashion because observation data $\mathbf{y}$ offers a valuable hint for clustering. Indeed, deterministic annealing (DA) — often known as graduated non-convexity (GNC) in the vision literature [29] — has been proposed as a statistical physics-inspired approach toward data clustering [28, 72].

The basic idea behind DA is simple and appealing; that is, we can modify the non-convex cost function $J(\mathbf{x})$ in such a way that the global optimum can be approximated through a sequence of convex cost functions $J_p^*(\mathbf{x})$ (the auxiliary variable p parameterizes the annealing process) as shown in Figure 5.4(a). When the iteration starts from a small parameter favoring a smooth cost function (i.e., high temperature), it is relatively easy to find a favorable local optimum. As the cost function becomes more jagged after several iterations, the local optimum will be gradually driven toward the global minimum as the temperature decreases [29]. In addition to the above standard interpretation, we add a twist from saddle-point thinking here; that is, $J_p^*(\mathbf{x})$'s do not need to be convex *everywhere* but only locally around the point of interest. In other words, the strategy of convex approximation can be made *data dependent*; therefore, as the point of interest $\mathbf{x}$ moves to $\mathbf{x}'$, we would have a different sequence of locally convex cost functions in action.

To illustrate how DA works, we have designed a Monte Carlo experiment as follows. Starting with random noise realization, we iteratively filter it by $\mathbf{x}^{n+1} = P_\theta \mathbf{x}^n$, where P_θ denotes a nonlocal filter with a "temperature" parameter θ (it is the threshold value T in nonlinear shrinkage or the Lagrangian multiplier λ in nonlocal regularization [73]). As long as the P_θ is nonexpansive [22], we can observe that the $lim_{n \to \infty} \mathbf{x}^n$ would converge to a fixed point — without any observation constraint, such fixed point could wander in the phase space depending on the order parameter θ. By varying the θ value, we could observe varying structures in physical systems (e.g., crystal versus glass); analogously in image restoration, we discover the constellations of an image manifold: *smooth regions, regular edges, and textures* as shown in Figure 5.4(b) where the hard thresholding stage of BM3D denoising [18] is used as P_T. It is also worth noting that the phenomenon of *phase transition* (as the temperature varies) has not been observed for conventional local image models such as Markov-Random-Field (MRF) [74]. In other words, clustering has played a subtle role in connecting nonlocal sparsity with nonconvex optimization, even though such connection has not been fully understood yet.

5.5 Reproducible Experimental Results

In alignment with the principle of reproducible research [75, 76], we have made the experimental results of this chapter fully reproducible. Due to space limitation, we will only discuss the potential of NSR and DA in three image restoration applications: deblurring, super resolution, and compressed sensing (more applications such as denoising can be found in another chapter of this book). For each of the three applications, we will selectly review one representative algorithm from our previous works and compare it with other competing approaches. Both subjective and objective quality comparison results will be reported for a pair of test images (one with abundant edges and the other with abundant textures), which reflects the equal importance of edges and textures in photographic images. The key take-home message is that *NSR-based iterative projection, when combined with DA-based nonconvex optimization, offers a class of powerful regularization techniques for various image restoration problems.*

5.5.1 Image Deblurring

Image deblurring refers to the restoration of an image from a noisy blurred image $\mathbf{y} = \mathbf{Hx} + \mathbf{w}$, where $\mathbf{H}$ denotes the blurring kernel. We assume that $\mathbf{H}$ is known (so-called nonblind deconvolution) in this experiment. A classical benchmark experiment that has been performed in many deblurring papers deals with the *cameraman* image and a 9×9 uniform blurring kernel at the noise level of $BSNR = 40$ dB ($BSNR = 10log_{10}\frac{\sigma_z}{\sigma_w^2}$, where $\mathbf{z} = \mathbf{Hx}$). Interestingly, the performance of various competing algorithms (e.g., total-variation based versus transform-based), when measured by improved SNR (ISNR), became "stuck" at around 8.4–8.6 dB for a long time. Only recently, we have shown that BM3D-based deblurring, when coupled with the idea of fine-granularity regularization and DA, can achieve ISNR of > 10 dB [22]. Here, we opt to compare four leading image deblurring algorithms with reproducible source codes: TVMM [77], shape-adaptive DCT (SADCT) [78] and iterative shrinkage/thresholding (IST) [11], and centralized sparse representation (CSR) [79]. The first three are leading deblurring algorithms based on local models; the last one is our latest work based on nonlocal models.

In addition to the popular test image *cameraman* (edge-class, 256×256), we have also worked with a *fingerprint* image (texture-class, 512×512). Figures 5.5 and 5.6 include a comparison of both subjective and objective qualities of deblurred images by different algorithms. For the *cameraman* image, we note that the most impressive visual quality improvement can be observed for the tripod region where conventional deblurring

(a)   (b)   (c)

(d)   (e)   (f)

Figure 5.5 Comparison of original *cameraman* image, noisy blurred image ($BSNR =$ 40 dB, 9×9 uniform blur) and deblurred images by different algorithms): (a) original; (b) noisy blurred; (c) TVMM [77] ($ISNR = 8.42$ dB); (d) SADCT [78] ($ISNR = 8.57$ dB); (e) TwIST [11] ($ISNR = 8.65$ dB); f) CSR [79] ($ISNR = 10.60$ dB).

techniques based on local models suffer from unpleasant ringing artifacts. The suppression of the Gibbs phenomenon by nonlocal sparse representation contributes to both a dramatic ISNR gain and visual quality improvement for this specific example. For the *fingerprint* image, it is not surprising to see that the TV-based method becomes less preferred because the total-variation model is a poor match for textures. When compared with DCT-based and wavelet-based techniques, CSR deblurring is capable of achieving over 1.6 dB ISNR gain and better preserving fine-detailed ridge patterns in fingerprints.

5.5.2   Super Resolution

In super resolution, a low-resolution (LR) image is generated by first applying a blurring kernel and then downsampling by a scale factor. In this experiment, a 7×7 Gaussian blurring kernel of standard deviation 1.6 is used to simulate the out-of-focus blur, and a downsampling ratio of 3 is used in both horizontal and vertical directions to generate LR images. The benchmark SR algorithms adopted in our experiment include (1) the softcut method (denoted by softcut) [80]; (2) the TV-based super resolution method (denoted by TV) [81], and (3) the sparse representation based method (denoted by Sparsity) [82]. In

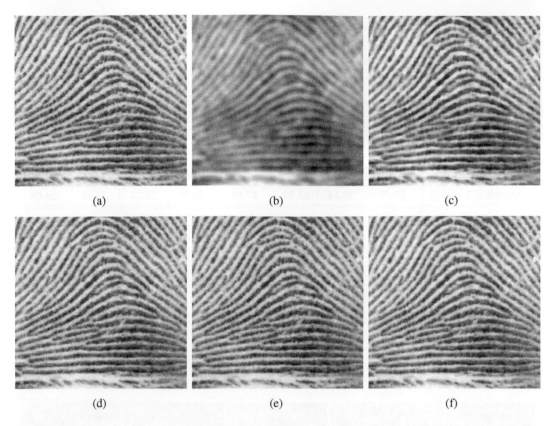

Figure 5.6 Comparison of 256 portion among original $fingerprint$ image, noisy blurred image ($BSNR$ = 40 dB, 9×9 uniform blur) and deblurred images by different algorithms): (a) original; (b) noisy blurred; (c) TVMM [77] ($ISNR$ = 8.30 dB); (d) SADCT [78] ($ISNR$ = 10.70 dB); (e) IST [11] ($ISNR$ = 10.46 dB); (f) CSR [79] ($ISNR$ = 12.38 dB).

our recent work [83], we have developed two sparsity-based SR algorithms: **Algorithm 1** (without nonlocal regularization) and **Algorithm 2** (with nonlocal regularization).

The PSNR results of reconstructed images by different algorithms are listed in Table 5.1. From the table we can see that **Algorithm 2** — empowered by local adaptation and nonlocal regularization — outperforms other competing methods. Subjective quality comparison of the reconstructed HR images by the test methods are shown in Figures 5.7 through 5.9. It can be seen that **Algorithm 2** is capable of reconstructing sharper and clearer edges/textures in SR-resolved images. The color version of these images can be found in [83].

5.5.3   Compressed Sensing

In an influential paper on compressed sensing [4], the authors reported a "puzzling numerical experiment": perfect reconstruction of a phantom image from 22 radial lines in the Fourier domain. Since [4], several new algorithms (e.g., nonconvex l_p-optimization where $0 < p < 1$ [84], spatially adaptive filtering [85], Bregmanized nonlocal regularization [86])

Table 5.1 PSNR (dB) Results (luminance components) of the Reconstructed HR Images

Images	\multicolumn{8}{c}{Noiseless, $\sigma_n = 0$}							
	Butterfly	*flower*	*Girl*	*Pathenon*	*Parrot*	*Leaves*	*Plants*	*Average*
TV [81]	26.56	27.51	31.24	26.00	27.85	24.51	31.34	27.86
Softcut [80]	24.74	27.31	31.82	25.95	27.99	24.34	31.19	27.72
Sparsity [82]	24.70	27.87	32.87	26.27	28.70	24.14	31.55	28.01
Algorithm 1 in [83]	26.96	28.90	33.59	26.95	33.26	26.32	33.26	29.46
Algorithm 2 in [83]	**27.29**	**29.14**	**33.59**	**27.04**	**30.58**	**26.77**	**33.42**	**29.69**
	\multicolumn{8}{c}{Noisy, $\sigma_n = 5$}							
TV [81]	25.49	26.57	29.86	25.35	27.01	23.75	29.70	26.82
Softcut [80]	24.53	26.98	31.30	25.72	27.69	23.17	30.57	27.37
Sparsity [82]	23.61	26.60	30.71	25.40	27.15	22.94	29.57	26.57
Algorithm 1 in [83]	25.79	27.62	31.52	26.17	28.95	25.20	30.92	28.02
Algorithm 2 in [83]	**26.03**	**27.74**	**31.84**	**26.30**	**29.39**	**25.49**	**31.27**	**28.30**

Figure 5.7 Reconstructed HR *Plants* images ($\sigma_n = 0$). From left to right: input LR image; softcut [80] (PSNR = 31.19 dB); TV [81] (PSNR = 31.34 dB); Sparsity [82] (PSNR = 31.55 dB); **Algorithm 2** in [83] (PSNR = 33.42 dB)

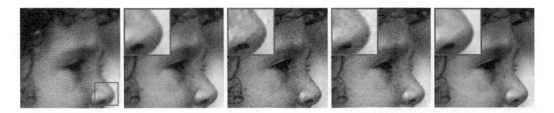

Figure 5.8 Reconstructed HR *Girl* images ($\sigma_n = 5$). From left to right: input LR image; softcut [80] (PSNR = 31.30 dB); TV [81] (PSNR = 29.86 dB); Sparsity [82] (PSNR = 30.71 dB); **Algorithm 2** in [83] (PSNR = 31.85 dB)

Figure 5.9 Reconstructed HR *Parrots* images ($\sigma_n = 0$). From left to right: input LR image; softcut [80] (PSNR = 27.99 dB); TV [81] (PSNR = 27.85 dB); Sparsity [82] (PSNR = 28.70 dB); **Algorithm 2** in [83] (PSNR = 30.46 dB)

have been developed; some of them have shown that perfect reconstruction of the phantom image can be achieved at a sampling rate much lower than twenty two radial lines (e.g., perfect reconstruction from ten and eleven radial lines was reported in [87] and [85] respectively). Since the release of source codes [85] in March 2011, it becomes trivial to verify that the BM3D-based approach [85] can achieve perfect reconstruction from as few as nine radial lines.

In our most recent work [88], we presented a novel approach toward compressed sensing via nonlocal Perona-Malik diffusion (PMD) [89]. The basic idea is to view the discrete implementation of PMD as a nonexpansive map and incorporate it into an alternative projection framework [90]. Despite the lack of rigorous proof about the convergence of PMD [91], experimental studies have shown its numerically stable behavior, which is also observed from our compressed sensing experiments. Two small tricks were invented to enhance the power of PMD under the context of compressed sensing: one is nonlocal extension [16] — based on the observation that the object of biomedical imaging is often bilateral symmetric, we propose to generalize the gradient/difference operator by including a nonlocal neighbor (the pixel at the location of mirrored position); the other is graduated non-nonconvexity [29] — the edge-stopping constant K in PMD is turned into a "temperature" variable that periodically decreases as the iteration proceeds.

In this experiment, we can show that perfect reconstruction of the 256×256 *phantom* image can be achieved from only eight radial lines. In our current implementation, we have adopted the following parameter setting for Algorithm 1 - $c(x) = c_1(x)$, $\lambda = 0.0125$, $K_0 = 4$, $t_{max} = 10^5$. The original 256×256 *phantom* image and its eight radial lines in the Fourier domain are shown in Figure 5.10. Figure 5.10 also shows the PSNR profile of Algorithm 1 — it is interesting to note that the phase transition behavior as the temperature parameter changes (after the half number of iterations). The convergent behavior of Algorithm 1 can be easily seen close to the end of iterations. The total running time of our MATLAB implementation (around twenty lines) on a typical PC (2.33 GHz dual-core processer, 4G memory) is around 1 hour; by contrast, it took the C-coded MEX-based implementation of BM3D-based approach [85] around 6 hours for the same task.

5.6 Conclusions and Connections

In this chapter, we have reviewed the history of sparse representations and revisited two closely related issues: *locality* and *convexity*. From local to nonlocal sparse representations,

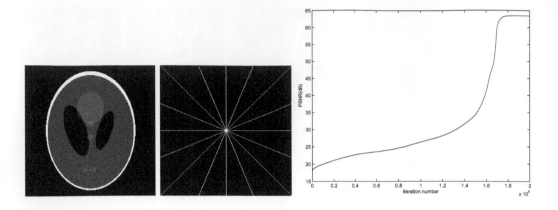

Figure 5.10 The original 256×256 *phantom* image (left), the eight radial lines in the Fourier domain (middle), and the PSNR result of nonlocal Perona–Malik diffusion.

we have discussed two exemplar ideas: dictionary learning (from signal-independent basis functions to signal-dependent ones) and structural clustering (from domain to range in the definition of local neighborhood). We have adopted a manifold-based geometric perspective to compare and understand these two lines of thoughts, which respectively correspond to global and local views toward discovering the topology underlying the image manifold. We have also challenged the popularity of compressed sensing (or l_1-based optimization) and argued that over-dependency on the convexity-blessed computational tools could blind us from gaining more fundamental insights into the organizational principles underlying photographic images. Based on an analogy between nonconvex optimization and statistical mechanics, we have advocated a deterministic annealing-based approach toward optimizing NSR-based image restoration. We have demonstrated the potential of nonlocal sparsity and deterministic annealing in four typical image restoration applications. All experimental results reported in this chapter are fully reproducible, which we hope can jump-start the research of young minds entering the field.

Readers with advanced mathematical training might find the writing style of this chapter in many aspects rigor-lacking. This is because we target engineering students who are not familiar with the calculus of variation. More rigorous treatment of this subject can be found in recent works dealing with nonlocal regularization [73, 92] as well as a research monograph devoted to sparse and redundant representations [93]. Meanwhile, we would argue that mentally reproducible research is not sufficient for computational science because the build-up of mathematical models is only one part of the story. The other equally important part is *the test of theoretic model against real-world observation data*. From this perspective, we have strived hard to make this research experimentally reproducible. Diligent readers might choose to rerun the four sets of experiments on other photographic images and develop better experimentally reproducible algorithms to advance the state-of-the-art. Only through "standing upon each other's shoulders" can we expedite the progress in our field and reach a higher impact in the venue of modern scientific research. More specifically, here is a list of issues/questions begging for answers:

• Locality or representation related: In addition to k-means and kNN, many other data clustering techniques (e.g., graph theoretic [94], spectral [95], mean-shift [96], EM-based [63]) have been developed by various technical communities. Even though many of them are intrinsically connected, how does the choice of clustering affect the nonlocal sparsity and its related image restoration algorithms? At the heart of all data clustering techniques is the definition of the *similarity/dissimilarity* metric. How do we learn the geodesic distance of a manifold from finite training data without running into the curse of dimensionality? Can such a manifold perspective turn useful to a nonlocal extension of MRF?

• Convexity or optimization related: Deterministic annealing has been proposed for clustering in the literature (e.g., central [72] versus pairwise [97]). Is it possible that clustering could lead us to connect NSR with nonlocal Monte Carlo methods [71]? To the best of our knowledge, Monte Carlo methods are the most powerful computational tool for high-dimensional data analysis. So maybe sampling from a nonconvex image manifold (i.e., the prior model) could benefit from the observation data, especially from a Bayesian point of view (prior becomes posterior after data are given). The physical implication of studying nonconvexity lies in its connection with bistability [98], hysteresis [99], and mechanism of biological memory [100].

The long-term objective along these lines of research seems to connect with the following scientific question: Where does *sparsity* come from? We acknowledge that this question is as deep as human's constant pursuit about how nature works. As another Nobel Laureate Abdus Salam once said, "Nature is not economical of structures but of principles." Is sparsity a manifest of a more general variational principle that has helped explain the mechanism of many physical and chemical systems (e.g., reaction-diffusion [101, 102])? Can sparsity help us understand the machinery of visual perception as the first step toward *what is intelligence*? If NSR can help us think out-of-the-box (e.g., Hilbert space), we might have a better chance of probing into the organizational principles underlying photographic images before our hands are tied to artificial objects created by ourselves (e.g., inner product or basis function). Once more, we can never deny the importance of mathematics but we had better also keep in mind that "imagination tires before nature" (Blaise Pascal).

Bibliography

[1] L. Rudin, S. Osher, and E. Fatemi, "Nonlinear total variation based noise removal algorithms," *Physica D*, vol. 60, pp. 259–268, 1992.

[2] L. Rudin and S. Osher, "Total variation based image restoration with free local constraints," in *IEEE International Conference on Image Processing*, pp. 31–35, 1994.

[3] D. Donoho, "Compressed sensing," *IEEE Transactions on Information Theory*, vol. 52, no. 4, pp. 1289–1306, 2006.

[4] E. J. Candès, J. K. Romberg, and T. Tao, "Robust uncertainty principles: Exact signal reconstruction from highly incomplete frequency information.," *IEEE Transactions on Information Theory*, vol. 52, no. 2, pp. 489–509, 2006.

[5] S. Mallat, *A Wavelet Tour of Signal Processing*. New York: Academic Press, 2nd ed., 1999.

[6] R. A. DeVore, B. Jawerth, and B. J. Lucier, "Image compression through wavelet transform coding," *IEEE Transactions on Information Theory*, vol. 38, pp. 719–746, Mar. 1992.

[7] D. Donoho and I. Johnstone, "Ideal spatial adaptation by wavelet shrinkage," *Biometrika*, vol. 81, pp. 425–455, 1994.

[8] D. Donoho, "De-noising by soft-thresholding," *IEEE Transactions on Information Theory*, vol. 41, pp. 613–627, 1995.

[9] Z. Xiong, K. Ramchandran, and M. Orchard, "Inverse halftoning using wavelets," *IEEE Transactions on Image Processing*, vol. 7, pp. 1479–1483, 1999.

[10] O. G. Guleryuz, "Nonlinear approximation based image recovery using adaptive sparse reconstructions and iterated denoising. Part I: Theory," *IEEE Transactions on Image Processing*, vol. 15, no. 3, pp. 539–554, 2006.

[11] J. Bioucas-Dias and M. Figueiredo, "A new TWIST: Two-step iterative shrinkage/thresholding algorithms for image restoration," *IEEE Transactions on Image Processing*, vol. 16, pp. 2992–3004, Dec. 2007. (http://www.lx.it.pt/ bioucas/code/TwIST_v1.zip)

[12] L. Mancera and J. Portilla, "Non-convex sparse optimization through determinisitic annealing and applications," *International Conference on Image Processing*, pp. 917–920, 2008.

[13] A. Efros and T. Leung, "Texture synthesis by non-parametric sampling," in *International Conference on Computer Vision*, pp. 1033–1038, 1999.

[14] A. A. Efros and W. T. Freeman, "Image quilting for texture synthesis and transfer," in *Proceedings of the 28th Annual Conference on Computer Graphics and Interactive Techniques*, pp. 341–346, 2001.

[15] A. Criminisi, P. Perez, and K. Toyama, "Region filling and object removal by exemplar-based image inpainting," *IEEE Transactions on Image Processing*, vol. 13, pp. 1200–1212, September 2004.

[16] I. Drori, D. Cohen-Or, and H. Yeshurun, "Fragment-based image completion," in *Proceedings of the 30th Annual Conference on Computer Graphics and Interactive Techniques*, pp. 303–312, 2003.

[17] A. Buades, B. Coll, and J.-M. Morel, "A non-local algorithm for image denoising," *IEEE Conference Computer Vision and Pattern Recognition*, vol. 2, pp. 60–65, 2005.

[18] K. Dabov, A. Foi, V. Katkovnik, and K. Egiazarian, "Image denoising by sparse 3-D transform-domain collaborative filtering," *IEEE Transactions on Image Processing*, vol. 16, pp. 2080–2095, Aug. 2007.

[19] V. Cheung, B. J. Frey, and N. Jojic, "Video epitomes," in *Proc. IEEE Conference Computer Vision and Pattern Recognition*, pp. 42–49, 2005.

[20] J. Yang, J. Wright, T. Huang, and Y. Ma, "Image super resolution as sparse representation of raw image patches," *IEEE Conference on Computer Vision and Pattern Recognition*, 2008.

[21] J. Mairal, F. Bach, J. Ponce, G. Sapiro, and A. Zisserman, "Non-local sparse models for image restoration," in *2009 IEEE 12th International Conference on Computer Vision*, pp. 2272–2279, 2009.

[22] X. Li, "Fine-granularity and spatially-adaptive regularization for projection-based image deblurring," *IEEE Transactions on Image Processing*, vol. 20, no. 4, pp. 971–983, 2011.

[23] I. Gelfand and S. Fomin, *Calculus of Variations*. Englewood Cliffs, NJ: Prentice Hall, 1963.

[24] J. Huang and D. Mumford, "Statistics of natural images and models," in *Proceedings of the IEEE Computer Society Conference on Computer Vision and Pattern Recognition*, vol. 1, pp. 541–547, 1999.

[25] A. Srivastava, A. Lee, E. Simoncelli, and S. Zhu, "On advances in statistical modeling of natural images," *Journal of Mathematical Imaging and Vision*, vol. 18, pp. 17–33, January 2003.

[26] W. T. Freeman, T. R. Jones, and E. C. Pasztor, "Example-based super resolution," *IEEE Computer Graphics and Applications*, vol. 22, pp. 56–65, 2002.

[27] I. Ekeland, "Nonconvex minimization problems," *Bulletin of the American Mathematical Society*, vol. 1, no. 3, pp. 443–474, 1979.

[28] K. Rose, "Deterministic annealing for clustering, compression, classification, regression, and related optimization problems," *Proceedings of the IEEE*, vol. 86, pp. 2210–2239, Nov. 1998.

[29] A. Blake and A. Zisserman, *Visual Reconstruction*. Cambridge, MA: MIT Press, 1987.

[30] D. Amit and V. Martin-Mayor, *Field Theory, The Renormalization Group, and Critical Phenomena*. Singapore: World Scientific, 1984.

[31] B. B. Mandelbrot, *The Fractal Geometry of Nature*. San Francisco: W.H. Freeman, 1982, Revised edition of: Fractals (1977), 1977.

[32] J. M. Shapiro, "Embedded image coding using zerotrees of wavelet coefficients," *IEEE Transactions on Acoustic Speech and Signal Processing*, vol. 41, no. 12, pp. 3445–3462, 1993.

[33] C. Bouman and K. Sauer, "A generalized Gaussian image model for edge-preserving MAP estimation," *IEEE Transactions on Image Processing*, vol. 2, no. 3, pp. 296–310, 1993.

[34] D. J. Field, "What is the goal of sensory coding?" *Neural Computation*, vol. 6, no. 4, pp. 559–601, 1994.

[35] B. Olshausen and D. Field, "Emergence of simple-cell receptive field properties by learning a sparse code for natural images," *Nature*, vol. 381, pp. 607–609, 1996.

[36] E. Candes and D. Donoho, "Curvelets: A surprisingly effective non-adaptive representation for objects with edges," in *Curve and Surface Fitting* (A. C. et al., Ed.), Nashville, TN: Vanderbilt University Press, 1999.

[37] M. N. Do and M. Vetterli, "The contourlet transform: An efficient directional multiresolution image representation," *IEEE Transactions on Image Processing*, vol. 14, pp. 2091–2106, Dec. 2005.

[38] E. Candes, *Ridgelets: Theory and Applications.* Ph.D. thesis, Stanford University, 1998. Department of Statistics.

[39] E. LePennec and S. Mallat, "Sparse geometric image representation with bandelets," *IEEE Transactions on Image Processing*, vol. 14, no. 4, pp. 423–438, 2005.

[40] M. Bertalmio, G. Sapiro, V. Caselles, and C. Ballester, "Image inpainting," in *Proceedings of SIGGRAPH*, (New Orleans, LA), pp. 417– 424, 2000.

[41] M. Elad and M. Aharon, "Image denoising via sparse and redundant representations over learned dictionaries," *IEEE Transactions on Image Processing*, vol. 15, pp. 3736–3745, December 2006.

[42] X. Li, "Variational bayesian image processing on graphical probability models," in *Proc. of International Conference on Image Processing*, 2008.

[43] P. Chatterjee and P. Milanfar, "Clustering-based denoising with locally learned dictionaries," *IEEE Transactions on Image Processing*, vol. 18, no. 7, pp. 1438–1451, 2009.

[44] D. L. Donoho and M. Elad, "Optimally sparse representation in general (nonorthogonal) dictionaries via 1 minimization," *Proceedings of the National Academy of Science*, vol. 100, pp. 2197–2202, Mar. 2003.

[45] M. Zibulevsky and M. Elad, "L1-l2 optimization in signal and image processing," *IEEE Signal Processing Magazine*, vol. 27, pp. 76 –88, May 2010.

[46] K. Wilson, "The renormalization group: Critical phenomena and the Kondo problem," *Reviews of Modern Physics*, vol. 47, no. 4, pp. 773–840, 1975.

[47] R. W. R. Gonzalez and S. Eddins, *Digital Image Processing Using MATLAB.* Englewood Cliffs, NJ: Prentice-Hall, 2004.

[48] D. Marr, *Vision— A Computational Approach.* New York: W.H. Freeman, 1982.

[49] T. Lindeberg, *Scale-space Theory in Computer Vision.* Dordrecht: Kluwer Academic Publishers, 1994.

[50] S. Mallat, "Multiresolution approximations and wavelet orthonormal bases of $l^2(\mathbf{r})$," *Transactions of the American Mathematical Society*, vol. 315, pp. 69–87, 1989.

[51] S. Mallat and W. Hwang, "Singularity detection and processing with wavelets," *IEEE Transactions on Information Theory*, vol. 8, pp. 617–643, 1992.

[52] R. Tibshirani, "Regression shrinkage and selection via the lasso," *Journal of the Royal Statistical Society, Series B*, vol. 58, pp. 267–288, 1996.

[53] Z. Xiong, M. Orchard, and Y. Zhang, "A deblocking algorithm for jpeg compressed images using overcomplete wavelet representations," *IEEE Transactions on Circuit and Systems for Video Technology*, vol. 7, pp. 433–437, 1997.

[54] I. Daubechies, "Where do wavelets come from? A personal point of view," *Proceedings of the IEEE*, vol. 84, no. 4, pp. 510–513, 1996.

[55] D. Bohm, "A suggested interpretation of the quantum theory in terms of hidden variables," *Physical Review*, vol. 85, no. 2, pp. 166–179, 1952.

[56] P. W. Anderson, "More is different," *Science*, vol. 177, pp. 393–396, Aug. 1972.

[57] C. Tomasi and R. Manduchi, "Bilateral filtering for gray and color images," in *International Conference on Computer Vision*, pp. 839–846, 1998.

[58] J. B. Tenenbaum, V. de Silva, and J. C. Langford, "A Global Geometric Framework for Nonlinear Dimensionality Reduction," *Science*, vol. 290, no. 5500, pp. 2319–2323, 2000.

[59] S. T. Roweis and L. K. Saul, "Nonlinear dimensionality reduction by locally linear embedding," *Science*, vol. 290, no. 5500, pp. 2323–2326, 2000.

[60] M. Elad, "Optimized projections for compressed sensing," *IEEE Transactions on Signal Processing*, vol. 55, no. 12, pp. 5695–5702, 2007.

[61] F. Girosi, "An equivalence between sparse approximation and support vector machines," *Neural Computation*, vol. 10, no. 6, pp. 1455–1480, 1998.

[62] R. Duda, P. Hart, and D. Stork, *Pattern Classification*. New York: Wiley, 2nd ed., 2001.

[63] M. A. T. Figueiredo and A. K. Jain, "Unsupervised learning of finite mixture models," *IEEE Transactions on Pattern Analysis and Machine Intelligence*, vol. 24, no. 3, pp. 381–396, 2002.

[64] K. Weinberger and L. Saul, "Unsupervised learning of image manifolds by semidefinite programming," *International Journal of Computer Vision*, vol. 70, no. 1, pp. 77–90, 2006.

[65] N. Kambhatla and T. K. Leen, "Dimension reduction by local principal component analysis," *Neural Computation*, vol. 9, no. 7, pp. 1493–1516, 1997.

[66] X. Li and Y. Zheng, "Patch-based video processing: a variational bayesian approach," *IEEE Transactions on Circuits and Systems for Video Technology* vol. 19, no. 1, pp. 27–40, 2009.

[67] W. Dong, X. Li, L. Zhang, and G. Shi, "Sparsity-based image via dictionary learning and structural clustering," *IEEE Conference on Computer Vision and Pattern Recognition*, 2011.

[68] S. Kirkpatrick, C. D. Gelatt, and M. P. Vecchi, "Optimization by simulated annealing," *Science*, vol. 220, pp. 671–680, 1983.

[69] R. Swendsen and J. Wang, "Nonuniversal critical dynamics in Monte Carlo simulations," *Physical Review Letters*, vol. 58, no. 2, pp. 86–88, 1987.

[70] U. Wolff, "Collective Monte Carlo updating for spin systems," *Physical Review Letters*, vol. 62, no. 4, pp. 361–364, 1989.

[71] J. S. Liu, *Monte Carlo Strategies in Scientific Computing*. Berlin: Springer Series in Statistics, 2001.

[72] K. Rose, E. Gurewwitz, and G. Fox, "A deterministic annealing approach to clustering," *Pattern Recognition Letters*, vol. 11, no. 9, pp. 589–594, 1990.

[73] A. Elmoataz, O. Lezoray, and S. Bougleux, "Nonlocal discrete regularization on weighted graphs: A framework for image and manifold processing," *IEEE Transactions on Image Processing*, vol. 17, pp. 1047–1060, July 2008.

[74] S. Geman and D. Geman, "Stochastic relaxation, Gibbs distributions, and the Bayesian restoration of images," *IEEE Transactions on Pattern Analysis and Machine Intelligence*, vol. 6, pp. 721–741, Nov. 1984.

[75] J. B. Buckheit and D. L. Donoho, "Wavelab and reproducible research," in *Lecture Notes Statistics*, pp. 55–81, Berlin: Springer-Verlag, 1995.

[76] P. Vandewalle, J. Kovacevic, and M. Vetterli, "Reproducible research in signal processing — What, why, and how," *IEEE Signal Processing Magazine*, vol. 26, pp. 37–47, May 2009.

[77] J. M. Bioucas-Dias, M. A. T. Figueiredo, and J. P. Oliveira, "Total variation-based image deconvolution: A majorization-minimization approach," in *International Conference on Acoustics, Speech and Signal Processing*, vol. 2, pp. 861–864, May 2006. (http://www.lx.it.pt/ bioucas/code/adaptive TVMM demo.zip)

[78] A. Foi, V. Katkovnik, and K. Egiazarian, "Pointwise shape-adaptive DCT for high-quality denoising and deblocking of grayscale and color images," *IEEE Transactions on Image Processing*, vol. 16, pp. 1395–1411, May 2007. (www.cs.tut.fi/ foi/SA-DCT/)

[79] W. Dong, L. Zhang, and G. Shi, "Centralized sparse representation for image restoration," *IEEE International Conference on Computer Vision (ICCV)*, 2011.

[80] S. Dai, M. Han, W. Xu, Y. Wu, Y. Gong, and A. Katsaggelos, "Softcuts: A soft edge smoothness prior for color image super resolution," *IEEE Transactions on Image Processing*, vol. 18, no. 5, pp. 969–981, 2009.

[81] A. Marquina and S. Osher, "Image super resolution by TV-regularization and breg-man iteration," *Journal of Scientific Computing*, vol. 37, no. 3, pp. 367–382, 2008.

[82] J. Yang, J. Wright, T. Huang, and Y. Ma, "Image super resolution via sparse repre-sentation," *IEEE Transactions on Image Processing*, vol. 19, no. 11, pp. 2861–2873, 2010.

[83] G. S. X. W. W. Dong, X. Li, and L. Zhang, "Image reconstruction with locally adap-tive sparsity and nonlocal robust regularization," *Inverse Problems*, to be submitted, 2011.

[84] R. Chartrand, "Exact reconstruction of sparse signals via nonconvex minimization," *IEEE Signal Processing Letters*, vol. 14, pp. 707–710, Oct. 2007.

[85] K. Egiazarian, A. Foi, and V. Katkovnik, "Compressed sensing image reconstruc-tion via recursive spatially adaptive filtering," in *IEEE International Conference on Image Processing*, vol. 1, (San Antonio, TX, USA), Sept. 2007.

[86] X. Zhang, M. Burger, X. Bresson, and S. Osher, "Bregmanized nonlocal regulariza-tion for deconvolution and sparse reconstruction," *UCLA CAM Report*, pp. 09–03, 2009.

[87] J. Trzasko and A. Manduca, "Highly undersampled magnetic resonance image re-construction via homotopic l_{0}-minimization," *IEEE Transactions on Medical Imaging*, vol. 28, no. 1, pp. 106–121, 2009.

[88] X. Li, "The magic of nonlocal Perona–Malik diffusion," *IEEE Signal Processing Letters*, vol. 18, no. 9, pp. 533–534.

[89] P. Perona and J. Malik, "Scale space and edge detection using anisotropic diffusion," *IEEE Transactions on Pattern Analysis and Machine Intelligence*, vol. 12, no. 7, pp. 629–639, 1990.

[90] D. Youla, "Generalized image restoration by the method of alternating orthogonal projections," *IEEE Transactions on Circuits and System*, vol. 9, pp. 694–702, Sept. 1978.

[91] S. Kichenassamy, "The Perona-Malik paradox," *SIAM Journal on Applied Mathe-matics*, vol. 57, no. 5, pp. 1328–1342, 1997.

[92] G. Gilboa and S. Osher, "Nonlocal operators with applications to image processing," *Multiscale Modeling and Simulation*, vol. 7, no. 3, pp. 1005–1028, 2008.

[93] M. Elad, *Sparse and Redundant Representations: From Theory to Applications in Signal and Image Processing*. Berlin: Springer, 2010.

[94] Z. Wu and R. Leahy, "An optimal graph theoretic approach to data clustering: Theory and its application to image segmentation," *IEEE Transactions on Pattern Analysis and Machine Intelligence*, vol. 15, no. 11, pp. 1101–1113, 1993.

[95] A. Ng, M. Jordan, and Y. Weiss, "On spectral clustering: Analysis and an algorithm," *Advances in Neural Information Processing Systems (NIPS)*, vol. 2, pp. 849–856, 2002.

[96] D. Comaniciu and P. Meer, "Mean shift: A robust approach toward feature space analysis," *IEEE Transactions on Pattern Analysis and Machine Intelligence*, vol. 24, pp. 603–619, May 2002.

[97] T. Hofmann and J. M. Buhmann, "Pairwise data clustering by deterministic annealing," *IEEE Transactions on Pattern Analysis and Machine Intelligence*, vol. 19, no. 1, pp. 1–14, 1997.

[98] L. Gammaitoni, P. Hanggi, P. Jung, and F. Marchesoni, "Stochastic resonance," *Reviews of Modern Physics*, vol. 70, no. 1, pp. 223–287, 1998.

[99] I. Mayergoyz, "Mathematical models of hysteresis," *Physical Review Letters*, vol. 56, no. 15, pp. 1518–1521, 1986.

[100] L. Mekler, "Mechanism of biological memory," *Nature*, vol. 215, pp. 481–484, 1967.

[101] A. Turing, "The chemical basis of morphogenesis," *Philosophical Transactions of the Royal Society of London. Series B, Biological Sciences*, vol. 237, no. 641, pp. 37–72, 1952.

[102] J. Smoller, *Shock Waves and Reaction-Diffusion Equations*. Berlin: Springer, 1994.

Chapter 6

Resolution Enhancement Using Prior Information

HSIN M. SHIEH
Feng Chia University

CHARLES L. BYRNE
University of Massachusetts

MICHAEL A. FIDDY
University of North Carolina

6.1 Introduction

Many image reconstruction techniques have been remarkably successful in applications such as medical imaging, remote sensing, nondestructive testing, and radar. The focus here is on a general problem that arises in many such applications: the reconstruction of a compactly supported function f from a limited number of measurements such as its image values. In reconstructing a function f from its finitely-many linear functional values, noisy or not, it is not possible to specify f uniquely. Estimates based on finite data may succeed in recovering broad features of f, but may fail to resolve important detail. While there has been considerable effort in the development of inverse or estimation algorithms, this problem has generally proved difficult.

For the typical example of reconstructing a real function (i.e., object irradiance) from a finite number of Fourier values, there will mathematically exist an infinite number of potential solutions [1,2]. A large number of inverse problems, associated with many applications, can be described as computing the image of an object from samples of its Fourier transformation. This includes computerized tomography, diffraction tomography, many variants of radar-imaging applications, and the estimate of spectra from a finite time series. Among all data-consistent estimates, the well-known minimum-norm (MN) estimation finds the one that has the smallest energy. The special MN estimation for the Fourier estimation problem, referred to as the discrete Fourier transform (DFT), can be easily and efficiently implemented with a finite-length data vector of uniformly sampled Fourier-transform values by the fast Fourier transform (FFT). However, this usually fails to achieve an acceptable resolution for a fine-structured object. It is well understood that the degrees of freedom from finitely sampled Fourier data limit the DFT resolution [2, 3].

To single out one particular data-consistent solution of f and combat the image degradation due to the limited nature of the data, one can require that some functional value of the image, such as its entropy, be optimized [2, 4, 5], or that the solution be closest to some other appropriate prior estimate according to a given distance criterion [6–9]. Linear and nonlinear, model-based data extrapolation procedures can be used to improve resolution, but at the cost of increased sensitivity to noise. For some applications the main requirement may be that the reconstruction algorithm be easily implemented and rapidly calculated. Best results are achieved when the criteria chosen force the reconstructed image to incorporate features of the true function that are known a priori.

The improvement in estimating linear-functional values of f that have not been measured from those that have been, represents to prior information about the function f, such as support information or, more generally, estimates of the overall profile of f. One way to do this is through minimum-weighted-norm (MWN) estimation, with the prior information used to determine the weights. Based on a weight function p characterizing the support limitation of f and the choice of minimum weighted L^2 norm among all data-consistent solutions, the prior discrete Fourier transform (PDFT) [2, 6–13] algorithm has proved its great potential in resolution enhancement. This MWN approach was presented in [6], where it was called the PDFT method. The PDFT was based on an earlier but noniterative version of the Gerchberg-Papoulis band-limited extrapolation procedure [14]. The PDFT was then applied to image reconstruction problems in [10], and an application of the PDFT was presented in [15]. In [7], the PDFT was extended to a nonlinear version, the indirect PDFT (IPDFT), that generalizes Burg's maximum entropy spectrum estimation method. The PDFT was applied to the phase problem in [8], and in [11] both the PDFT and IPDFT were examined in the context of Wiener filter approximation. More recent work on these topics is discussed in the book [2].

Apart from image reconstruction from Fourier-transform data, the PDFT has been extended to the imaging problem based on Radon-transform values. While computerized tomography and B-scan ultrasonic imaging have been the first successful examples, other extensions can be developed in a similar manner. In addition, difficulties with implementing the PDFT on a large data set (tomographic problem) have been overcome through the use of the discrete PDFT (DPDFT) approximation method. The DPDFT involves the calculation of a minimum-weighted-norm solution of a large system of linear equations, for which the iterative scheme of the algebraic reconstructive techniques (ART) method has been chosen to improve the computational efficiency.

Compressed sensing (CS) is a sampling mechanism that reconstructs signals/images of interest using a small number of measurements, and has attracted much attention in various applications. This issue becomes more important when success of digital data acquisition is increasing pressure on signal/image processing hardware and software to support, or the measurements are limited by either one or a combination of the factors such as instrument constraints, limited observation time, nonstationary observation process and the ill-posed nature of the problem. The treatment of this problem can resort to finding a sparse solution, with nonzero entries only on a minimal support of the solution domain, for which a maximally sparse representation is typically the solution to be expected and can often be accomplished by one-norm minimization. We specifically provide a discussion about the roles of prior knowledge and norm, in particular to develop a superior data acquisition system and an optimal, possibly most sparse representation of the function we make measurements of.

The PDFT has been successfully applied for resolution enhancement of the image reconstruction of a compactly supported function from only finitely-many data values, but typically fails to obtain acceptable details over regions containing small-scale features of interest. It is also often the goal to recover accurately some small subregions of the object. This critical issue is much harder, especially when the small-scale portions involve only a small amount of the total energy from the object. For improving the image estimate, a new approach has been developed to extend the PDFT by allowing different weight functions to modulate the different spatial frequency components of the reconstructed image. The resolution enhancement of this new approach comes from the incorporation of prior knowledge of the domain of the object as well as the locations of the small-scale subregions of interest. The simple strategy in the choice of weights takes only two different window functions, while the improved one allows more flexible window functions to be chosen as weights. More flexibility in the choice of weights can provide a superior image estimate, but typically at the cost of a time-consuming search process.

6.2 Fourier Transform Estimation and Minimum L^2-Norm Solution

The image reconstruction problem is to reconstruct a (possibly complex-valued) function $f : S \subseteq R^D \to C$ from measurements pertaining to that function. The subset S is the support of the object function, which is almost always assumed to be finite, but which may not be known precisely. The common practical case is that of finitely many measurements d_n, $n = 1, ..., N$, that are related linearly to the distribution function f. For such problems, Hilbert space techniques will clearly play an important role. One important issue, which we discuss in detail subsequently, is the choice of the Hilbert space within which to formulate the problem. Having selected the Hilbert space, our approach will be to seek the estimate of f of minimum norm that is consistent with the measured data.

6.2.1 Hilbert Space Reconstruction Methods

To model the operator that transforms f into the data values, we need to select an ambient space containing f. A typical approach is to choose a Hilbert space. The inner product induces a norm, and the reconstruction is that function, consistent with the data, for which this norm is minimized. What follows is a brief introduction of the well-known Hilbert space minimum norm solution method of reconstruction; for the moment, the Hilbert space is $L^2(S)$.

6.2.2   Minimum L^2-Norm Solutions

The estimation problem concerned is highly under-determined; there are infinitely many functions in $L^2(S)$ that are consistent with the data and might be the right answer. We proceed conservatively, selecting as the estimate that function consistent with the data that has the smallest norm.

Assume that the data vector $\mathbf{d} = (d_1, d_2, ..., d_N)^T$ is in C^N and the linear operator $\mathcal{H}$ from $L^2(S)$ to C^N takes f to $\mathbf{d}$; so we have

$$\mathbf{d} = \mathcal{H}f. \tag{6.1}$$

Typically, the data d_n, $n = 1, ..., N$, are associated with some linearly independent functions $h_n, n = 1, 2, \ldots, N$, and

$$d_n = \langle f, h_n \rangle_2 = \int_S f(\mathbf{x}) \, \overline{h_n(\mathbf{x})} \, d\mathbf{x}, \tag{6.2}$$

where the overline indicates the conjugate operator. Associated with the mapping $\mathcal{H}$ is its adjoint operator, $\mathcal{H}^\dagger$, going from C^N to $L^2(S)$ and given, for each vector $\mathbf{a} = (a_1, a_2, ..., a_N)^T$, by

$$\mathcal{H}^\dagger \mathbf{a} = a_1 h_1 + ... + a_N h_N. \tag{6.3}$$

The operator from C^N to C^N defined by $\mathcal{H}\mathcal{H}^\dagger$ corresponds to an $N \times N$ matrix, which we also denote by $\mathcal{H}\mathcal{H}^\dagger$. If the functions h_n are linearly independent, then this matrix is positive-definite, and therefore invertible.

Given the data vector $\mathbf{d}$, we can solve the system of linear equations

$$\mathbf{d} = \mathcal{H}\mathcal{H}^\dagger \mathbf{a} \tag{6.4}$$

for the vector $\mathbf{a}$. Then the function

$$f_{\text{M2N}} = \mathcal{H}^\dagger \mathbf{a} \tag{6.5}$$

is consistent with the measured data and is the function in $L^2(S)$ of least norm for which this is true. The function $u = f - f_{\text{M2N}}$ has the property $\mathcal{H}u = 0$. It is easy to see that

$$\|f\|_2^2 = \|f_{\text{M2N}}\|_2^2 + \|u\|_2^2, \tag{6.6}$$

where $\|\cdot\|_2$ indicates the operator of L^2 norm. The estimate $f_{\text{M2N}}(r)$ is the minimum-norm solution, with respect to the $L^2(S)$ norm.

6.2.3   Case of Fourier-Transform Data

To illustrate the minimum-norm solution, we consider the case in which the data d_n, $n = 1, ..., N$, are values of the Fourier transform of f. Specifically, suppose that

$$d_n = \int_S f(x) \, \frac{e^{-ik_n x}}{2\pi} \, dx, \tag{6.7}$$

for arbitrary values k_n. For simplicity, we consider the one-dimensional case.

6.2.3.1 The $L^2(-\pi, \pi)$ Case

Assume that $f(x) = 0$, for $|x| > \pi$. Then $S = [-\pi, \pi]$. The minimum L^2-norm solution in $L^2(S)$ has the form

$$f_{\text{M2N}}(x) = \sum_{m=1}^{N} a_m \, e^{ik_m x}, \tag{6.8}$$

with

$$d_n = \sum_{m=1}^{N} a_m \int_{-\pi}^{\pi} \frac{e^{i(k_m - k_n)x}}{2\pi} \, dx. \tag{6.9}$$

For the equispaced values $k_n = n$, we have $a_m = d_m$, and the minimum-norm solution is the discrete Fourier transform (DFT)

$$f_{\text{M2N}}(x) = \sum_{n=1}^{N} d_n e^{inx}. \tag{6.10}$$

6.2.3.2 Over-Sampled Case

Suppose now that $f(x) = 0$ for $|x| > X$, where $0 < X < \pi$. If we know such an X, we can use $L^2(-X, X)$ as the Hilbert space; clearly, $S \subseteq [-X, X]$, but S need not be all of $[-X, X]$. Denote by $\chi_X(x)$ the characteristic function of the interval $[-X, X]$, which is one for $|x| \leq X$ and zero otherwise.

For equispaced data at $k_n = n$, it will have

$$d_n = \int_{-\pi}^{\pi} f(x)\chi_X(x)\frac{e^{-inx}}{2\pi} \, dx, \tag{6.11}$$

so that the minimum-norm solution has the form

$$f_{\text{M2N}}(x) = \chi_X(x) \sum_{m=1}^{N} a_m \, e^{imx}, \tag{6.12}$$

with

$$d_n = \sum_{m=1}^{N} a_m \frac{\sin X(m - n)}{\pi(m - n)}. \tag{6.13}$$

The minimum-norm solution is supported on $[-X, X]$ and consistent with the Fourier-transform data.

6.2.4 Case of Under-Determined Systems of Linear Equations

Given a system of linear equations, with M unknowns and N equations, the system is under-determined if $N < M$ and also involves the ambiguity of nonunique solutions. This is a typical case when the discrete solution is solved alternately in the approximation sense for realistic applications, with the aid of digital computing processors. For explanation, let's assume that

$$\mathbf{d} = \mathbf{H}\mathbf{f}, \tag{6.14}$$

with $\mathbf{d} = (d_1, d_2, ..., d_N)^T$ is in C^N, $\mathbf{f} = (f_1, f_2, ..., f_M)^T$ is in C^M and the linear operator $\mathbf{H}$ from C^M to C^N. For Equation (6.7), it can be discretized approximately as

$$d_n = \sum_{m=1}^{M} f(x_m) \frac{e^{-ik_n x_m} \triangle x}{2\pi}, \tag{6.15}$$

$n = 1, 2, \ldots, N$. $\triangle x$ indicates the sampling interval of the function f and $\mathbf{H}$ is the $M \times N$ matrix with entries $\mathbf{H}_{mn} = \exp(-ik_n x_m)\triangle x/2\pi$. Then it typically contains the situation that $\mathbf{d} = \mathbf{H}\mathbf{f} = \mathbf{H}\mathbf{z}$ and $\mathbf{f} \neq \mathbf{z}$. In such cases, one can select one solution out of the infinitely many possibilities by requiring that the solution also satisfy some additional constraints. It is reasonable to ask for a solution $\mathbf{f}$ of smallest two-norm

$$||\mathbf{f}||_2 = \sqrt{\sum_{n=1}^{N} |\mathbf{f}_n|^2}. \tag{6.16}$$

The minimum two-norm solution of $\mathbf{d} = \mathbf{H}\mathbf{f}$ is a vector of the form

$$\mathbf{f}_{\mathrm{M2N}} = \mathbf{H}^{\ddagger}\mathbf{z}, \tag{6.17}$$

where $\ddagger$ indicates the operator of the conjugate transpose. Then $\mathbf{d} = \mathbf{H}\mathbf{f}_{\mathrm{M2N}}$ becomes $\mathbf{d} = \mathbf{H}\mathbf{H}^{\ddagger}\mathbf{z}$. Typically, $(\mathbf{H}\mathbf{H}^{\ddagger})^{-1}$ will exist, and $\mathbf{z} = (\mathbf{H}\mathbf{H}^{\ddagger})^{-1}\mathbf{d}$, from which it follows that the minimum two-norm solution is

$$\mathbf{f}_{\mathrm{M2N}} = \mathbf{H}^{\ddagger}(\mathbf{H}\mathbf{H}^{\ddagger})^{-1}\mathbf{d}. \tag{6.18}$$

When M and N are not too large, forming the matrix $\mathbf{H}\mathbf{H}^{\ddagger}$ and solving for $\mathbf{z}$ is not prohibitively expensive or time-consuming. Otherwise, when M and N are large, one can turn to iterative algorithms to find the minimum two-norm solution. Both the ART and the Landweber algorithm converge to that solution closest to the starting vector $\mathbf{f}^0$, in the two-norm sense. Therefore, beginning with $\mathbf{f}^0 = \mathbf{0}$, these algorithms essentially give the minimum two-norm solution.

If C is a closed convex set in R^N, the projected Landweber algorithm converges to that solution $\mathbf{f}$ in C closest to $\mathbf{f}^0$, in the two-norm sense. Again, if taking $\mathbf{f}^0 = \mathbf{0}$, the projected Landweber algorithm converges to that solution $\mathbf{f}$ in C having the smallest two-norm.

6.3   Minimum Weighted L^2-Norm Solution

Typically, the minimum L^2-norm solution is a conservative approach to combat the ambiguity of nonunique data-consistent solutions in a $L^2(S)$. At the same time, however, there often exists some prior information about f that one can incorporate in the estimate by some means. One way to achieve both of these goals is to select the norm to incorporate prior information about f, and then to take as the estimate of f the function consistent with the data, for which the chosen norm is minimized.

If one changes the norm on $L^2(S)$, or, equivalently, the inner product, then the minimum-norm solution will change. For any continuous linear operator $\mathcal{T}$ on $L^2(S)$, the adjoint operator, denoted $\mathcal{T}^\dagger$, is defined by

$$\langle \mathcal{T}f, h \rangle_2 = \langle f, \mathcal{T}^\dagger h \rangle_2. \tag{6.19}$$

Thus, the adjoint operator will change as the inner product varies.

6.3.1 Class of Inner Products

Let $\mathcal{T}$ be a continuous, linear, and invertible operator on $L^2(S)$, and define the $\mathcal{T}$-inner product to be

$$\langle f, h \rangle_\mathcal{T} = \langle \mathcal{T}^{-1}f, \mathcal{T}^{-1}h \rangle_2. \tag{6.20}$$

Associated with the $\mathcal{T}$-inner product, the $\mathcal{T}$-norm is defined by

$$\|f\|_\mathcal{T} = \sqrt{\langle \mathcal{T}^{-1}f, \mathcal{T}^{-1}f \rangle_2}. \tag{6.21}$$

If we use the $\mathcal{T}$-inner product to define the problem to be solved, we will have that

$$d_n = \langle f, t_n \rangle_\mathcal{T} \tag{6.22}$$

for known functions t_n, $n = 1, 2, \ldots, N$. And thus, with the definition of this $\mathcal{T}$-inner product, we find that

$$\begin{aligned} d_n &= \langle f, h_n \rangle_2 \\ &= \langle \mathcal{T}f, \mathcal{T}h_n \rangle_\mathcal{T}. \end{aligned} \tag{6.23}$$

The adjoint operator for $\mathcal{T}$, with respect to the $\mathcal{T}$-norm, is denoted $\mathcal{T}^*$, and is defined by

$$\langle \mathcal{T}f, h \rangle_\mathcal{T} = \langle f, \mathcal{T}^*h \rangle_\mathcal{T}. \tag{6.24}$$

Therefore,

$$d_n = \langle f, \mathcal{T}^*\mathcal{T}h_n \rangle_\mathcal{T}. \tag{6.25}$$

Lemma 6.3.1

$$\mathcal{T}^*\mathcal{T} = \mathcal{T}\mathcal{T}^\dagger. \tag{6.26}$$

Consequently,

$$d_n = \langle f, \mathcal{T}\mathcal{T}^\dagger h_n \rangle_\mathcal{T}. \tag{6.27}$$

6.3.2   Minimum $\mathcal{T}$-Norm Solutions

The function f_{MTN} that is both data consistent and has the smallest $\mathcal{T}$-norm has the algebraic form

$$f_{\mathrm{MTN}} = \sum_{m=1}^{N} c_m \mathcal{T}\mathcal{T}^\dagger h_m \, . \tag{6.28}$$

Applying the $\mathcal{T}$-inner product to both sides of Equation (6.28), we get

$$
\begin{aligned}
d_n &= \langle f_{\mathrm{MTN}}, \mathcal{T}\mathcal{T}^\dagger h_n \rangle_{\mathcal{T}} & (6.29)\\
&= \sum_{m=1}^{N} c_m \langle \mathcal{T}\mathcal{T}^\dagger h_m, \mathcal{T}\mathcal{T}^\dagger h_n \rangle_{\mathcal{T}} \, . & (6.30)
\end{aligned}
$$

Therefore,

$$d_n = \sum_{m=1}^{N} c_m \langle \mathcal{T}^\dagger h_m, \mathcal{T}^\dagger h_n \rangle_2 \, . \tag{6.31}$$

Our reconstruction procedure is to solve this system for the c_m and insert them into Equation (6.28). The Gram matrix that appears in Equation (6.31) is positive-definite, but is often ill-conditioned; increasing the main diagonal by a percent or so usually is sufficient regularization.

6.3.3   Case of Fourier-Transform Data

To illustrate these minimum $\mathcal{T}$-norm solutions, we return now to the important case in which the measured data are values of the Fourier transform of the function f.

6.3.3.1   Using a Prior Estimate of f

Suppose that $f(x) = 0$ for $|x| > \pi$ again, and that $p(x)$ satisfies

$$0 < \epsilon \le p(x) \le E < +\infty \tag{6.32}$$

for all x in $[-\pi, \pi]$. Defining the operator $\mathcal{T}$ by $(\mathcal{T}f)(x) = \sqrt{p(x)}f(x)$, the $\mathcal{T}$-norm is then

$$\langle f, h \rangle_{\mathcal{T}} = \int_{-\pi}^{\pi} f(x)\overline{h(x)}p(x)^{-1}dx \, . \tag{6.33}$$

It follows that

$$d_n = \int_{-\pi}^{\pi} f(x)p(x)\frac{e^{-inx}}{2\pi}\, p(x)^{-1}dx \, , \tag{6.34}$$

so that the minimum $\mathcal{T}$-norm solution is

$$f_{\mathrm{MTN}}(x) = p(x) \sum_{m=1}^{N} a_m e^{imx} \, , \tag{6.35}$$

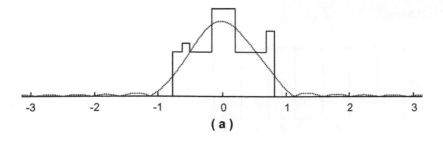

(a)

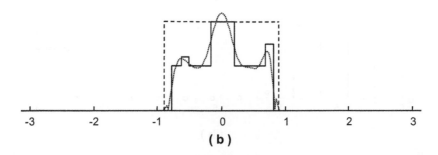

(b)

Figure 6.1 Reconstruction of a compactly supported function from fifteen sampled Fourier values: (a) DFT estimate; (b) PDFT estimate with $p = \chi_{0.9}$.

where

$$d_n = \sum_{m=1}^{N} a_m \int_{-\pi}^{\pi} p(x) \frac{e^{i(m-n)x}}{2\pi} \, dx . \qquad (6.36)$$

The estimate $f_{MTN}(x)$ is $f_{MW2N}(x)$ in the terminology of a previous section, and also is f_{PDFT} (the PDFT estimate) in more official terminology. If we have prior knowledge about the support of f, or some idea of its shape, one can incorporate that prior knowledge into the reconstruction through the choice of $p(x)$.

6.3.4 Case of $p(x) = \chi_X(x)$

If the function f is known to be compactly supported on a domain S equal to or in $[-X, X]$, $f(x) = 0$ at least for $x \notin [-X, X]$, then the prior estimate of f can be chosen to be $\chi_X(x)$ and $f_{MW2N}(x)$ can be written in the form of Equation (6.12), with the coefficients determined by Equation (6.13). While the mathematical expressions of $f_{MW2N}(x)$ are the same as that of $f_{M2N}(x)$ in Equation (6.12), $f_{MW2N}(x)$ is the function of minimum weighted norm with respect to $L^2(-\pi, \pi)$, and $f_{M2N}(x)$ in Equation (6.12) is the one of minimum $L^2(-X, X)$ norm. This simply explains that the resolution enhancement of f_{MW2N} (PDFT) superior to the usual $f_{M2N}(x)$ is driven by the use of p, with over-sampled data values. Based on a prior knowledge ($[-X, X]$) more accurately accommodating the true domain of f, it makes better use of the degrees of freedom over the domain f does not reside, as shown in Figure 6.1.

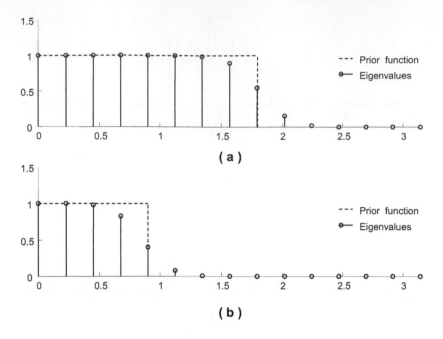

Figure 6.2 Eigenvalues of the matrix $\mathbf{P}$ (a) corresponding to $\chi_{1.8}$; (b) corresponding to $\chi_{0.9}$.

6.3.5   Regularization

The resolution enhancement of the PDFT essentially depends on over-sampled data and a decent estimate of the true support of the function f. The more over-sampled the data and the more accurately the true support is known, the greater the improvement in resolution, but also the greater the sensitivity to noise and model error. To clarify this point, we convert Equation (6.36) in the matrix equation form into

$$\mathbf{d} = \mathbf{P}\,\mathbf{a}\,, \tag{6.37}$$

with $\mathbf{d} = (d_1, d_2, \ldots, d_N)^T$, $\mathbf{a} = (a_1, a_2, \ldots, a_N)^T$, and the matrix $\mathbf{P}$ having the entries

$$\mathbf{P}_{mn} = \int_{-\pi}^{\pi} p(x) \frac{e^{i(n-m)x}}{2\pi}\, dx\,. \tag{6.38}$$

If $p(x) = \chi_X(x)$, then $\mathbf{P}_{mn} = \sin[X(m-n)]/[\pi(m-n)]$, with $\mathbf{P}_{mm} = X/\pi$. Typically, $\mathbf{P}$ will have $(X/\pi)N$ eigenvalues near one and the remaining ones near zero, as shown in Figure 6.2. Solving the equations in (6.37) by means of the inverse of $\mathbf{P}$ is therefore an ill-conditioned problem, as X grows smaller.

For the remainder of this section the eigenvalues of $\mathbf{P}$ are denoted as $\lambda_1 > \lambda_2 > \cdots > \lambda_N > 0$, with associated orthonormal eigenvectors $\mathbf{u}_1, \mathbf{u}_2, \ldots, \mathbf{u}_N$. The matrix $\mathbf{P}$ then has the form

$$\mathbf{P} = \sum_{n=1}^{N} \lambda_n \mathbf{u}_n (\mathbf{u}_n)^{\ddagger}\,, \tag{6.39}$$

so that

$$\mathbf{P}^{-1} = \sum_{n=1}^{N} \lambda_n^{-1} \mathbf{u}_n (\mathbf{u}_n)^{\ddagger}. \tag{6.40}$$

Let's denote the functions $U_n(x)$, $n = 1, 2, \ldots, N$, by

$$U_n(x) = \sum_{m=1}^{N} (\mathbf{u}_n)_m e^{-imx}. \tag{6.41}$$

With the orthonormality of these eigenvectors,

$$\int_{-\pi}^{\pi} U_k(x) \overline{U_n(x)} dx = 0, \tag{6.42}$$

for $k \neq n$ and

$$\int_{-\pi}^{\pi} U_n(x) \overline{U_n(x)} dx = \int_{-\pi}^{\pi} |U_n(x)|^2 dx = 1. \tag{6.43}$$

Also

$$\int_{-X}^{X} |U_n(x)|^2 dr = (\mathbf{u}_n)^{\ddagger} Q \, \mathbf{u}_n = \lambda_n. \tag{6.44}$$

Therefore, λ_n is the proportion of the energy of $U_n(x)$ for x in the interval $[-X, X]$. The function $U_1(x)$ is the most concentrated in that interval, while $U_N(x)$ is the least. In Figure 6.3 the typical behavior of $U_n(x)$ for an example with fifteen data values is demonstrated.

The function $U_1(x)$ has a single large main lobe and no zeros within $[-X, X]$. The function $U_2(x)$, being orthogonal to $U_1(x)$ but still largely concentrated within $[-X, X]$, has a single zero in that interval. Each succeeding function has one more zero within $[-X, X]$ and is somewhat less concentrated there than its predecessors. At the other extreme, the function $U_N(x)$ has $N-1$ zeros in $[-X, X]$, is near zero throughout that interval, and is concentrated mainly outside the interval.

If $X = \pi$, then the DFT estimate in Equation 6.10 can be written in terms of $U_1(x), U_2(x), \ldots, U_N(x)$,

$$f_{\mathrm{DFT}}(x) = \sum_{n=1}^{N} \left[(\mathbf{u}_n)^{\ddagger} \mathbf{d} \right] U_n(x). \tag{6.45}$$

Similarly, using Equation (6.40), the PDFT estimate in Equation (6.35) can be written as

$$f_{\mathrm{PDFT}}(x) = \sum_{n=1}^{N} \lambda_n^{-1} \left[(\mathbf{u}_n)^{\ddagger} \mathbf{d} \right] U_n(x). \tag{6.46}$$

Comparing Equations (6.45) and (6.46), the PDFT places greater emphasis on those $U_n(x)$ corresponding to larger values of n. These are the functions least concentrated within the interval $[-X, X]$, but they are also those with the greater number of zeros within that interval. That means that these functions are much more oscillatory within $[-X, X]$ and better

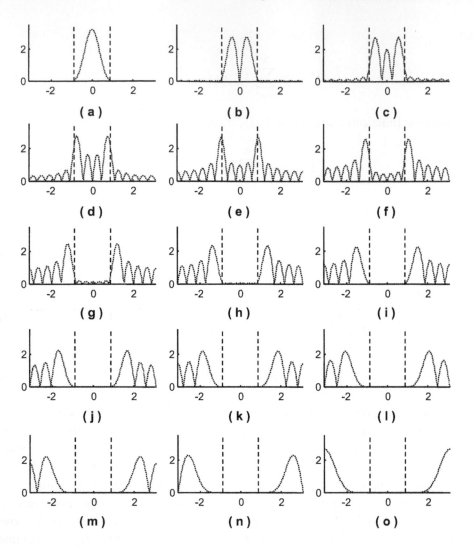

Figure 6.3 Example showing $|U_n|$ for $n = 1, 2, \ldots, 15$, where their corresponding eigenvalues are (a) 1.00e+00; (b) 9.99e-01; (c) 9.82e-01; (d) 8.29e-01; (e) 4.02e-01; (f) 7.82e-02; (g) 6.83e-03; (h) 3.52e-04; (i) 1.21e-05; (j) 2.91e-07; (k) 4.85e-09; (l) 5.53e-11; (m) 4.12e-13; (n) 1.72e-15; (o) 4.28e-18. Two vertical dashed lines indicate the boundaries of $p = \chi_{0.9}$.

suited to resolve closely spaced peaks in $f(x)$. Because the inner product $(\mathbf{u}_n)^{\ddagger}\mathbf{d}$ can be written as

$$(\mathbf{u}_n)^{\ddagger}\mathbf{d} = \frac{1}{2\pi} \int_{-\pi}^{\pi} f(x) \, \overline{U_n(x)} dx \,, \tag{6.47}$$

this term will be relatively small for the larger values of n and so the product $\lambda_n^{-1} \left[(\mathbf{u}_n)^{\ddagger}\mathbf{d} \right]$ will not be excessively large, provided that we have selected X properly. If X is too small and the support of $f(x)$ extends beyond the interval $[X, X]$, then the term $(\mathbf{u}_n)^{\ddagger}\mathbf{d}$ will not be as small and the product $\lambda_n^{-1} \left[(\mathbf{u}_n)^{\ddagger}\mathbf{d} \right]$ will be too large. This is what happens when there is noise in the data; the object function corresponding to the noisy data is not simply

$f(x)$ but contains a component that can be viewed as extending throughout the interval $[-\pi, \pi]$.

To reduce the sensitivity to noise, while not sacrificing resolution, one can resort to regularization. The simplest way to do this is to add a small positive quantity, $\epsilon > 0$, to each of the diagonal elements of the matrix $\mathbf{P}$. This is equivalent to having the prior $p(x)$ consist of two components, one the original $\chi_X(x)$ and the second a small positive $\chi_\pi(x)$. The effect of this regularization is to increase each of the λ_n by ϵ, without altering the eigenvectors or the $U_n(x)$. Because we now have $1/(\lambda_n + \epsilon)$ instead of the (potentially) much larger $1/\lambda_n$ in Equation (6.46), the sensitivity to noise and to poor selection of the Ω is reduced. At the same time, however, we have reduced the importance for the PDFT of the $U_n(r)$ for larger values of n; this will lead to a loss of resolution and PDFT that behaves like the DFT if ϵ is too large. Selecting the proper ϵ is somewhat of an art; it will certainly depend on what the eigenvalues are and on the signal-to-noise ratio. The eigenvalues, in turn, will depend on the ratio X/π.

This section has only focused on the particular case in which the variable x is one-dimensional and the prior $p(x) = \chi_X(x)$ describes only the support of the object function $f(x)$. However, for other choices of $p(x)$, the eigenvalues of the corresponding matrix $\mathbf{P}$ are similarly behaved and ill-conditioning is still an important issue, although the distribution of the eigenvalues may be somewhat different. For two-dimensional vector $\mathbf{x}$, the support of $f(\mathbf{x})$ can be described using a prior $p(\mathbf{x})$ that is nonzero on a rectangle, on a circle, on an ellipse, or on a more general region. For each of those choices, the corresponding matrix will have eigenvalues that decay toward zero, but perhaps at different rates.

6.3.6 Multidimensional Problem

The extension of the PDFT to the multidimensional case is straightforward. For a multidimensional vector $\mathbf{x} = (x_1, x_2, \ldots, x_M)$, let the object function $f(\mathbf{x})$ be compactly supported in $S = \{(x_1, x_2, \ldots, x_M) \mid |x_k| \leq \pi, k = 1, 2, \ldots, M\}$, and, for simplicity, have its sampled M-dimensional Fourier-transform values given by

$$F(\mathbf{n}) = \int_S f(\mathbf{x})e^{-i\mathbf{n}\cdot\mathbf{x}}\, d\mathbf{x}, \qquad (6.48)$$

where $\mathbf{n} = (n_1, n_2, \ldots, n_M)$ and $n_1 = 1, 2, \ldots, N_1, \ldots, n_M = 1, 2, \ldots, N_M$. The M-dimensional algorithm is formulated as

$$f_{\text{PDFT}}(\mathbf{x}) = p(\mathbf{x}) \sum_{n_1=1}^{N_1} \cdots \sum_{n_M=1}^{N_M} a_{n_1, n_2, \ldots, n_M}\, e^{i\mathbf{n}\cdot\mathbf{x}}, \qquad (6.49)$$

and the coefficients $a_{n_1, n_2, \ldots, n_M}$ for $n_1 = 1, 2, \ldots, N_1, \ldots, n_M = 1, 2, \ldots, N_M$ can be determined by

$$F(\mathbf{m}) = \sum_{n_1=1}^{N_1} \cdots \sum_{n_M=1}^{N_M} a_{n_1, n_2, \ldots, n_M}\, P(\mathbf{m} - \mathbf{n}), \qquad (6.50)$$

where $\mathbf{m} = (m_1, m_2, \ldots, m_M)$ for $m_1 = 1, 2, \ldots, N_1, \ldots, m_M = 1, 2, \ldots, N_M$ and

$$P(\mathbf{m} - \mathbf{n}) = \int_S p(\mathbf{x})e^{-i(\mathbf{m}-\mathbf{n})\cdot\mathbf{x}}\, d\mathbf{x}. \qquad (6.51)$$

The determination of the coefficients $a_{n_1,n_2,...,n_M}$ for $n_1 = 1, 2, ..., N_1, ..., n_M = 1, 2, ..., N_M$ in Equation (6.50) is done by vectorization, which turns multiple sums into a single sum, and results in a matrix equation as Equation (6.37).

6.3.7 Case of Radon-Transform Data: Tomographic Data

In some applications the measurements are related to the object function through the Radon transform,

$$Rf(\mathbf{x}) = \int_L f(\mathbf{x})|d\mathbf{x}|, \tag{6.52}$$

where L indicates the path along which the line integral is evaluated. For transmission tomography, the idealized mathematical model is that the data values are line integrals of the attenuation function f along the many different travel paths of the electromagnetic rays. In practice, the measurements will not be precisely line integrals, but more like strip integrals, and, in any case, there will be finitely many of them. For simplicity, we consider here only the case $D = 2$; that is, f is a function of two real variables. Therefore, Equation (6.52) can be represented as

$$d_n = \int f(\mathbf{x}) \, h_n(\mathbf{x}) \, d\mathbf{x}, \tag{6.53}$$

for $n = 1, 2, ..., N$. Each $h_n(\mathbf{x})$ indicates the characteristic function of a strip, that is, $h_n(\mathbf{x}) = 1$ for points $\mathbf{x}$ within the strip, and $h_n(\mathbf{x}) = 0$ for $\mathbf{x}$ outside the strip. By a strip we mean the set of all points that lie within $\delta > 0$ of a line segment through the domain of $f(\mathbf{x})$, where δ is fixed and small. These problems tend to be ill-posed and the data noisy. To reduce instability, we replace any zero values of the prior with an $\epsilon > 0$ to introduce a small amount of regularization and to make the reconstruction less sensitive to noise. In addition, the computation can be costly with the closed-form PDFT estimate when the data set is large, as is the case for multidimensional problems such as tomography. The function $f(\mathbf{x})$ can be discretized into finitely many pixel values, the PDFT approximated by the the discrete PDFT (DPDFT) and ART used to find the minimum-weighted-norm solution.

For some other applications, such as ultrasonic and seismic image reconstruction, the data values can be also formulated as Equation (6.52) based on the echo imaging model, in which all the points on L correspond to the coherent contribution of received signals. Therefore, the PDFT estimate can be obtained from Equation (6.53) in a similar sense. Successful implementational examples regarding this aspect can be seen in [2, 6, 16].

6.3.8 Under-Determined Systems of Linear Equations

To illustrate the minimum-weighted-norm solution, we consider the inverse problem of recovering the function f from the data d_n, $n = 1, ..., N$, in which the data are values of the Fourier transform in a Riemann sum approximation, as shown in Equation (6.15). In particular, solving this problem in the matrix equation form of $\mathbf{d} = \mathbf{H}\mathbf{f}$, this will be typically the case to solve an under-determined system of linear equations. A common choice is to select the minimum-norm solution in Equation (6.18).

Suppose there exists some prior information about the shape of the function f, such as it is zero outside some interval $[a, b]$ contained within S, or, more generally, that $|f(x)|$ can be approximated by some nonnegative function $p(x) \geq 0$. Then let $\mathbf{p} = (p_1, p_2, \ldots, p_N)^T$ be the vector of discrete samples of the function $p(x)$ and $w_n = p_n^{-1/2}$ whenever $p_n > 0$; let $w_n = \alpha > 0$ for some small $\alpha > 0$ otherwise. Let $\mathbf{W}$ be the diagonal matrix with entries w_n. The minimum-weighted-norm solution of $\mathbf{d} = \mathbf{H}\mathbf{f}$ is

$$\mathbf{f}_{\text{MW2N}} = \mathbf{W}^{-1}\mathbf{H}^{\ddagger}(\mathbf{H}\mathbf{W}^{-1}\mathbf{H}^{\ddagger})^{-1}\mathbf{d}\,. \tag{6.54}$$

This minimum-weighted-norm solution can be obtained from the minimum norm solution of a related system of linear equations. Let

$$\mathbf{B} = \mathbf{H}\mathbf{W}^{-1/2} \tag{6.55}$$

and

$$\mathbf{g} = \mathbf{W}^{1/2}\mathbf{f}\,. \tag{6.56}$$

Then $\mathbf{d} = \mathbf{H}\mathbf{f} = \mathbf{B}\mathbf{g}$. The minimum norm solution of $\mathbf{d} = \mathbf{B}\mathbf{g}$ is

$$\begin{aligned} \mathbf{g}_{\text{M2N}} &= \mathbf{B}^{\ddagger}(\mathbf{B}\mathbf{B}^{\ddagger})^{-1}\mathbf{d} \\ &= \mathbf{W}^{-1/2}\mathbf{H}^{\ddagger}(\mathbf{H}\mathbf{W}^{-1}\mathbf{H}^{\ddagger})^{-1}\mathbf{d} \end{aligned} \tag{6.57}$$

and

$$\mathbf{f}_{\text{MW2N}} = \mathbf{W}^{-1/2}\mathbf{g}_{\text{M2N}}\,. \tag{6.58}$$

When the data are noisy, an exact solution of $\mathbf{d} = \mathbf{H}\mathbf{f}$ is not reliable. One can regularize by taking the approximate solution that minimizes the functional value of

$$\psi(\mathbf{g}) = ||\mathbf{B}\mathbf{g} - \mathbf{d}||_2^2 + \epsilon^2||\mathbf{g}||_2^2\,, \tag{6.59}$$

for some $\epsilon > 0$. Thus, with Equation (6.55) and Equation (6.56), the regularized solution can be given by

$$\mathbf{f}_{\text{RMW2N}} = \mathbf{W}^{-1}\mathbf{H}^{\ddagger}(\mathbf{H}\mathbf{W}^{-1}\mathbf{H}^{\ddagger} + \epsilon^2\mathbf{I})^{-1}\mathbf{d}\,, \tag{6.60}$$

where $\mathbf{I}$ is the identity matrix.

6.3.9 Discrete PDFT

Instead of solving the PDFT estimate for Equation (6.2) in the continuous form by Equation (6.28), the basic idea in the discrete PDFT (DPDFT) approach is to discretize Equation (6.2) into a matrix equation, and then solve it by Equation (6.54). Thus, it discretizes the functions f, h_n, and $p > 0$, replacing the function f with the $J \times 1$ vector $\mathbf{f}$, h_n with the $J \times 1$ vector $\mathbf{h}_n$, and p with the $J \times 1$ vector $\mathbf{p}$. Then the data values d_n are related to the vector $\mathbf{f}$ by

$$d_n = \sum_{j=1}^{J} \mathbf{f}_j \, \overline{(\mathbf{h}_n)_j}\,. \tag{6.61}$$

In matrix notation, we can write this system as $\mathbf{H}\mathbf{f} = \mathbf{d}$, where $\mathbf{H}$ has the entries $\mathbf{H}_{nj} = \overline{(\mathbf{h}_n)_j}$, and the entries of $\mathbf{d}$ are the d_n. It is typically the case that $J > N$, so that this system of equations is under-determined. By analogy with the PDFT, it now seeks the vector $\mathbf{f}$, satisfying the system in Equation (6.61), for which the norm

$$||\mathbf{f}||_{\mathbf{p}} = \sqrt{\sum_{j=1}^{J} |f_j|^2 \, \mathbf{p}_j^{-1}} \tag{6.62}$$

is minimized.

Following the treatment in Section 6.3.8, $\mathbf{g}_j = \mathbf{f}_j / \sqrt{\mathbf{p}_j}$ and $\mathbf{B}_{nj} = \mathbf{H}_{nj} \sqrt{\mathbf{p}_j}$, we have $\mathbf{Bg} = \mathbf{d}$. The minimum norm solution of the system $\mathbf{Bg} = \mathbf{d}$ is the solution we seek. Iterative algorithms such as the algebraic reconstruction technique (ART) produce this minimum norm solution when the system is under-determined and the initial vector is zero. Thus, the iterative step of the DPDFT with the ART can be written in the form of [17]

$$\mathbf{f}_j^{k+1} = \mathbf{f}_j^k + \mathbf{p}_j \frac{\overline{\mathbf{H}}_{nj} \left(d_n - \sum_{m=1}^{J} \mathbf{H}_{nm} \mathbf{f}_m^k \right)}{\sum_{m=1}^{J} \mathbf{p}_m |\mathbf{H}_{nm}|^2} , \tag{6.63}$$

where $n = \mathrm{mod}(k, N) + 1$. Notice that we no longer need to calculate the Gram matrix and the solution to the system is approximated iteratively, which is computationally feasible, even for large N.

When the data are noisy, even this minimum norm solution can have a large norm and does not correspond to a useful reconstruction. We need a regularizing scheme that essentially avoids the exact solutions of $\mathbf{H}\mathbf{f} = \mathbf{d}$, but seeks instead a solution of $\mathbf{g}$ (Equation (6.56)) that minimizes the functional value of ψ in Equation (6.59). Due to the regularization, this method uses the ART to solve the system of equations given in the matrix equation by

$$\begin{bmatrix} \mathbf{B} & \epsilon\mathbf{I} \end{bmatrix} \begin{bmatrix} \mathbf{g} \\ \mathbf{v} \end{bmatrix} = \mathbf{d} . \tag{6.64}$$

For the ART iterative process of Equation (6.64), we begin at $\mathbf{g}^0 = 0$ and $\mathbf{v}^0 = 0$, then the limit for its upper component $\mathbf{g}^\infty = \hat{\mathbf{g}}$. The iterative step can be represented as

$$\mathbf{f}_j^{k+1} = \mathbf{f}_j^k + \mathbf{p}_j \overline{\mathbf{H}}_{nj} \left(\frac{d_n - \sum_{m=1}^{J} \mathbf{H}_{nm} \mathbf{f}_m^k - \epsilon \mathbf{v}_n^k}{\epsilon^2 + \sum_{m=1}^{J} \mathbf{p}_m |\mathbf{H}_{nm}|^2} \right) \tag{6.65}$$

and

$$\mathbf{v}_n^{k+1} = \mathbf{v}_n^k + \epsilon \left(\frac{d_n - \sum_{m=1}^{J} \mathbf{H}_{nm} \mathbf{f}_m^k - \epsilon \mathbf{v}_n^k}{\epsilon^2 + \sum_{m=1}^{J} \mathbf{p}_m |\mathbf{H}_{nm}|^2} \right) , \tag{6.66}$$

where

$$\mathbf{v}_m^{k+1} = \mathbf{v}_m^k \qquad \text{for} \qquad m \neq \mathrm{mod}(k, N) + 1 . \tag{6.67}$$

Then the reconstructed image is the limit of the sequence $\{\mathbf{f}^k\}$.

6.4   Solution Sparsity and Data Sampling

The problem of finding the localized energy solution from a set of limited measurements, a problem of increasing importance to both the industrial and research communities, arises in many applications, such as band-limited extrapolation, spectral estimation, signal reconstruction, signal classification, direction-of-arrival estimation, image restoration, biomagnetic inverse problems, channel equalization, and echo cancellation. The limitation of measurements typically comes from a combination of instrument constraints, limited observation time, nonstationary observation processes, and the ill-posed nature of the problem. One way to treat this underlying problem is to define a sparse signal as the signal of interest for which a maximally sparse representation is typically the solution to be expected. Therefore, the inverse problem of reconstructing the signal of interest amounts to seeking a minimum number of vectors to represent this signal of interest.

One area that has attracted much attention lately is compressed sensing or compressed sampling (CS) [18]. For applications such as medical imaging, CS may provide a means of reducing radiation dosage to the patient without sacrificing image quality. An important aspect of CS is finding sparse solutions of under-determined systems of linear equations, which can often be accomplished by one-norm minimization, [19].

6.4.1   Compressed Sensing

The objective in CS is to exploit sparseness to reconstruct a vector $\mathbf{f}$ in R^J from relatively few linear functional measurements [18].

Let $U = \{\mathbf{u}_1, \mathbf{u}_2, ..., \mathbf{u}_J\}$ and $V = \{\mathbf{v}_1, \mathbf{v}_2, ..., \mathbf{v}_J\}$ be two orthonormal bases for R^J, with all members of R^J represented as column vectors. For $i = 1, 2, ..., J$, let

$$\mu_i = \max_{1 \leq j \leq J} \{ |\langle \mathbf{u}_i, \mathbf{v}_j \rangle| \} \tag{6.68}$$

and

$$\mu(U, V) = \max\{ \mu_i \,|\, i = 1, ..., I \} . \tag{6.69}$$

From Cauchy's Inequality

$$|\langle \mathbf{u}_i, \mathbf{v}_j \rangle| \leq 1, \tag{6.70}$$

and from Parseval's Equation

$$\sum_{j=1}^{J} |\langle \mathbf{u}_i, \mathbf{v}_j \rangle|^2 = ||\mathbf{u}_i||^2 = 1, \tag{6.71}$$

giving

$$\frac{1}{\sqrt{J}} \leq \mu(U, V) \leq 1 . \tag{6.72}$$

The quantity $\mu(U, V)$ is the coherence measure of the two bases; the closer $\mu(U, V)$ is to the lower bound of $1/\sqrt{J}$, the more incoherent the two bases are.

Let $\mathbf{f}$ be a fixed member of R^J and expanded in the V basis as

$$\mathbf{f} = x_1 \mathbf{v}_1 + x_2 \mathbf{v}_2 + ... + x_J \mathbf{v}_J . \tag{6.73}$$

The coefficient vector $\mathbf{x} = (x_1, ..., x_J)^T$ is said to be s-sparse if s is the number of nonzero x_j.

If s is small, most of the x_j are zero, but because there is no information about which ones these are, we would have to compute all the linear functional values

$$x_j = \langle \mathbf{f}, \mathbf{v}_j \rangle \tag{6.74}$$

to recover $\mathbf{f}$ exactly. In fact, the smaller s is, the harder it would be to learn anything from randomly selected x_j, as most would be zero. The idea in CS is to obtain measurements of $\mathbf{f}$ with members of a different orthonormal basis, which is called the U basis. If the members of U are very much like the members of V, then nothing is gained. But if the members of U are quite unlike the members of V, then each inner product measurement

$$y_i = \langle \mathbf{f}, \mathbf{u}_i \rangle = \mathbf{f}^T \mathbf{u}_i \tag{6.75}$$

can provide information about $\mathbf{f}$. If the two bases are sufficiently incoherent, then relatively few y_i values should tell us a lot about $\mathbf{f}$. Specifically, we have the following result due to Candès and Romberg [20]: suppose the coefficient vector $\mathbf{x}$ for representing $\mathbf{f}$ in the V basis is s-sparse. Select uniformly random $M \leq J$ members of the U basis and compute the measurements $y_i = \langle \mathbf{f}, \mathbf{u}_i \rangle$. Then, if M is sufficiently large, it is highly probable that $\mathbf{z} = (z_1, ..., z_J)^T = \mathbf{x}$ also solves the problem of minimizing the one-norm

$$\|\mathbf{z}\|_1 = |z_1| + |z_2| + ... + |z_J|, \tag{6.76}$$

subject to the conditions

$$y_i = \langle \mathbf{g}, \mathbf{u}_i \rangle = \mathbf{g}^T \mathbf{u}_i, \tag{6.77}$$

for those M randomly selected $\mathbf{u}_i$ where

$$\mathbf{g} = z_1 \mathbf{v}_1 + z_2 \mathbf{v}_2 + ... + z_J \mathbf{v}_J. \tag{6.78}$$

The smaller $\mu(U, V)$ is, the smaller the M is permitted to be without reducing the probability of perfect reconstruction.

6.4.2 Sparse Solutions

Suppose that $\mathbf{A}$ is a real $M \times N$ matrix, with $M < N$, and that the linear system $\mathbf{Ax} = \mathbf{b}$ has infinitely many solutions. For any vector $\mathbf{x} = (x_1, \ldots, x_N)^T$, let's define the support of $\mathbf{x}$ to be the subset S of $\{1, 2, ..., N\}$ consisting of those n for which the entries $x_n \neq 0$. For any under-determined system $\mathbf{Ax} = \mathbf{b}$, there will, of course, be at least one solution of minimum support, that is, for which $s = |S|$, the size of the support set S, is minimum. However, finding such a maximally sparse solution requires combinatorial optimization, and is known to be computationally difficult. It is important, therefore, to have a computationally tractable method for finding maximally sparse solutions.

6.4.3 Why Sparseness?

One obvious reason for wanting sparse solutions is that one has prior knowledge that the desired solution is sparse. Such a problem arises in signal analysis from Fourier-transform data. In other cases, such as in the reconstruction of locally constant signals, it is not the signal itself, but its discrete derivative that is sparse.

6.4.3.1 Signal Analysis

Suppose that a signal $f(t)$ is known to consist of a small number of complex exponentials, so that $f(t)$ has the form

$$f(t) = \sum_{j=1}^{J} a_j e^{i\omega_j t}, \qquad (6.79)$$

for some small number of frequencies ω_j in the interval $[0, 2\pi)$. For $n = 0, 1, ..., N - 1$, let $f_n = f(n)$, and let $\mathbf{f}$ be the N-vector with entries f_n; let's assume that J is much smaller than N. The discrete (vector) Fourier transform of $\mathbf{f}$ is the vector $\hat{\mathbf{f}}$ having the entries

$$\hat{\mathbf{f}}_k = \frac{1}{\sqrt{N}} \sum_{n=0}^{N-1} f_n\, e^{2\pi i k n / N}, \qquad (6.80)$$

for $k = 0, 1, ..., N - 1$; it can be written in the form of $\hat{\mathbf{f}} = \mathbf{E}\mathbf{f}$, where $\mathbf{E}$ is the $N \times N$ matrix with entries $\mathbf{E}_{kn} = (1/\sqrt{N})\, e^{2\pi i k n / N}$. If N is large enough, one can safely assume that each of the ω_j is equal to one of the frequencies $2\pi i k$ and that the vector $\hat{\mathbf{f}}$ is J-sparse. The question now is: How many values of $f(n)$ does one need to calculate in order to be sure that $f(t)$ can be recaptured exactly? We have the following theorem

Theorem 6.4.1 *Let N be prime. Let S be any subset of $\{0, 1, ..., N - 1\}$ with $|S| \geq 2J$. Then the vector $\hat{\mathbf{f}}$ can be uniquely determined from the measurements f_n for n in S.*

It is known that

$$\mathbf{f} = \mathbf{E}^{\ddagger}\hat{\mathbf{f}} \qquad (6.81)$$

The point here is that, for any matrix $\mathbf{R}$ obtained from the identity matrix $\mathbf{I}$ by deleting $N - |S|$ rows, one can recover the vector $\hat{\mathbf{f}}$ from the measurements $\mathbf{R}\mathbf{f}$.

If N is not prime, then the assertion of the theorem may not hold, as $n = \mathrm{mod}(0, N)$, without $n = 0$. However, the assertion remains valid for most sets of J frequencies and most subsets S of indices; therefore, with high probability, it can recover the vector $\hat{\mathbf{f}}$ from $\mathbf{R}\mathbf{f}$.

Note that the matrix $\mathbf{E}$ is unitary, that is, $\mathbf{E}^{\ddagger}\mathbf{E} = \mathbf{I}$, and, equivalently, the columns of $\mathbf{E}$ form an orthonormal basis for C^N. The data vector is

$$\mathbf{b} = \mathbf{R}\mathbf{f} = \mathbf{R}\mathbf{E}^{\ddagger}\hat{\mathbf{f}}. \qquad (6.82)$$

In this example, the vector $\mathbf{f}$ is not sparse, but can be represented sparsely in a particular orthonormal basis, namely as $\mathbf{f} = \mathbf{E}^{\ddagger}\hat{\mathbf{f}}$, using a sparse vector $\hat{\mathbf{f}}$ of coefficients. The representing basis then consists of the columns of the matrix $\mathbf{E}^{\ddagger}$. The measurements pertaining to the vector $\mathbf{f}$ are the values f_n, for n in S. Because f_n can be viewed as the inner product of $\mathbf{f}$ with δ^n, the nth column of the identity matrix $\mathbf{I}$, that is,

$$f_n = \langle \delta^n, \mathbf{f} \rangle, \qquad (6.83)$$

the columns of $\mathbf{I}$ provide the so-called sampling basis. With $\mathbf{A} = \mathbf{R}\mathbf{E}^{\ddagger}$ and $\mathbf{x} = \hat{\mathbf{f}}$, it then has

$$\mathbf{A}\mathbf{x} = \mathbf{b}, \qquad (6.84)$$

with the vector $\mathbf{x}$ sparse. It is important for what follows to note that the matrix $\mathbf{A}$ is random, in the sense that we choose which rows of $\mathbf{I}$ to use to form $\mathbf{R}$.

6.4.3.2   Locally Constant Signals

Suppose now that the function $f(t)$ is locally constant, consisting of some number of distinct regions. Let's discretize the function $f(t)$ to get the vector $\mathbf{f} = (f(0), f(1), ..., f(N))^T$. The discrete derivative vector is $\mathbf{g}$, with

$$\mathbf{g}_n = f(n) - f(n-1) . \tag{6.85}$$

Because $f(t)$ is locally constant, the vector $\mathbf{g}$ is sparse. Typically, the data will not be values $f(n)$. The goal will be to recover $\mathbf{f}$ from M linear functional values pertaining to $\mathbf{f}$, where M is much smaller than N. For explanation, let's assume that the value $f(0)$ can be known either by measurement or by estimation.

The data vector $\mathbf{d} = (d_1, \ldots, d_M)^T$ consists of measurements pertaining to the vector $\mathbf{f}$:

$$d_m = \sum_{n=0}^{N} \mathbf{H}_{mn} \mathbf{f}_n , \tag{6.86}$$

for $m = 1, ..., M$, where the $\mathbf{H}_{mn}$ are known. One can then write

$$d_m = f(0) \left(\sum_{n=0}^{N} \mathbf{H}_{mn} \right) + \sum_{k=1}^{N} \left(\sum_{j=k}^{N} \mathbf{H}_{mj} \right) \mathbf{g}_k . \tag{6.87}$$

Because $f(0)$ is known, we have

$$b_m = d_m - f(0) \left(\sum_{n=0}^{N} \mathbf{H}_{mn} \right) = \sum_{k=1}^{N} \mathbf{A}_{mk} \, \mathbf{g}_k , \tag{6.88}$$

where

$$\mathbf{A}_{mk} = \sum_{j=k}^{N} \mathbf{H}_{mj} . \tag{6.89}$$

The problem is then to find a sparse solution of $\mathbf{A}\mathbf{x} = \mathbf{g}$. As in the previous example, one has the freedom to select the linear functions, that is, the values $\mathbf{H}_{mn}$, so the matrix $\mathbf{A}$ can be viewed as random.

6.4.4   Tomographic Imaging

The reconstruction of tomographic images is an important aspect of medical diagnosis, and one that combines aspects of both of the previous examples. The data one obtains from the scanning process can often be interpreted as values of the Fourier transform of the desired image; this is precisely the case in magnetic resonance imaging, and approximately true for x-ray transmission tomography, positron emission tomography (PET), and single-photon emission computed tomography (SPECT). The images one encounters in medical diagnosis are often approximately locally constant, so the associated array of discrete partial derivatives will be sparse. If this sparse derivative array can be recovered from relatively few Fourier-transform values, then the scanning time can be reduced.

6.4.5 Compressed Sampling

The goal is to recover the vector $\mathbf{f} = (f_1, ..., f_N)^T$ from M linear functional values of $\mathbf{f}$, where M is much less than N. In general, this is not possible without prior information about the vector $\mathbf{f}$. In compressed sampling, the prior information concerns the sparseness of either $\mathbf{f}$ itself, or another vector linearly related to $\mathbf{f}$.

Let $\mathbf{U}$ and $\mathbf{V}$ be unitary $N \times N$ matrices, so that the column vectors of both $\mathbf{U}$ and $\mathbf{V}$ form orthonormal bases for C^N. The bases associated with $\mathbf{U}$ and $\mathbf{V}$ can be referred to as the sampling basis and the representing basis, respectively. The first objective is to find a unitary matrix $\mathbf{V}$ so that $\mathbf{f} = \mathbf{Vx}$, where $\mathbf{x}$ is sparse. It then finds a second unitary matrix $\mathbf{U}$ such that, when an $M \times N$ matrix $\mathbf{R}$ is obtained from $\mathbf{U}$ by deleting rows, the sparse vector $\mathbf{x}$ can be determined from the data $\mathbf{b} = \mathbf{RVx} = \mathbf{Ax}$. Theorems in compressed sensing describe properties of the matrices $\mathbf{U}$ and $\mathbf{V}$ such that, when $\mathbf{R}$ is obtained from $\mathbf{U}$ by a random selection of the rows of $\mathbf{U}$, the vector $\mathbf{x}$ will be uniquely determined, with high probability, as the unique solution that minimizes the one-norm.

6.5 Minimum L^1-Norm and Minimum Weighted L^1-Norm Solutions

Consider the problem P_0: among all solutions $\mathbf{x}$ of the consistent system $\mathbf{b} = \mathbf{Ax}$, find one, call it $\hat{\mathbf{x}}$, that is maximally sparse, that is, has the minimum number of nonzero entries. Obviously, there will be at least one such solution having minimal support, but finding one, however, is a combinatorial optimization problem and is generally NP-hard. One way to solve this problem is the Hilbert space method associated with the minimization of some chosen Hilbert space norm. This minimum norm solution is the best known example, in which the functional value of

$$\sqrt{\langle \mathbf{x}, \mathbf{x} \rangle} \tag{6.90}$$

is minimized.

6.5.1 Minimum L^1-Norm Solutions

Instead of the minimum two-norm solution, we seek a minimum one-norm solution $\mathbf{x}_{\mathrm{M1N}}$, that is, solve the problem P_1: minimize

$$||\mathbf{x}||_1 = \sum_{n=1}^{N} |\mathbf{x}_n|, \tag{6.91}$$

subject to $\mathbf{Ax} = \mathbf{b}$. Problem P_1 can be formulated as a linear programming problem and so is more easily solved. It is important to know when P_1 has a unique solution $\mathbf{x}_{\mathrm{M1N}}$ and when $\mathbf{x}_{\mathrm{M1N}} = \hat{\mathbf{x}}$. The problem P_1 will have a unique solution if and only if $\mathbf{A}$ is such that the one-norm satisfies

$$||\mathbf{x}_{\mathrm{M1N}}||_1 < ||\mathbf{x}_{\mathrm{M1N}} + \mathbf{v}||_1, \tag{6.92}$$

for all nonzero $\mathbf{v}$ in the null space of $\mathbf{A}$.

The entries of $\mathbf{x}$ need not be nonnegative, so the problem is not yet a linear programming problem. Let

$$\mathbf{B} = [\mathbf{A} \quad -\mathbf{A}] \tag{6.93}$$

and consider the linear programming problem of minimizing the function

$$\mathbf{c}^T \mathbf{z} = \sum_{n=1}^{2N} \mathbf{z}_n, \tag{6.94}$$

subject to the constraints $\mathbf{z}_n \geq 0$, and $\mathbf{Bz} = \mathbf{b}$. Let $\mathbf{z}^*$ be the solution given by

$$\mathbf{z}^* = \begin{bmatrix} \mathbf{u}^* \\ \mathbf{v}^* \end{bmatrix}. \tag{6.95}$$

Then, $\mathbf{x}^* = \mathbf{u}^* - \mathbf{v}^*$ minimizes the one-norm, subject to $\mathbf{Ax} = \mathbf{b}$.

It is true that $\mathbf{u}_n^* \mathbf{v}_n^* = 0$ for $n = 1, 2, \ldots, N$. If this were not the case and there is an n such that $0 < \mathbf{v}_n^* < \mathbf{u}_n^*$, then a new vector $\mathbf{z}$ can be created by replacing the old $\mathbf{u}_n^*$ with $\mathbf{u}_n^* - \mathbf{v}_n^*$ and the old $\mathbf{v}_n^*$ with zero, while maintaining $\mathbf{Bz} = \mathbf{b}$. But then, because $\mathbf{u}_n^* - \mathbf{v}_n^* < \mathbf{u}_n^* + \mathbf{v}_n^*$, it follows that $\mathbf{c}^T \mathbf{z} < \mathbf{c}^T \mathbf{z}^*$, which is a contradiction. Consequently, $||\mathbf{x}^*||_1 = \mathbf{c}^T \mathbf{z}^*$.

We next select any $\mathbf{x}$ with $\mathbf{Ax} = \mathbf{b}$ and write $\mathbf{u}_n = \mathbf{x}_n$, if $\mathbf{x}_n \geq 0$ and $\mathbf{u}_n = 0$ otherwise. Let $\mathbf{v}_n = \mathbf{u}_n - \mathbf{x}_n$, so that $\mathbf{x} = \mathbf{u} - \mathbf{v}$, and let

$$\mathbf{z} = \begin{bmatrix} \mathbf{u} \\ \mathbf{v} \end{bmatrix}. \tag{6.96}$$

Then $\mathbf{b} = \mathbf{Ax} = \mathbf{Bz}$ and $\mathbf{c}^T \mathbf{z} = ||\mathbf{x}||_1$, giving

$$\begin{aligned} ||\mathbf{x}^*||_1 &= \mathbf{c}^T \mathbf{z}^* \\ &\leq \mathbf{c}^T \mathbf{z} \\ &= ||\mathbf{x}||_1, \end{aligned} \tag{6.97}$$

and $\mathbf{x}^*$ must be a minimum one-norm solution.

6.5.2   Why the One-Norm?

When a system of linear equations $\mathbf{Ax} = \mathbf{b}$ is under-determined, one can find the minimum two-norm solution that minimizes the square of the two-norm,

$$||\mathbf{x}||_2^2 = \sum_{n=1}^{N} |\mathbf{x}_n|^2, \tag{6.98}$$

subject to $\mathbf{Ax} = \mathbf{b}$. One drawback to this approach is that the two-norm penalizes relatively large values of $\mathbf{x}_n$ much more than the smaller ones, and so tends to provide nonsparse solutions. Alternatively, we seek the solution for which the one-norm,

$$||\mathbf{x}||_1 = \sum_{n=1}^{N} |\mathbf{x}_n|, \tag{6.99}$$

is minimized. The one-norm still penalizes relatively large entries $\mathbf{x}_n$ more than the smaller ones, but much less than the two-norm does. As a result, it often happens that the minimum one-norm solution actually solves P_0 as well.

6.5.3 Comparison with the PDFT

The PDFT approach to solving the under-determined system $\mathbf{Ax} = \mathbf{b}$ is to select weights $\mathbf{w}_n > 0$ and then to find the solution $\tilde{\mathbf{x}}$ that minimizes the weighted two-norm given by

$$\sum_{n=1}^{N} |\mathbf{x}_n|^2 \mathbf{w}_n .\tag{6.100}$$

The intention is to select weights $\mathbf{w}_n$ so that $\mathbf{w}_n^{-1}$ is reasonably close to the absolute value of the corresponding entry of the minimum one-norm solution $|(\mathbf{x}_{\mathrm{M1N}})_n|$; consider, therefore, what happens when $\mathbf{w}_n^{-1} = |(\mathbf{x}_{\mathrm{M1N}})_n|$. We claim that $\tilde{\mathbf{x}}$ is also a minimum one-norm solution.

To see why this is true, note that, for any $\mathbf{x}$, we have

$$\sum_{n=1}^{N} |\mathbf{x}_n| = \sum_{n=1}^{N} \frac{|\mathbf{x}_n|}{\sqrt{|(\mathbf{x}_{\mathrm{M1N}})_n|}} \sqrt{|(\mathbf{x}_{\mathrm{M1N}})_n|}$$

$$\leq \sqrt{\sum_{n=1}^{N} \frac{|\mathbf{x}_n|^2}{|(\mathbf{x}_{\mathrm{M1N}})_n|}} \sqrt{\sum_{n=1}^{N} |(\mathbf{x}_{\mathrm{M1N}})_n|} .$$

Therefore,

$$\sum_{n=1}^{N} |\tilde{\mathbf{x}}_n| \leq \sqrt{\sum_{n=1}^{N} \frac{|\tilde{\mathbf{x}}_n|^2}{|(\mathbf{x}_{\mathrm{M1N}})_n|}} \sqrt{\sum_{n=1}^{N} |(\mathbf{x}_{\mathrm{M1N}})_n|}$$

$$\leq \sqrt{\sum_{n=1}^{N} \frac{|(\mathbf{x}_{\mathrm{M1N}})_n|^2}{|(\mathbf{x}_{\mathrm{M1N}})_n|}} \sqrt{\sum_{n=1}^{N} |(\mathbf{x}_{\mathrm{M1N}})_n|}$$

$$= \sum_{n=1}^{N} |(\mathbf{x}_{\mathrm{M1N}})_n| ,$$

thus, $\tilde{\mathbf{x}}$ also minimizes the one-norm.

6.5.4 Iterative Reweighting

Let $\mathbf{x}$ be the truth. Generally, each weight $\mathbf{w}_n$ is required to be a good prior estimate of the reciprocal of $|\mathbf{x}_n|$. Because $\mathbf{x}$ is not known typically, we can take a sequential-optimization approach, beginning with weights $\mathbf{w}_n^0 > 0$, finding the PDFT solution using these weights, then using this PDFT solution to get a better choice for the weights, and so on. This sequential approach was successfully implemented in the early 1980s by Michael Fiddy and his students [12].

In [21], the same approach is taken, but with respect to the one-norm. Because the one-norm still penalizes larger values disproportionately, balance can be achieved by minimizing a weighted one-norm, with weights close to the reciprocals of the $|\mathbf{x}_n|$. Again, not yet knowing $\mathbf{x}$, they employ a sequential approach, using the previous minimum-weighted-one-norm solution to obtain the new set of weights for the next minimization. At each step of the sequential procedure, the previous reconstruction is used to estimate the true support of the desired solution.

It is interesting to note that an ongoing debate among users of the PDFT concerns the nature of the prior weighting. Does w_n approximate $|\mathbf{x}_n|^{-1}$ or $|\mathbf{x}_n|^{-2}$? This is close to the issue treated in [21], the use of a weight in the minimum one-norm approach.

It should be noted again that finding a sparse solution is not usually the goal in the use of the PDFT, but the use of the weights has much the same effect as using the one-norm to find sparse solutions: to the extent that the weights approximate the entries of $\hat{\mathbf{x}}$, their use reduces the penalty associated with the larger entries of an estimated solution.

6.6   Modification with Nonuniform Weights

In many applications it is often the goal to image accurately some small subregions of the object. This critical issue is much harder to address, especially when the small-scale portions generate only a small amount of the total energy received from the object. Even with accurate prior information about the true object support, the PDFT may not reconstruct small-scale features with the necessary accuracy. To achieve improved resolution of such small-scale subregions, the modification of the PDFT reconstruction changes the estimate of $f(x)$ into

$$\hat{f}(x) = \sum_{m=1}^{N} a_m w_m(x) e^{imx} , \qquad (6.101)$$

where the $w_m(x)$ are selected nonnegative window functions. Data consistency is achieved when the a_m satisfy the equations

$$d_n = \sum_{m=1}^{N} a_m \int_{-\pi}^{\pi} w_m(x) e^{i(m-n)x} dx . \qquad (6.102)$$

6.6.1   Selection of Windows

Suppose that the support of $f(x)$, which is taken to equal Ω, can be divided into two disjoint regions, a larger one, denoted L, in which there are no small-scale features, and a smaller one, denoted S, containing small-scale features of interest to us. Suppose also that the set $\{k_n\}$ of "spatial frequencies" can be divided into two sets, LF and HF, with $LF = \{k_n | n = 1, ..., K\}$ consisting of lower values of k_n and $HF = \{k_n | n = K + 1, ..., N\}$ consisting of higher values.

Our PDFT reconstructed image is

$$f_{\text{PDFT}}(x) = \chi_L(x) \sum_{n=1}^{N} a_n \exp(jxk_n) + \chi_S(x) \sum_{n=1}^{N} a_n \exp(jxk_n) , \qquad (6.103)$$

using

$$\chi_\Omega(x) = \chi_L(x) + \chi_S(x) .$$

If it can be safely assumed that within the set L, the function $f(x)$ involves no high spatial frequencies, and within the set S no low spatial frequencies, then it makes sense to modify Equation (6.103) and write

$$\hat{f}(x) = \chi_L(x) \sum_{n|k_n \in LF} a_n \exp(jxk_n) + \chi_S(x) \sum_{n|k_n \in HF} a_n \exp(jxk_n). \quad (6.104)$$

For a theoretical understanding, we make the simplifying assumption that

$$\int_L \exp(jxk_n) \exp(-jxk_m) dx = \int_S \exp(jxk_m) \exp(-jxk_n) dx = 0 \quad (6.105)$$

whenever $k_n \in LF$ and $k_m \in HF$. Then the matrix involved in the system of equations to be solved is block-diagonal, with P_L in the northwest block and P_S in the southeast block. As a result, the reconstruction given by Equation (6.104) will be a sum of the $U_n(x)$ for the matrix P_L, for x in L, and the $U_n(x)$ for the matrix P_S, for x in S. Because $\chi_S(x)$ has the small set S for its support, the eigenfunctions $U_n(x)$ associated with P_S oscillate rapidly within the set S, allowing the reconstruction to reveal small-scale details within that region.

The reconstruction in Equation (6.101) simply extends this idea, allowing each spatial frequency to have its own region, and has developed successful demonstrations in [22].

6.6.2 Multidimensional Case

The extension of the method in Equation (6.101) to the multidimensional problem is straightforward. This can be easily understood from Section 6.3.6, although minor differences exist. Let's take the same assumptions for the object function, $f(\mathbf{x})$ and its sampled M-dimensional Fourier-transform values as that in Section 6.3.6. Thus, the two-dimensional algorithm is formulated as

$$\hat{f}(\mathbf{x}) = \sum_{n_1=1}^{N_1} \cdots \sum_{n_M=1}^{N_M} a_{n_1,n_2,\ldots,n_M} w_{n_1,n_2,\ldots,n_M}(\mathbf{x}) e^{i\mathbf{n}\cdot\mathbf{x}}, \quad (6.106)$$

and the coefficients $a_{n_1,n_2,\ldots,n_M}$ for $n_1 = 1, 2, \ldots, N_1, \ldots, n_M = 1, 2, \ldots, N_M$ can be determined by

$$F(\mathbf{m}) = \sum_{n_1=1}^{N_1} \cdots \sum_{n_M=1}^{N_M} a_{n_1,n_2,\ldots,n_M} W_{n_1,n_2,\ldots,n_M}(\mathbf{m} - \mathbf{n}), \quad (6.107)$$

where $\mathbf{m} = (m_1, m_2, \ldots, m_M)$ for $m_1 = 1, 2, \ldots, N_1, \ldots, m_M = 1, 2, \ldots, N_M$ and

$$W_{n_1,n_2,\ldots,n_M}(\mathbf{m} - \mathbf{n}) = \int_S w_{n_1,n_2,\ldots,n_M}(\mathbf{x}) e^{-i(\mathbf{m}-\mathbf{n})\cdot\mathbf{x}} dx. \quad (6.108)$$

6.6.3 Challenge of the Modified PDFT for Realistic Applications

For the purpose of this discussion, let $f(x)$ be an object function supported on the interval $[-\pi, \pi]$ of the real line, and its sampled Fourier-transform values at frequencies of

$k_1, k_2, \ldots, k_N$ can be represented as

$$F(k_n) = \int_{-\pi}^{\pi} f(x) \exp(-jxk_n) dx . \tag{6.109}$$

In [22], the proposed extension to the PDFT takes the linear estimation form

$$\hat{f}(x) = \sum_{n=1}^{N} a_n w_n(x) \exp(jxk_n) \tag{6.110}$$

with

$$F(k_m) = \sum_{n=1}^{N} a_n W_n(k_m - k_n) , \tag{6.111}$$

where $w_n(x)$ for $n = 1, 2, \ldots, N$ are selected nonnegative window functions and their Fourier transforms are $W_n(k)$ for $n = 1, 2, \ldots, N$, respectively. Suppose that the support of $f(x)$, which we take to equal Ω, can be divided into P disjoint regions $S_1, S_2, \ldots, S_P$:

$$\chi_{\Omega}(x) = \chi_{S_1}(x) + \chi_{S_2}(x) + \cdots + \chi_{S_p}(x) . \tag{6.112}$$

S_1 is assumed to be the area in which there are no small-scale subregions, and for $S_2, S_3, \ldots, S_P$, each includes an individual subregion of small-scale features. For complex exponential bases $\exp(jxk_n), n = 1, 2, \ldots, N$, a simple strategy in the choice of their corresponding window functions $w_n(x)$, proposed in [22], can be written in the form of

$$w_n(x) = \sum_{p=1}^{P} \sigma_{np} \chi_{S_p}(x) , \tag{6.113}$$

with two options in Equation (6.114) for binary-valued weights $\sigma_{n1}, \sigma_{n2}, \sigma_{n3}, \cdots, \sigma_{nP}$:

$$\begin{cases} \text{(i) } \sigma_{n1} = 1 \text{ and } \sigma_{n2} = \sigma_{n3} = \cdots = \sigma_{nP} = 0 , & \text{for } k_n \in LF \\ \text{(ii) } \sigma_{n1} = 0 \text{ and } \sigma_{n2} = \sigma_{n3} = \cdots = \sigma_{nP} = 1 , & \text{for } k_n \in HF. \end{cases} \tag{6.114}$$

Sampled spatial frequencies, $\{k_1, k_2, \ldots, k_N\}$, are divided into two sets, LF and HF, with LF consisting of lower values of k_n and HF consisting of higher values. According to Equations (6.110), (6.113), and (6.114), complex exponential bases having their spatial frequencies within the set HF contribute the image reconstruction over all subregions of small-scale features, and on the other hand those having their spatial frequencies within the set LF contribute the image reconstruction for the rest of the object function's support.

In practice, a priori information about $S_2, S_3, \ldots, S_P$ might not be correct but the PDFT estimate with a tight prior (χ_{Ω}) can be used to acquire this information. A wider support for each subregion ($S_2, S_3, \ldots, S_P$) is recommended in the first step, and the image quality can be increased by narrowing the support widths until the presence of significant delta-like artifacts appear around any subregion's boundaries.

In general, the method from Equations (6.110), (6.111), (6.113), and (6.114) has the potential to provide a high-resolution image superior to the classical PDFT, as seen by the

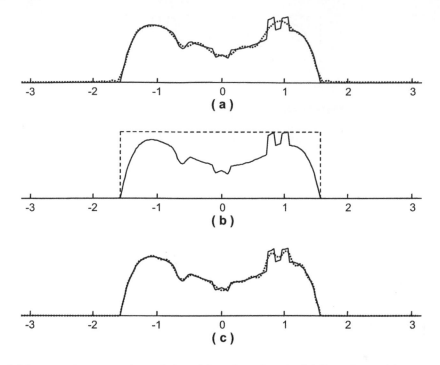

Figure 6.4 Image reconstruction of the object function (solid line) from thirty-seven sampled Fourier transform values: (a) the DFT estimate (dotted line); (b) the prior function (dashed line); (c) the PDFT estimate (dotted line) with the prior in (b).

comparison between Figure 6.4 and Figure 6.5. However, there usually exist some delta-like shapes around the boundaries of $S_2, S_3, \ldots, S_P$ in the image estimate, as seen in Figure 6.5(c). These artifacts are typical when the prior function is over-constrained and the variation in the spatial-frequency content among $S_2, S_3, \ldots, S_P$ is not consistent with the data. To resolve this, the choice of weighted windows for each complex exponential basis function must rely on a robust strategy for selection. A new technique based on the genetic algorithms (GA) search engine is presented in the next section, it being a statistical-searching approach for optimizing the choice of weighted windows.

6.6.4 Modified Strategy in the Choice of Weighted Windows

Thus, to accommodate more realistic signals and images, instead of taking the simple steps in Equation (6.114), the selection of weighted windows $\{w_1, w_2, \ldots, w_N\}$ is modified in the sense of

$$\begin{cases} \text{(i) } \sigma_{n1} = 1 \text{ and } \sigma_{n2} = \sigma_{n3} = \cdots = \sigma_{nP} = 0\,, & \text{for } k_n \in RF \\ \text{(ii) } \sigma_{n1} = 0\,, & \text{for } k_n \in SF \end{cases} \qquad (6.115)$$

for $n = 1, 2, \ldots, N$. For option (ii) in Equation (6.115), $\sigma_{n2}, \sigma_{n3}, \ldots, \sigma_{nP}$ for $n = 1, 2, \ldots, N$ can be either 1 or 0, but the cases of $\sigma_{n2} = \sigma_{n3} = \cdots = \sigma_{nP} = 0$ should be excluded. Sampled spatial frequencies, $\{k_1, k_2, \ldots, k_N\}$, are divided into two sets, SF and RF, with $k_n \in SF$ essentially contributing to the image reconstruction over regions

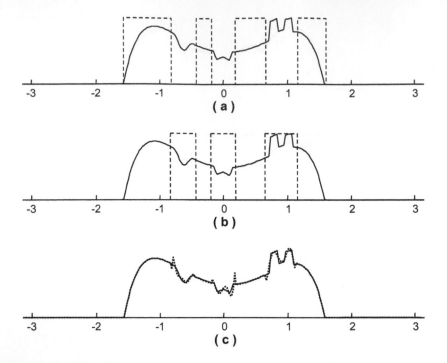

Figure 6.5 The same image reconstruction problem as that in Figure 6.4 but different estimation method: (a) the weighted window χ_{S_1} (dashed line) associated with low-frequency
profile; (b) the weighted windows χ_{S_2}, χ_{S_3}, and χ_{S_4} (dashed line) associated with small-
scale subregions of interest; (c) the estimate by Equations (6.110), (6.111), (6.113), and
(6.114) (dotted line) with weighted windows in (a) and (b).

containing small-scale features of interest and $k_n \in RF$ for the rest of the object function's
support. For easy understanding, Equation (6.113) can be replaced with

$$w_n(x) = \mathbf{G}_n^T \mathbf{X}, \tag{6.116}$$

where $\mathbf{G}_n = (\sigma_{n1}, \sigma_{n2}, \dots, \sigma_{nP})^T$ and $\mathbf{X} = (\chi_{S_1}, \chi_{S_2}, \dots, \chi_{S_P})^T$. The two options
in Equation (6.115) indicate that for $\mathbf{G}_1, \mathbf{G}_2, \dots, \mathbf{G}_N$, each has 2^{P-1} different possibilities, $(0, 0, \dots, 0, 0, 1)^T$, $(0, 0, \dots, 0, 1, 0)^T$, $(0, 0, \dots, 0, 1, 1)^T$, $(0, 0, \dots, 1, 0, 0)^T$, $\dots$,
$(0, 1, \dots, 1, 1, 1)^T$, $(1, 0, \dots, 0, 0, 0)^T$, and essentially provides a flexibility in choosing
w_n, as compared to that in Equation (6.114) (only $(0, 1, \dots, 1, 1)^T$ and $(1, 0, \dots, 0, 0)^T$
available). To better accommodate the complexity of the spatial-frequency profile in small-
scale features of interest, it particularly resorts to the GA search engine for the purpose
of optimization in the determination of $\mathbf{G}_1, \mathbf{G}_2, \dots, \mathbf{G}_N$. For the same image reconstruction problem demonstrated in Figure 6.4 and 6.5, the approach with the implementation
of GA into Equation (6.110), (6.111), (6.113), and (6.115) provides a further resolution
improvement superior to that in Figure 6.5, as seen in Figure 6.6.

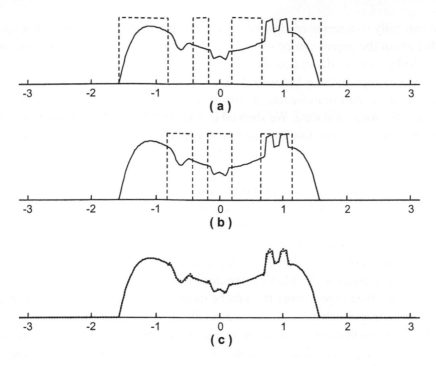

Figure 6.6 The same image reconstruction problem as that in Figure 6.4 but different esti-mation method: (a) the weighted window χ_{S_1} (dashed line) associated with low-frequency profile; (b) the weighted windows χ_{S_2}, χ_{S_3} and χ_{S_4} (dashed line) associated with small-scale features of interest; (c) the estimate by applying the GA into Equation 6.110, 6.111, 6.113 and 6.115 (dotted line) with weighted windows in (a) and (b).

6.7 Summary and Conclusions

In this chapter we have described the ubiquitous problem of estimating an image from a limited number of noisy measured samples, acquired somehow from an object of interest. The assumption throughout has been that these data can be described by a linear functional of the object function, and that the object is well approximated by a function of compact support. Given the inherent ill-posedness of this problem, an inevitable underlying theme to what was presented is that an infinity of possible reconstructed images of that object can be found. Consequently, we adopt an appropriate model that is both consistent with the measured data as well as any prior information about the object that we might have. In the simple case of measured Fourier data, it was clear that a DFT of these data spreads energy in the image domain over intervals that might be much larger than the prior knowledge we might have regarding the extent of the object's compact support. In addition, the model creates a framework that allows us to select a unique image estimate, which has minimum energy and that minimizes the energy different between that chosen estimate and the true object energy. We are accepting that these seem to be reasonable requirements. Regardless, we do not forget that we are engaged in a modeling exercise that, from the start, admits to our ignorance when finite noisy data are all we (inevitably) have to work from.

The Hilbert space formalism, which focused on the L^2-norm as a distance measure, led quite naturally to a series of algorithms, labeled the PDFT, P denoting that some prior knowledge about the object can be incorporated. It is not always clear what one means by prior knowledge but we illustrated this with several examples. These examples were not intended to be comprehensive but were given to provide some insight as to the procedure one could adopt in order to make use of whatever kind of information one might have prior to processing the measured data. We showed that knowledge of the compact support of the object, or at least a conservative estimate of it, can lead to dramatic improvements in the corresponding minimum norm image estimate. One can think of this as redistributing the available degrees of freedom, embodied in the measured data, as being redistributed to find the coefficients of a much more appropriate basis set to represent the object of interest. It is somewhat obvious that for an object of compact support, a representation of infinitely periodic sines and cosines might not be the efficient or most compact representation for an object known to reside somewhere in the interval $[-X, X]$.

The minimum norm methods presented are easily regularized and provide a data-consistent optimal estimate. In addition, we described how these methods that incorporate prior knowledge about object support could be modified to include information within that support about subregions that are known to have different spatial frequency content, or even to weight subregions that are simply of greater interest to the user. With increased noise and the need for larger regularization parameters, we described how the (Miller-Tikhonov) regularization procedure can be viewed as a controlled relaxation of the incorporated prior knowledge support function. This provided an intuitive understanding about the necessary trade-offs between resolution and noise. However, it also points to a way to update a sequence of prior estimates and explore iteratively their compatibility with the energies of the corresponding minimum norm estimates, thereby providing a kind of search tool that can lead to signatures or task specific information that might be more useful than simply computing a nice-looking image.

The minimum norm approach, through the choice of a suitable Hilbert space, defines distances and hence metrics for assessing the quality of the reconstructed image. A key feature of this is that an improved set of basis functions is defined and can be improved in the sense that we exploit prior knowledge to design them. We are very familiar with successful techniques for image compression, and accept that a smaller number of coefficients associated with a basis set such as wavelets, can dramatically reduce the number of data to be stored or transmitted, while still yielding a high-quality image. One can also imagine a data acquisition scheme whereby those ideal coefficients are the data measured directly, rather than being computed using a minimum norm estimation procedure. The simple modified basis function used in the PDFT example, which incorporates object support information, leads to a higher-resolution image from a fixed number of measured data. Equivalently, we can think of those new coefficients calculated in the PDFT algorithm as actually "better" data for our image estimation problem. This idea of what constitutes "better" data that allows image information we require to be recovered from some small measured data set, has been labeled compressive or random sampling in recent years. This is an approach that makes the very best use of what might be a small number of data or degrees of freedom to recover image information one might otherwise have expected to need much larger data sets. The key to this approach was to think about the basis set for the object representation and then the basis for the domain in which data are measured. These are preferably quite different and preferably incoherent; incoherence was explicitly defined and the smaller that parameter, $\mu(U, V)$ (i.e., the closer it is to $1/\sqrt{J}$), the better. Selecting U and V is to a large extent in our hands, as we can choose the representation we want for the image estimate and we can to a large extent choose a basis appropriate to a measurement procedure. The prior knowledge that drives this modeling approach is that of sparseness, meaning that the object can be well represented by a small number of coefficients in some appropriate Hilbert space. The weighted L^2 space for the PDFT is but one example of such a representation. A simpler example would be an object consisting of a few delta-like features. This is sparse in the object domain while in the measurement space, such as its associated Fourier domain, basis functions are continuous and extend throughout that space. Prior knowledge of sparseness permits relatively few random measurements in the Fourier domain to be sufficient for an excellent image reconstruction of the object, based on an L^1-norm minimization and far fewer measurements than one might expect, based on Nyquist sampling criteria [20]. We connected the minimum L^2-norm methods described earlier with this minimum L^1 approach specifically through the example of the PDFT.

In conclusion, the way in which we view image estimation has undergone a transition, as we move away from dealing with the imperfections of our existing modeling methods, that typically involve algorithms for deconvolution, enhancements, and somewhat cosmetic processing, to one grounded in degrees of freedom and Hilbert space metrics. By regarding the limited number of noisy measurements we have available as being derived from a customized model for our object that were projected onto a basis set of our choosing, we can generate a better image estimate. More precisely, we may choose not to form an image at all, but more effectively better reveal information about the object that we need. Of course, with too little data and too much noise, there will always be limitations to what can be done; and if we choose our prior knowledge very badly, we must accept the consequences!

Bibliography

[1] M. Bertero and P. Boccacci, *Introduction to Inverse Problems in Imaging*. Bristol and Philadelphia: IOP Publishing, 1998.

[2] C. L. Byrne, *Signal Processing: A Mathematical Approach*. Wellesley, MA: AK Peters, Ltd., 2005.

[3] M. Bertero, *Inverse problems in scattering and imaging*. Malvern Physics Series, Adam Hilger, Bristol: IOP Publishing, 1992.

[4] J. P. Burg, "Maximum entropy spectral analysis," in *the 37th Annual Society of Exploration Geophysicists Meeting*, (Oklahoma City, Oklaoma), 1967.

[5] J. P. Burg, "The relationship between maximum entropy spectra and maximum likelihood spectra," *Geophysics*, vol. 37, pp. 375–376, 1972.

[6] C. L. Byrne and R. M. Fitzgerald, "Reconstruction from partial information, with applications to tomography," *SIAM Journal of Applied Mathematics*, vol. 42, pp. 933–940, 1982.

[7] C. L. Byrne and R. M. Fitzgerald, "Spectral estimators that extend the maximum entropy and maximum likelihood methods," *SIAM Journal of Applied Mathematics*, vol. 44, pp. 425–442, 1984.

[8] C. L. Byrne and M. A. Fiddy, "Estimation of continuous object distributions from limited Fourier magnitude measurements," *Journal of the Optical Society of America A*, vol. 4, pp. 112–117, 1987.

[9] H. M. Shieh, C. L. Byrne, and M. A. Fiddy, "Image reconstruction: a unifying model for resolution enhancement and data extrapolation. Tutorial," *Journal of the Optical Society of America A*, vol. 23, pp. 258–266, 2006.

[10] C. L. Byrne, R. M. Fitzgerald, M. A. Fiddy, T. J. Hall, and A. M. Darling, "Image restoration and resolution enhancement," *Journal of the Optical Society of America A*, vol. 73, pp. 1481–1487, 1983.

[11] C. L. Byrne and M. A. Fiddy, "Image as power spectral; reconstruction as a Wiener filter approximation," *Inverse Problems*, vol. 4, pp. 399–409, 1988.

[12] T. J. Hall, A. M. Darling, and M. A. Fiddy, "Image compression and restoration incorporating prior knowledge," *Optics Letters*, vol. 7, pp. 467–468, 1982.

[13] H. M. Shieh and M. A. Fiddy, "Accuracy of extrapolated data as a function of prior knowledge and regularization," *Applied Optics*, vol. 45, pp. 3283–3288, 2006.

[14] C. L. Byrne and R. M. Fitzgerald, "A unifying model for spectrum estimation," in *Proceedings of the RADC Workshop on Spectrum Estimation*, (Griffiss AFB, Rome, NY), October, 1979.

[15] C. L. Byrne, B. M. Levine, and J. Dainty, "Stable estimation of the probability density function of intensity from photon frequency counts," *Journal of the Optical Society of America A*, vol. 1, pp. 1132–1135, 1984.

[16] H. M. Shieh, C.-H. Chung, and C. L. Byrne, "Resolution enhancement in computerized tomographic imaging," *Applies Optics*, vol. 47, pp. 4116–4120, 2008.

[17] H. M. Shieh, C. L. Byrne, M. E. Testorf, and M. A. Fiddy, "Iterative image reconstruction using prior knowledge," *Journal of the Optical Society of America A*, vol. 23, pp. 1292–1300, 2006.

[18] D. L. Donoho, "Compressed sampling," *IEEE Transactions on Information Theory*, vol. 52, pp. 1289–1306, 2006.

[19] A. M. Bruckstein, D. L. Donoho, and M. Elad, "From sparse solutions of systems of equations to sparse modeling of signals and images," *SIAM Review*, vol. 51, pp. 34–81, 2009.

[20] E. Candès and J. Romberg, "Sparsity and incoherence in compressive sampling," *Inverse Problems*, vol. 23, pp. 969–985, 2007.

[21] E. J. Candès, M. B. Wakin, and S. P. Boyd, "Enhancing sparsity by reweighted l_1 minimization," *Journal of Fourier Analysis and Applications*, vol. 14, pp. 877–905, 2007.

[22] H. M. Shieh, Y.-C. Hsu, C. L. Byrne, and M. A. Fiddy, "Resolution enhancement of imaging small-scale portions in a compactly supported function," *Journal of the Optical Society of America A*, vol. 27, pp. 141–150, 2010.

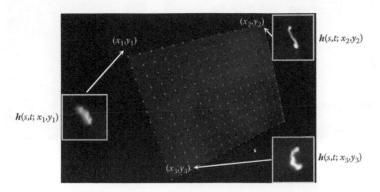

FIGURE 3.1 Spatially varying PSF for motion blur caused by camera shake. The image was acquired in a dark room by taking a picture of an LCD displaying regularly spaced white dots.

FIGURE 3.2 Defocus PSF acquired by deliberately defocusing an LCD covered by regularly spaced white dots. Vignetting clips the polygonal shape of the PSF, especially on the periphery of the field of view.

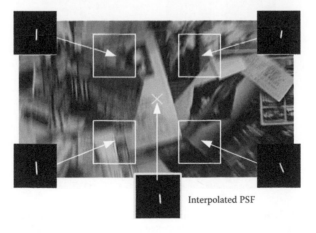

FIGURE 3.3 If the PSF varies slowly, we can estimate convolution kernels on a grid of positions and approximate the PSF in the rest of the image by interpolation of four adjacent kernels.

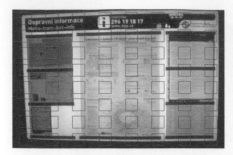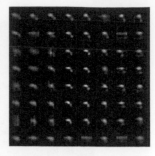

FIGURE 3.12 First of six input images (1700 × 1130 pixels) used for super resolution (left) and 8 × 8 local convolution kernels computed by the SR algorithm [39] (right). Spatial variance of the PSF is obvious. Squares in the image depict patches, in which the kernels were estimated.

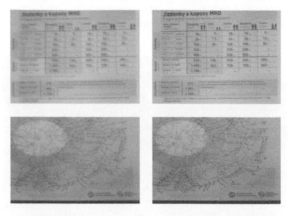

FIGURE 3.13 Two details of the blurred low-resolution image from Figure 3.12 (left column) and the resulting high-resolution image of 3400 × 2260 pixels (right column).

FIGURE 4.1 Comparison of weight distribution of different methods for image *Houston*: (a) original image; (b) enlarged portion corresponding to the dotted rectangle in (a), with small solid rectangles show the center patch and examples of similar patches; (c) weight distribution corresponding to the bilateral filter; (d) weight distribution corresponding to the NLM filter.

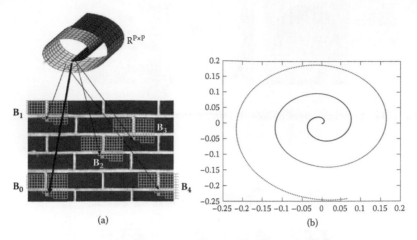

FIGURE 5.1   Illustration of manifold: (a) two adjacent points on a manifold could be spatially distant (e.g., B_i vs. B_j ($0 \le i \ne j \le 4$)); (b) spiral data (a 1D manifold embedded in R^2).

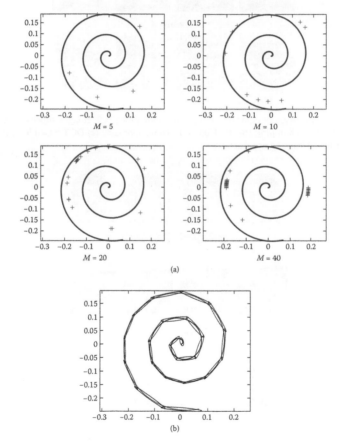

FIGURE 5.2   Two ways of discovering the manifold structure underlying a signal: (a) dictionary learning (atoms in dictionary **A** are highlighted in '+'); (b) structural clustering (each segment denotes a different cluster $p(x|\theta_m)$).

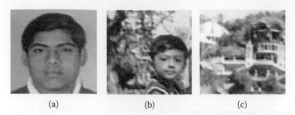

(a)   (b)   (c)

FIGURE 7.14   Color test images. (a) *face*; (b) *child*; and (c) *palace*.

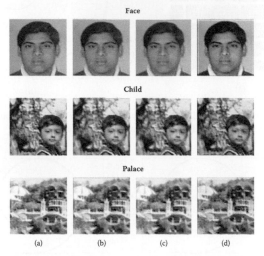

(a)   (b)   (c)   (d)

FIGURE 7.15   Comparison of color image super resolution for $q = 2$. Images super-resolved using; (a) Kim and Kwon approach in [45]; (b) DWT-based learning approach; (c) DCT-based learning approach; (d) Contourlet-based learning approach.

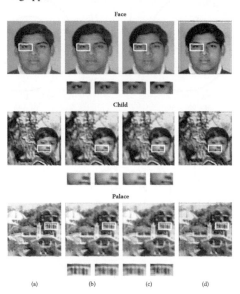

(a)   (b)   (c)   (d)

FIGURE 7.16   Comparison of color image super resolution for $q = 4$. Images super-resolved using; (a) Kim and Kwon approach in [45]; (b) DWT-based learning approach; (c) DCT-based learning approach; (d) Contourlet-based learning approach. Below each image a zoomed-in version of the rectangle region is shown for easier visual inspection.

FIGURE 8.5 Unsupervised classification results of MODIS Band 01 and Band 02. (a) MODIS 250m data clustering; (b) MODIS 500m data clustering; (c) clustering of bilinear interpolated images; (d) clustering of MAP-uHMT super resolved images.

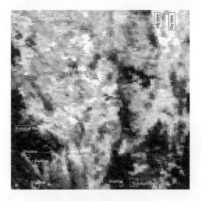

FIGURE 8.6 Training fields in data set (a) of MODIS.

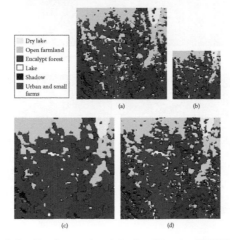

FIGURE 8.7 Supervised classification results of the first 7 bands of MODIS. (a) MODIS 500m data classification; (b) MODIS 1km data classification; (c) classification of bilinear interpolated images; (d) classification of MAP-uHMT super resolved images.

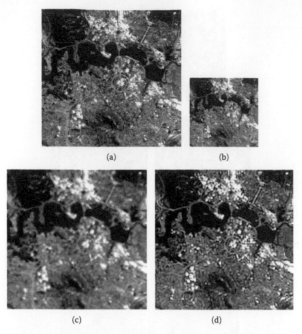

FIGURE 8.8 Five bands of ETM+ data displayed in color composite with Bands 1 (blue), 2 (green), 3 (red). (a) original High Resolution (HR) images; (b) sub-sampled LR images; (c) interpolated images; (d) MAP-uHMT super resolved images.

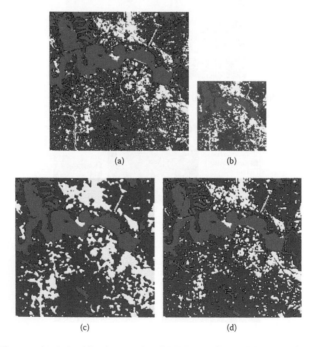

FIGURE 8.9 Unsupervised classification results of ETM+ Band 1 to 5. (a) clustering of original high resolution data; (b) clustering of LR data; (c) clustering of bilinear interpolated data; (d) clustering of MAP-uHMT super resolved data.

FIGURE 8.10 Training fields in data set (a) of ETM+.

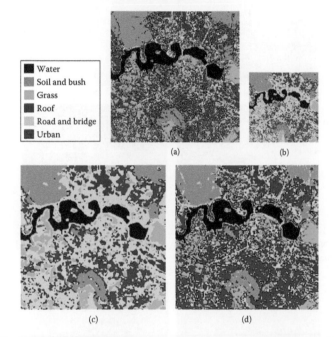

Water
Soil and bush
Grass
Roof
Road and bridge
Urban

(a)

(b)

(c)

(d)

FIGURE 8.11 Supervised classification results of the 5 bands of Landsat7. (a) classification results of the 5 bands of the original HR images; (b) classification results of the 5 bands of the LR images; (c) classification of bilinear interpolated images; (d) classification of MAP-uHMT super resolved images.

(a)

(b)

FIGURE 9.1 Additive color mixing. (a) image obtained by shifting the three grayscale images on the left, top and bottom of the figure which correspond to the red, green, and blue color channels, respectively; (b) original RGB image.

(a)   (b)

FIGURE 9.3 Chrominance representation of the image shown in Figure 9.1b using (a) Equation 9.2 and (b) Equation 9.4.

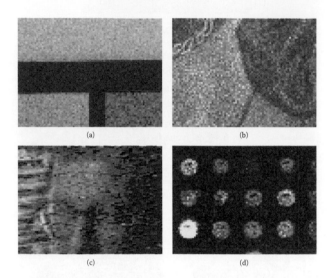

(a)   (b)

(c)   (d)

FIGURE 9.4 Noise present in real-life images: (a) digital camera image; (b) digitized artwork image; (c) television image; (d) microarray image.

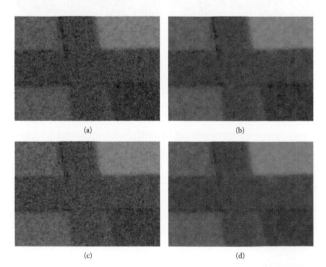

(a)   (b)

(c)   (d)

FIGURE 9.7 Noise reduction in a color image: (a) captured noisy image; (b) luminance noise suppression; (c) chrominance noise suppression; (d) both luminance and chrominance noise suppression.

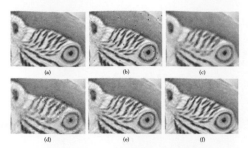

FIGURE 9.8 Impulsive noise suppression using vector order-statistics filters: (a) noise-free image; (b) noisy image; (c) vector median filter; (d) basic vector directional filter; (e) optimal weighted vector directional filter; (f) switching vector median filter. The filtered images are produced in a single iteration using a 3×3 supporting window.

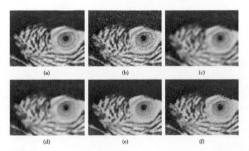

FIGURE 9.9 Additive Gaussian noise suppression using vector combination methods: (a) noise-free image; (b) noisy image; (c) fuzzy vector filter; (d) data-adaptive filter based on digital paths; (e) standard 4-point anisotropic diffusion filter after 20 iterations; (f) bilateral filter operating using an 11×11 supporting window. Unless otherwise stated, the filtered images are produced in a single iteration using a 3×3 supporting window.

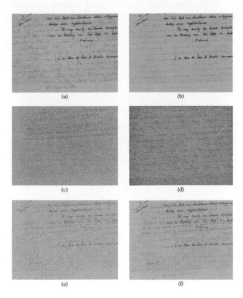

FIGURE 10.1 Example of removing show-through while enhancing a hidden pattern: (a) the original RGB document; (b) the document cleansed from interferences and colorized; (c) a cleansed map of the hidden printed text alone; (d) a cleansed map of the show-through pattern alone; (e) the handwritten main text plus the hidden printed text; (f) a pseudocolor image built using images in (b) (grayscale version) and (e).

(a)   (b)   (c)

FIGURE 10.2 Example of stain attenuation via symmetric whitening: (a) the original RGB document; (b) the grayscale version of the document; (c) first output of SW (note the enhancement of the bleed-through pattern).

FIGURE 10.5 Example of interference removal and recovery of the main text from an RGB manuscript: (a) the original image; (b) the clean main text obtained with the regularization technique based on a linear convolutional data model; (c) the best extraction of the main text obtainable with ICA (after histogram thresholding).

High-resolution
Low-frame-rate

Low-resolution
High-frame-rate

Time

FIGURE 11.1 Tradeoff between resolution and frame rates. Top: Image from a high-resolution, low-frame-rate camera. Bottom: Images from a low-resolution, high-frame-rate camera. (© 2010 IEEE)

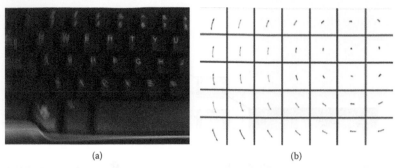

(a)   (b)

FIGURE 11.4 Spatially-varying blur kernel estimation using optical flows. (a) motion blur image; (b) estimated blur kernels of (a) from optical flows. (© 2010 IEEE)

(a)   (b)

(c)   (d)

FIGURE 11.7 Convolution with kernel decomposition. (a) convolution result without kernel decomposition, where full blur kernels are generated on-the-fly per pixel using optical flow integration; (b) convolution using 30 PCA-decomposed kernels; (c) convolution using a patch-based decomposition; (d) convolution using delta function decomposition of kernels, with at most 30 delta functions per pixel. (© 2010 IEEE)

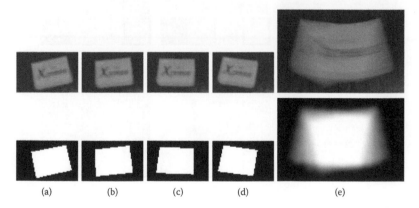

(a)   (b)   (c)   (d)   (e)

FIGURE 11.9 Layer separation using a hybrid camera: (a)-(d) low-resolution frames and their corresponding binary segmentation masks; (e) high-resolution frame and the matte estimated by compositing the low-resolution segmentation masks with smoothing. (© 2010 IEEE)

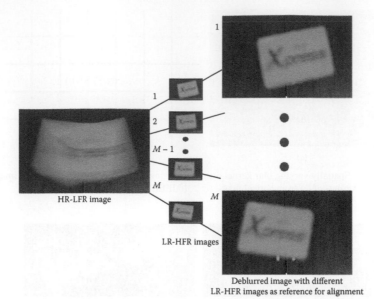

FIGURE 11.10 Relationship of high-resolution deblurred result to corresponding low-resolution frame. Any of the low-resolution frame can be selected as a reference frame for the deblurred result. This allows up to M deblurred solutions to be obtained. (© 2010 IEEE)

FIGURE 11.11 Image deblurring using globally invariant kernels. (a) input; (b) result generated with the method of [14], where the user-selected region is indicated by a black box; (c) result generated by [3]; (d) result generated by back-projection [34]; (e) our results; (f) the ground truth sharp image. Close-up views and the estimated global blur kernels are also shown. (© 2010 IEEE)

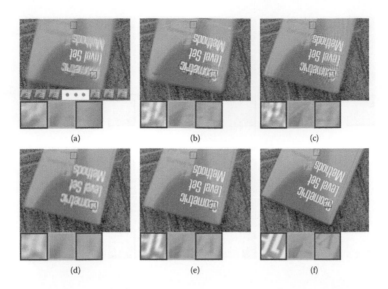

FIGURE 11.12 Image deblurring with spatial varying kernels from rotational motion. (a) input; (b) result generated with the method of [30] (obtained courtesy of the authors of [30]); (c) result generated by [3] using spatially-varying blur kernels estimated from optical flow; (d) result generated by back-projection [34]; (e) our results; (f) the ground truth sharp image. Close-ups are also shown. (© 2010 IEEE)

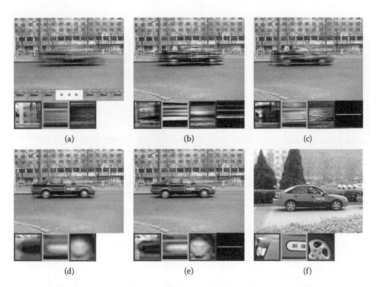

FIGURE 11.13 Image deblurring with translational motion. In this example, the moving object is a car moving horizontally. We assume that the motion blur within the car is globally invariant. (a) input; (b) result generated by [14], where the user-selected region is indicated by the black box; (c) result generated by [3]; (d) result generated by back-projection [34]; (e) our results; (f) the ground truth sharp image captured from another car of the same model. Close-up views and the estimated global blur kernels within the motion blur layer are also shown. (© 2010 IEEE)

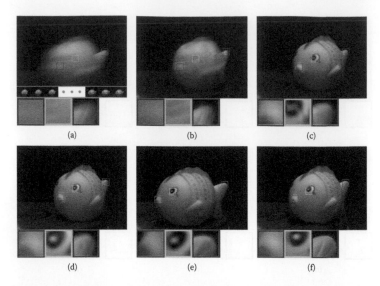

FIGURE 11.14 Image deblurring with spatially-varying kernels. In this example, the moving object contains out-of-plane rotation with both occlusion and disocclusion at the object boundary. (a) input; (b) result generated by [3]; (c) result generated by back projection [34]; (d) our results using the first low-resolution frame as the reference frame; (e) our results using the last low-resolution frame as the reference frame; (f) the ground truth sharp image. Close-ups are also shown. (© 2010 IEEE)

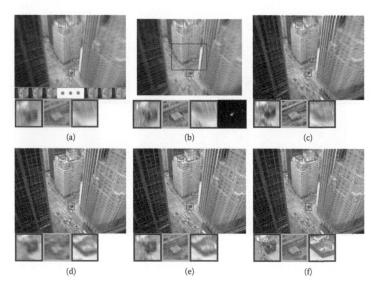

FIGURE 11.15 Image deblurring with spatially-varying kernels. In this example, the camera is zooming into the scene. (a) input; (b) result generated by [14]; (c) result generated by [3]; (d) result generated by back-projection [34]; (e) our results; (f) the ground truth sharp image. Close-ups are also shown. (© 2010 IEEE)

FIGURE 11.16 Deblurring with and without multiple high-resolution frames. (a) and (b) input images containing both translational and rotational motion blur; (c) deblurring using only (a) as input; (d) deblurring using only (b) as input; (e) deblurring of (a) using both (a) and (b) as inputs; (f) ground truth sharp image. Close-ups are also shown. (© 2010 IEEE)

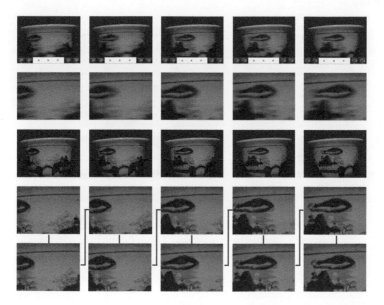

FIGURE 11.17 Video deblurring with out-of-plane rotational motion. The moving object is a vase with a center of rotation approximately aligned with the image center. First Row: Input video frames. Second Row: Close-ups of a motion blurred region. Third Row: Deblurred video. Fourth Row: Close-ups of deblurred video using the first low-resolution frames as the reference frames. Fifth Row: Close-ups of deblurred video frames using the fifth low-resolution frames as the reference frames. The final video sequence has higher temporal sampling than the original high-resolution video, and is played with frames ordered according to the red lines. (© 2010 IEEE)

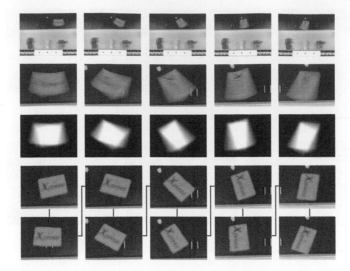

FIGURE 11.18 Video deblurring with a static background and a moving object. The moving object is a tossed box with arbitrary (in-plane) motion. First Row: Input video frames. Second Row: Close-up of the motion blurred moving object. Third Row: Extracted alpha mattes of the moving object. Fourth Row: The deblurred video frames using the first low-resolution frames as the reference frames. Fifth Row: The deblurred video frames using the third low-resolution frames as the reference frames. The final video with temporal super-resolution is played with frames ordered as indicated by the red lines. (© 2010 IEEE)

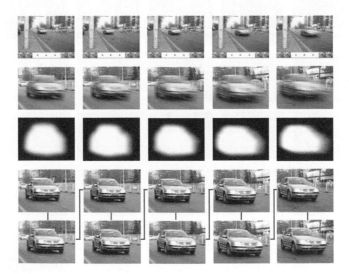

FIGURE 11.19 Video deblurring in an outdoor scene. The moving object is a car driving towards the camera, which produces both translation and zoom-in blur effects. First Row: Input video frames. Second Row: The extracted alpha mattes of the moving object. Third Row: The deblurred video frames using the first low-resolution frames as the reference frames. Fourth Row: The deblurred video frames using the third low-resolution frames as the reference frames. The final video consists of frames ordered as indicated by the red lines. By combining results from using different low-resolution frames as reference frames, we can increase the frame rate of the deblurred video. (© 2010 IEEE)

Chapter 7

Transform Domain-Based Learning for Super Resolution Restoration

PRAKASH P. GAJJAR
Dhirubhai Ambani - Institute of Information and Communication Technology

MANJUNATH V. JOSHI
Dhirubhai Ambani - Institute of Information and Communication Technology

KISHOR P. UPLA
Sardar Vallabhbhai National Institute of Technology

7.1   Introduction to Super Resolution

The term *resolution* refers to the smallest measurable physical quantity. The resolution of an imaging system is defined as the ability of the system to record finer details in a distinguishable manner [2]. The term *image resolution* can be defined as the smallest measurable detail in a visual presentation. In image processing, it is a measure of the amount of detail provided by an image or a video signal. The term *image resolution* is classified into different types: spatial resolution, brightness resolution, spectral resolution, and temporal resolution. In this chapter we address the problem of increasing the spatial resolution of given low spatial resolution images. A digital image is represented using a set of picture elements. These picture elements are called "pixels" or "pels." A pixel at any location in an image carries the information regarding the image intensity at that location in the image. An image represented using a large number of pixels conveys more information as compared to the same image when represented using fewer pixels. Spatial resolution refers to the spacing of the pixels in an image and is measured in pixels per inch (ppi). High spatial resolution allows for sharp details and fine-intensity transitions across all directions. The representation of an image having sharp edges and subtle intensity transition by less dense pixels gives rise to blocky effects. On the other hand, the images with a dense set of pixels gives the viewer the perception of finer details and offers a pleasing view. In the rest of this chapter, the term *resolution* is explicitly used to refer to *spatial resolution* unless specified otherwise.

7.1.1 Limitations of Imaging Systems

The sensor is the most important component of any digital imaging system. It converts optical energy into an electrical signal. In modern digital cameras, charge-coupled devices (CCD) and CMOS sensors are widely used to capture digital images. The CCD or CMOS sensors consist of an array of photodetectors. The spatial resolution of an image captured using a digital camera is determined by the number of photodetector elements in the sensor. A sensor with fewer photodetectors samples the scene with a low sampling frequency and causes an aliasing effect. Such sensors produce low-resolution images with blocky effects. The direct solution for enhancing the spatial resolution of an image is to increase the number of photodetectors in the sensor. As the number of photodetectors in the sensor chip is increased, the size of the chip also increases. This leads to an increase in capacitance [3], causing limitation on the charge transfer rate, and hence this approach is not considered effective. An alternative to this solution is to increase the photodetector density by reducing the size of photodetector elements. As the photodetector size decreases, the amount of light falling on each photodetector also decreases. Above some limit of the size of the detector, the amount of light collected reaches such a low level that the signal is no longer prominent as compared to the noise. This generates shot noise that degrades the image quality severely [4,5]. Thus, there exists a lower limit on reducing the size of the photodetector element. The optimal size photodetector to generate a light signal without suffering from the effects of shot noise is estimated at about $40~\mu m^2$ for a 0.35-μm CMOS process. Current image sensor technology has almost reached this level. Therefore, new approaches toward increasing spatial resolution are required to overcome the inherent limitations of sensors and optical imaging systems. The cost of high-precision optics and image sensors is also an important factor in many commercial applications. Hence, a promising approach is to use algorithms based on digital signal processing techniques to construct a high-resolution image from one or more available low-resolution observations.

7.1.2 Super Resolution Concept

In the process of capturing an image using a digital image acquisition system, there is a natural loss of spatial resolution. There are many factors that contribute to this effect. These include out of focus, blur, motion blur due to limited shutter speed or relative movement between camera and the object, undersampling, and the noise occurring within the sensor or during transmission and insufficient sensor density. The resulting image is usually degraded due to blur, noise, and aliasing effects. Figure 7.1 illustrates the degradations introduced in the image at various stages in the image capturing process. Aliasing occurs when the image is sampled at a low spatial sampling rate and that causes the distortions in the high-frequency contents of the image. Resolution improvement by applying tools from digital signal processing techniques has been a topic of great interest. The term *super resolution* refers to such signal processing techniques that reconstruct a high spatial resolution image

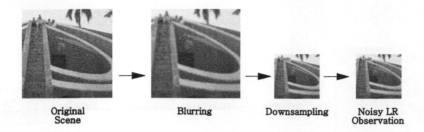

Figure 7.1 Information loss during image acquisition. A noisy low-resolution (LR) image of the scene is recorded.

from one or more low-resolution images. The goal of super resolution techniques is to recover the high-frequency content that is lost during the image acquisition process. The word *super* in super resolution represents very well the characteristics of the technique of overcoming the inherent resolution limitation of low-resolution imaging systems. The advantages of the super resolution approach are that the existing low-resolution imaging systems can still be utilized without any additional hardware, and it costs less and offers flexibility.

The super resolution reconstruction problem is closely related to the image restoration problem. The goal of image restoration is to recover an image from degradations such as blur and noise, but does not increase the size of the image. Thus, for an image restoration problem, the size of the restored image is the same as that of the observed image, while it is different in image super resolution, depending on the resolution factor of the super-resolved image. Image interpolation is another area referred to by the super resolution community. It increases the size of the image using a single aliased low-resolution observation. Because the single image can provide no more nonredundant information, the quality of the interpolated images is very much limited. Single image interpolation techniques cannot recover the high-frequency components lost or degraded during a low-resolution sampling process. For this reason, image interpolation methods are not considered super resolution techniques.

7.1.3 Super Resolution: Ill-Posed Inverse Problem

Super resolution algorithms attempt to reconstruct the high-resolution image corrupted by the limitations of the optical imaging systems. The algorithms estimate the high-resolution image from one or more degraded low-resolution images. This is an inverse problem, wherein the original information is retrieved from the observed data. A schematic representation of the same is shown in Figure 7.2. Solving the inverse problem requires inverting the effects of the forward model. It is difficult to invert without amplifying the noise in the observed data. Assuming a linear model, the forward model of high-resolution (HR) to low-resolution (LR) transformation is reduced to matrix manipulations, and hence it is logical to formulate the restoration problem as matrix inversion. The problem is worsened by the fact that we end up with an under-determined system and hence an infinite space of solutions. In other words, there exist an infinite number of high-resolution images that are consistent with the original data. Thus, the super resolution problem is an ill-posed inverse problem. Hence, some form of constraints to limit the space of solutions must be included.

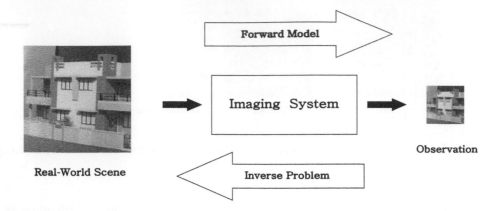

Figure 7.2 Schematic representation of inverse problem. The forward model is a mathematical description of the image degradation process. The inverse problem addresses the issue of reconstructing the original scene from one or more observations.

The regularization-based approach solves the ill-posed inverse problem by making it better-posed using the prior information about the solution. It is a systematic method for adding additional information to the solution. The Bayesian super resolution reconstruction approach is commonly employed for solving ill-posed inverse problems. This method is used when a posterior probability density function of the original image can be established. The major advantages of the Bayesian super resolution approach are its robustness and flexibility in modeling noise characteristics and a priori knowledge about the solution.

7.2   Related Work

Super resolution (SR) has attracted a growing interest as a purely computational means to increase imaging sensors performance [6–10]. The pioneer work of super resolution reconstruction goes back to 1984 by Tsai and Huang [11]. Since then, a variety of approaches for solving the super resolution problem have been proposed [12–14]. We categorize SR reconstruction methods into three main divisions: motion-based SR reconstruction, motion-free super resolution, and single frame super resolution. Motion based techniques use the relative motion between the low-resolution observations as a cue in estimating the high-resolution image, while motion-free super resolution techniques use cues such as blur, zoom, and defocus. Single-frame super resolution approaches attempt to reconstruct the super resolution using a single low resolution observation. In this case, because only a single undersampled and degraded input image is available, the task of obtaining a super-resolved image consists of recovering the additional spatial data from the available database of high-resolution images.

7.2.1   Motion-Based Super Resolution

Motion-based super resolution approaches obtain super resolution from several subsampled and misregisterd low-resolution (LR) images of a scene. The low-resolution images can be obtained either as a sequence taken over a time, or taken at the same time with different sensors. In particular, camera and scene motion lead to multiple observations containing similar, but not identical information. Because, the observations have subpixel shifts and are aliased, new information in each of the observations can be exploited to construct a high-resolution (HR) image. The frequency domain-based approaches for solving the SR problem use the relationship between LR images and the HR image in the frequency domain. Formulation of the system of equations relating the aliased DFT coefficients of the observed images to samples of the DFT of the unknown scene requires knowledge of the translational motion between frames to sub-pixel accuracy. Each observation image must contribute to independent equations, which places restrictions on the inter-frame motion that contributes useful data. Tsai and Huang [11] assume that the desired HR image is bandlimited and derive analytical relationships between observations and the desired HR image. They obtain super resolution using the shifting property of the Fourier transform and the aliasing relationship between the Fourier transform of HR and of the observed LR images. Other approaches using frequency domain techniques include [15–18]. Approaching the super resolution problem in the frequency domain makes a lot of sense because it is relatively simple and computationally efficient. The capability of parallel implementation of these techniques makes the hardware less complex. However, there are some problems with a frequency domain formulation. It restricts the inter-frame motion to being translational because the DFT assumes uniformly spaced samples. The observation model is restricted only to global translational motion. Another disadvantage is that prior knowledge that might be used to constrain or regularize the super resolution problem is often difficult to express in the frequency domain. In other words, the frequency domain approach makes it difficult to apply the spatial domain prior for regularization.

A variety of techniques exist for the super resolution problem in the spatial domain. The primary advantages to working in the spatial domain are support for unconstrained motion between frames and ease of incorporating prior knowledge into the solution. Stochastic methods, which treat SR reconstruction as a statistical estimation problem, have rapidly gained prominence because they provide a powerful theoretical framework for the inclusion of a priori constraints necessary for satisfactory solution of the ill-posed SR inverse problem. The statistical techniques explicitly handle prior information and noise. Inclusion of prior knowledge is usually more natural using a stochastic approach. The stochastic SR reconstruction using the Bayesian approach provides a flexible and convenient way to model a priori knowledge about the final solution. This method can be applied when an a posteriori probability density function of the original image can be estimated. The maximum a posteriori (MAP) approach to estimating the super resolution seeks the estimate for which the a posteriori probability is maximum. It is common to utilize Markov random field (MRF) image models as the prior term. Under typical assumptions of Gaussian noise, the prior may be chosen to ensure a convex optimization enabling the use of simple optimization procedures.

Schultz and Stevenson propose SR reconstruction from LR video frames using the MAP technique [19]. They employ a discontinuity-preserving Huber-Markov Gibbs prior model and constrained optimization. Farsiu et al. [20] propose a unified approach of demosaicing and super resolution of a set of low-resolution color images. They employ bilateral regularization of the luminance term for the reconstruction of sharp edges, and that of the chrominance term and intercolor dependencies term, to remove the color artifacts from the

HR estimate. Irani and Peleg [21] propose a super resolution algorithm based on iterative back-projection (IBP). The key idea here is that the error between the observed low-resolution images and the corresponding low-resolution images formed using an estimate of the SR image can be used to iteratively refine the estimated SR image. This approach begins by guessing an initial HR image. This initial HR image is then downsampled to simulate the observed LR images. The simulated LR images are subtracted from the observed LR image. If the initial HR image was the real observed HR image, then the simulated LR images and the observed LR images would be identical and their difference would be zero. Hence, the computed differences can be "back-projected" to improve the initial guess. The back-projecting process is repeated iteratively to minimize the difference between the simulated and the observed LR images, and subsequently produce a better HR image.

7.2.2   Motion-Free Super Resolution

All the above approaches use motion as a cue for solving the super resolution problem. The primary factor that controls the quality of the super resolution obtained using these approaches is the extremely precise alignment, that is, registration between the low-resolution observations. Park et al. [22] has shown by example that small error in registration can considerably affect the super resolution results. Most of the registration algorithms tend to be sensitive to illumination, blur variations, and noise. Approaches that use frequency domain processing to compute the registration parameters are relatively stable under various image artifacts. However, they are limited in the class of transformations that can be estimated between two images [23].

It is shown that super resolution is also possible from the observations captured without relative motion between them. There has been a substantial amount of work performed on the spatial resolution enhancement by using cues that do not involve a motion estimation among low-resolution observations. The new approaches based on cues other than motion such as blur, defocus [24], and zoom [25, 26] are known as motion-free super resolution approaches. In these approaches, the additional information of the super-resolved image is obtained from the multiple observations captured by varying camera parameters. In general, changes in these low-resolution images caused by the camera or scene motion, camera zoom, focus, and blur allow us to recover the missing information for reconstructing an output image at a resolution much higher than the original resolution of the camera. The authors in [27] describe an MAP-MRF-based super resolution technique for SR reconstruction from several blurred and noisy low-resolution observations. In [28], the authors recover both the high-resolution scene intensity and the depth fields simultaneously using a defocus cue. Rajagopalan and Kiran [24] propose a frequency domain approach for SR reconstruction using the defocus cue. They also show that the estimation of the HR image improves as the relative blur increases.

In [29, 30] the authors show the estimation of a super-resolved image and depth map using a photometric cue. They model the surface gradients and albedo as the Markov random fields and use line fields for discontinuity preservation. Because they use simulated annealing for minimization, the approach is computationally very taxing. Joshi et al. [25] demonstrate the use of zoom cue for super resolution. They capture the low-resolution observations by varying the zoom setting of a camera and obtain super resolution using a MAP-MRF framework. A learning-based approach for SR reconstruction from zoomed observations is proposed in [31]. The HR image is modeled as an autoregressive (AR) model where the model parameters are estimated using the most zoomed observation. The authors in [32] propose a shape from focus method to super-resolve the focused image of 3D objects. Using the observations in the shape from focus stack and the depth map of the object, they reconstruct super resolution by magnification factors of 2 or greater using a MAP-MRF technique.

7.2.3 Learning-Based Super Resolution

The super resolution approaches discussed thus far recover the super-resolved image using the nonredundant information available in multiple low-quality images. For these techniques to succeed, a sufficient number of low-resolution images are needed, so as to enable the recovery of the aliased pixels. Based on this reasoning, one might be led to the conclusion that SR based on a single LR image is impossible. Is it indeed so? The answer depends on the availability of the information in the LR observations that the reconstruction process needs for super-resolving. One fascinating and promising alternative is to use a database of high-resolution training images to recover the finer detail of the super-resolved image. The learning-based super resolution approaches attempt to obtain super resolution using a single observation. These approaches are also known as "single-frame super resolution" approaches. In comparison to the multiple-image case, this problem is more severely constrained as the available information about the scene is limited.

Given a database consisting of training-images, the learning-based algorithms learn or estimate the finer details corresponding to different image regions seen at a low resolution. A number of learning-based SR algorithms have been studied [33–42]. These algorithms use a learning scheme to capture the high-frequency details by determining the correspondence between LR and HR training images. Compared to traditional methods, which basically process images at the signal level, learning-based SR algorithms incorporate application-dependent priors to infer the unknown high-resolution image. The input LR image is split into either overlapping or non-overlapping patches. Then, for each LR patch from the input image, either one best-matched patch or a set of best-matched LR patches is selected from the training set. The corresponding HR patches are used to reconstruct the output HR image.

Freeman et al. [33] propose an example-based super resolution technique. They estimate missing high-frequency details by interpolating the input low-resolution image into the desired scale and then search the high spatial frequency patches from the database. They embed two matching conditions into a Markov network. The first condition is the similarity of the LR training image patch being similar to the test image patch, while the other condition is that the contents of the corresponding HR patch should be consistent with its neighbors. The super resolution is performed by the nearest-neighbor-based estimation of high-frequency patches based on the corresponding patches of input low-resolution image. A learning-based image hallucination technique is proposed in [43]. Here, the authors use primal sketch priors for primitive layers (edges and junctions) and employ patch-based learning using a large image database. In [41], Baker and Kanade propose a hallucination technique based on the recognition of generic local features. These local features are then used to predict a recognition-based prior, rather than a smoothness prior as is the case with most super resolution techniques. The authors in [44] present a learning-based method to super-resolve face images using a kernel principal component analysis (PCA) based prior model. They regularize the solution using prior probability based on the energy lying outside the span of principal components identified in a higher-dimensional feature space. Based on the framework of Freeman et al. [39], Kim and Kwon investigate a regression-based approach for single-image super resolution [45]. Here, the authors generate a set of candidates for each pixel using patch-wise regression and combine them based on the estimated confidence for each pixel. In the post-processing step, they employ a regularization technique using a discontinuity preserving prior. A novel method with manifold learning is proposed in [46]. In this paper, neighbor embedding with training images is adopted to recover the super resolution image. A disadvantage of the approach is that the recovery of super resolution image is easily affected by the training image, which needs to be selected manually within the related contents. Ni and Nguyen used support vector regression to learn the relationship of DCT coefficients between the low- and high-resolution images [47].

Various techniques are investigated in the literature to obtain smooth image boundaries, such as level-set [48], multiscale tensor voting [49], and snake-based vectorization [50] techniques. An edge smoothness prior is favored because it is able to suppress the jagged edge artifact effectively. However, it is difficult to obtain analytical forms for evaluating the smoothness of the soft edges with gradual intensity transitions. In [51], the authors propose a neighbor embedding-based super resolution through edge detection and feature selection (NeedFS). They propose a combination of appropriate features for preserving edges as well as smoothing the color regions. The training patches are learned with different neighborhood sizes depending on edge detection. Liu et al. [52] present a two-step hybrid approach for super-resolving a face image by combining Freeman's image primitive technique [33] and a PCA model-based approach. They propose a global parametric model called "global face image," carrying the common facial properties and a local nonparametric model called "local feature image" that records the local individualities. The high-resolution face image is obtained by composition of the global face image and the local feature image.

In [53], Glasner et al. obtain SR from as little as a single low-resolution image without any additional external information. Their approach combines the power of classical SR and example-based SR by exploiting the fact that patches in a single natural image tend to redundantly recur many times inside the image, both within the same scale as well as across different scales. Recurrence of patches within the same image scale forms the basis for applying the classical SR constraints to information from a single image. Recurrence of patches across different image scales provides examples of low-resolution and high-resolution pairs of patches, thus giving rise to example-based super resolution from a single image without any external database or any prior examples. Because the approach relies on the presence of redundantly occuring image patches, it may not work well for the natural images that do not have redundant information. In [54], the authors propose a single-image upscaling technique that exploits a local scale invariance in natural images. Observing that the small patches are very similar to themselves upon small scaling factors, they search for example patches at extremely localized regions in the input image and compare this localized search with other alternatives for obtaining example patches. The authors implement non-dyadic scalings using nearly-biorthogonal filter banks derived for

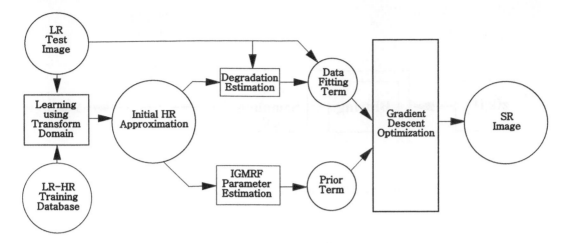

Figure 7.3 Schematic representation of proposed approach for image super resolution.

general $N + 1 : N$ upsampling and downsampling ratios. Yang et al. [55] demonstrate formulation and solution to the image super resolution problem based on sparse representations in terms of coupled dictionaries jointly trained from high- and low-resolution image patch pairs. Instead of working with high- and low-resolution training images, the authors learn a compact representation for patches from the input image patch pairs to capture the co-occurrence prior. However, an important question is to determine the optimal dictionary size for natural image patches in terms of super resolution tasks.

7.3 Description of the Proposed Approach

The basic idea of the proposed technique of learning-based super resolution is illustrated by the block diagram shown in Figure 7.3. Given a low-resolution observation (test image), we learn its high-frequency contents from a database consisting of a set of low-resolution images and their high-resolution versions. It may be noted that the LR images are not constructed by downsampling the HR images, as is done by most learning-based approaches. Instead, they are captured using a real camera comprised of various resolution settings and hence represent the true LR-HR versions of the scenes. In order to learn the high-frequency components, we consider a transform-based method. The transform coefficients corresponding to the high-frequency contents are learned from the database, and an initial estimate for the high-resolution version of the observation is obtained by taking the inverse transform. This initial HR estimate is then used for degradation estimation as well as for estimating the prior model parameters. The estimated degradation models the aliasing due to undersampling, and the model parameters inject the geometrical properties in a test image corresponding to the unknown high-resolution image. We then use a MAP estimation to arrive at a cost function consisting of a data fitting term and the prior term. A suitable optimization is exploited to minimize the cost function. The minimization leads to the final super-resolved image. We extend this method to color image super resolution where we super-resolve the luminance component using the proposed method and use the interpola-

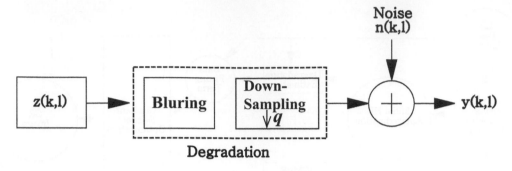

Figure 7.4 Image formation model. It includes blurring, downsampling by a factor of q, and additive noise.

tion in the wavelet domain for chrominance components. The luminance and chrominance components are then combined to obtain the super resolution.

7.3.1   Image Acquisition Model

We propose a super resolution algorithm that attempts to estimate the high-resolution image from a single low-resolution observation. This is an inverse problem. Solving such a problem needs a forward model that represents the image formation process. We represent the image formation process using a linear forward model. The pictorial representation of this observation model is shown in Figure 7.4. Let $\mathbf{y}$ and $\mathbf{z}$ represent the lexicographically ordered vectors of a low-resolution blurred observation of size $M \times M$ pixels and the corresponding HR image of size $qM \times qM$ pixels, respectively. Here, q is a decimation factor. The forward model can be written as

$$\mathbf{y} = A\mathbf{z} + \mathbf{n}, \tag{7.1}$$

Where $\mathbf{n}$ is the independent and identically distributed (i.i.d.) noise vector with zero mean and variance σ_n^2. It has the same size as $\mathbf{y}$. The degradation is represented using A, which takes care of aliasing and blur. It can be expressed as

$$A = DH, \tag{7.2}$$

where D is the decimation matrix of size $M^2 \times q^2 M^2$ and H is the blur matrix of size $q^2 M^2 \times q^2 M^2$. Many researchers have attempted to solve the super resolution problem assuming that the observations captured are not blurred. In other words, they consider the identity matrix for blur. In such a case, the degradation model to obtain the aliased pixel intensities from the high-resolution pixels, for a decimation factor of q, has the form [19]

$$A = D = \frac{1}{q^2} \begin{pmatrix} 1\ 1 \dots 1 & & \mathbf{0} \\ & 1\ 1 \dots 1 & \\ \mathbf{0} & & 1\ 1 \dots 1 \end{pmatrix}. \tag{7.3}$$

The multivariate noise probability density is given by

$$P(\mathbf{n}) = \frac{1}{(2\pi\sigma_n^2)^{\frac{M^2}{2}}} e^{-\frac{1}{2\sigma_n^2}\mathbf{n}^T\mathbf{n}}. \tag{7.4}$$

Our problem is to estimate $\mathbf{z}$ given $\mathbf{y}$ and the training database. Given an LR observation, the estimation of the HR image, maximizing any of the conditional distributions that describe the image acquisition model shown above, is a typical example of an ill-posed problem. Therefore, we have to regularize the solution by introducing a priori models on the HR image.

7.3.2 Learning the Initial HR Estimation

As mentioned earlier, the learning-based super resolution approaches use a database of high-resolution training images to recover the finer detail of the super-resolved image. Here, many researchers obtain the LR images in the database either by downsampling the high-resolution images (i.e., simulate the LR images or use an interpolated version of the LR image while searching). Such a database does not represent the true spatial relationship of features between LR-HR pairs, as they do not correspond to the images captured by a real camera. We use a database consisting of LR training images and the corresponding HR training images all captured using a real camera. Each LR-HR training image pair in this database exhibits the true spatial relationship of image features across the scales. Our database contains a large number of sets of LR and HR images covering indoor scenes as well as outdoor scenes taken at different times and with different lighting conditions. Because the construction of the database is a one-time and offline operation, we can use the computer memory for the storage of database images. This allows us to capture a large number of images, even when the memory of the camera is limited. We consider the discrete wavelet transform, the discrete cosine transform, and contourlet transform seperately for obtaining the initial HR estimate. We learn the transform coefficients that correspond to the high-frequency contents of the super-resolved image from HR training images in the database, and obtain the initial HR estimate of the super-resolved image by taking the inverse transform. We use this initial HR estimate to obtain the prior model parameters and entries in the degradation matrix. A detailed discussion on learning the initial HR estimate using different transforms is presented in Section 7.4.

7.3.3 Degradation Estimation

The degradation model given by equation (7.3) assumes that the observation is not blurred and indicates that a low-resolution pixel intensity $y(i, j)$ is obtained by integrating the intensities of q^2 pixels corresponding to the same scene in the high-resolution image and adding noise intensity $n(i, j)$. We cannot consider the aliasing effect as an averaging effect. We describe the estimation of the degradation matrix that takes care of both the blur and aliasing in the LR image formation. The degradation matrix A in Equation (7.1) is repeated here for the sake of simplicity,

$$A = DH, \tag{7.5}$$

where D is the decimation matrix of size $M^2 \times q^2 M^2$ and H is the blur matrix of size $q^2 M^2 \times q^2 M^2$. The blur matrix H can be expressed as

$$H = \begin{pmatrix} H_0 & H_{q^2 M-1} & H_{q^2 M-2} \cdots & H_1 \\ H_1 & H_0 & H_{q^2 M-1} \cdots & H_2 \\ & & & \\ H_{q^2 M-1} & H_{q^2 M-2} & H_{q^2 M-3} \cdots & H_0 \end{pmatrix}, \, . \tag{7.6}$$

Assuming a space-invariant blur, each entry H_j in the above matrix is a block matrix that can be written as

$$H_j = \begin{pmatrix} h_{j,0} & h_{j,q-1} & h_{j,q-2} \cdots & h_{j,1} \\ h_{j,1} & h_{j,0} & h_{j,q-1} \cdots & h_{j,2} \\ & & & \\ h_{j,q-1} & h_{j,q-2} & h_{j,q-3} \cdots & h_{j,0} \end{pmatrix}, \tag{7.7}$$

where $h_{.,.}$ are values of a point spread function (PSF). So, H is a block circulant matrix. Now, multiplication of D and H will result in the A matrix. A matrix can be written as

$$A = \begin{pmatrix} A_1 & A_2 & \dots & A_{q^2 M^2-1} & A_{q^2 M^2} \\ A_{q^2 M^2-q+1} & \dots & A_1 & A_2 \dots & A_{q^2 M^2-q} \\ & & & & \\ \dots & \dots & & \dots & A_1 \end{pmatrix}. \tag{7.8}$$

For $M = 2$ and $q = 2$, the structure of the A matrix with size 4×16 can be written as

$$A = \begin{pmatrix} A_1 & A_2 & A_3 & A_4 & A_5 & A_6 & A_7 & A_8 & A_9 & A_{10} & A_{11} & A_{12} & A_{13} & A_{14} & A_{15} & A_{16} \\ A_{15} & A_{16} & A_1 & A_2 & A_3 & A_4 & A_5 & A_6 & A_7 & A_8 & A_9 & A_{10} & A_{11} & A_{12} & A_{13} & A_{14} \\ A_9 & A_{10} & A_{11} & A_{12} & A_{13} & A_{14} & A_{15} & A_{16} & A_1 & A_2 & A_3 & A_4 & A_5 & A_6 & A_7 & A_8 \\ A_7 & A_8 & A_9 & A_{10} & A_{11} & A_{12} & A_{13} & A_{14} & A_{15} & A_{16} & A_1 & A_2 & A_3 & A_4 & A_5 & A_6 \end{pmatrix}. \tag{7.9}$$

In the above equation, A entries are coming from the corresponding multiplication terms of D and H. Because the initial HR estimate is already available, we obtain the entries in Equation (7.8) using available LR images and the initial HR estimate. A simple gradient descent approach has been used for estimation of the above values.

7.3.4 Image Field Model and MAP Estimation

As discussed in Section 7.1.3, super resolution is an ill-posed inverse problem. There are infinite solutions to equation (7.1). A reasonable assumption about the nature of the true image makes the ill-posed problem a better posed, and this leads to a better solution. What do we know a priori about the HR images? We expect the images to be smooth. In many computer vision problems, MRF is the most commonly used prior model. In MRF prior model-based HR reconstruction schemes, there is a fundamental trade-off between the smoothness of the super-resolved image and the amount of noise or visually unappealing artifacts. This occurs because the solution penalizes discontinuities in the image. Because the simplest Gaussian model tends to oversmooth reconstructions, it has been rejected in favor of various edge-preserving alternatives. The problem is not with the Gaussian family, but rather with the assumption of homogeneity. A more efficient model is one that considers that only homogeneous regions are smooth and that edges must remain sharp. This motivates us to consider an inhomogeneous prior that can adapt to the local structure of the image in order to provide a better reconstruction.

7.3.4.1 Inhomogeneous Gaussian Markov Random Field Prior Model

We propose an inhomogeneous Gaussian Markov random field as a prior model for super resolution. The advantage in using this model is that it is adaptive to the local structures in an image and hence eliminates the need for a separate edge-preserving prior. Inhomogeneous Gaussian random fields have been investigated by Aykroyd [56]. The simplicity of the Gaussian model allows rapid calculation, and the flexibility of the spatially varying prior parameter allows varying degrees of spatial smoothing. The inhomogeneous model allows greater flexibility; small features are not masked by the smoothing, and constant regions obtain sufficient smoothing to remove the effects of noise. This also helps to eliminate the need for separate priors to preserve edges as well as smoother regions in an image. Further, while using the edge-preserving IGMRF prior, we employ a simple gradient descent approach and thus avoid the use of computationally taxing optimization techniques such as simulated annealing.

The authors in [57] model the super-resolved image by an inhomogeneous Gaussian MRF with an energy function that allows us to adjust the amount of regularization locally. They define a corresponding energy function as

$$U(\mathbf{z}) = \sum_{i,j} \left[b_{i,j}^x (\mathcal{D}_x \mathbf{Z})_{i,j}^2 + b_{i,j}^y (\mathcal{D}_y \mathbf{Z})_{i,j}^2 \right], \tag{7.10}$$

where $\mathcal{D}_x$ and $\mathcal{D}_y$ are first-order derivative operators with respect to rows and columns and $\mathbf{Z}$ is the super-resolved image. Here, $b_{i,j}^x$ and $b_{i,j}^y$ are the IGMRF parameters at location (i,j) for the vertical and horizontal directions, respectively. In the above energy function, the authors model the spatial dependency at a pixel location by considering a first-order neighborhood and thus considering edges occurring in the horizontal and vertical directions only. However, in practice, there may be diagonal edges in the reconstructed image. In order to take care of these edges, we consider a second-order neighborhood and modify the energy function as follows:

$$U(\mathbf{z}) = \sum_{i,j} \left[b_{i,j}^x (\mathcal{D}_x \mathbf{Z})_{i,j}^2 + b_{i,j}^y (\mathcal{D}_y \mathbf{Z})_{i,j}^2 + b_{i,j}^g (\mathcal{D}_g \mathbf{Z})_{i,j}^2 + b_{i,j}^h (\mathcal{D}_h \mathbf{Z})_{i,j}^2 \right]. \tag{7.11}$$

Here, $b_{i,j}^g$ and $b_{i,j}^h$ are the IGMRF parameters at location (i,j) for diagonal directions. A low value of b indicates the presence of an edge between two pixels. These parameters help us obtain a solution that is less noisy in smooth areas and preserve sharp details in other areas. Now, in order to estimate the IGMRF parameters, we need the true super-resolved image, which is not available and must be estimated. Therefore, an approximation of $\mathbf{Z}$ must be accurately determined if we want the parameters obtained from it to be significant for regularization. This is why we choose to use the learning-based approach to compute the close approximation of $\mathbf{Z}$ that serves as the initial HR approximation $\mathbf{Z}_0$. Because the results of this approach exhibit sharp textures and are sufficiently close to the original image, it enables us to estimate the adaptive parameters from it.

7.3.4.2   Estimation of IGMRF Parameters

The maximum likelihood estimate on complete data with respect to the original image $\mathbf{Z}$ is

$$\hat{b}_{i,j}^{x,y,g,h} = \arg \max_{b_{i,j}^{x,y,g,h}} \left[log P(\mathbf{Z} | b^x, b^y, b^g, b^h) \right], \qquad (7.12)$$

and the log-likelihood derivatives are

$$\frac{\partial log P(\mathbf{Z} | b^x, b^y, b^g, b^h)}{\partial b_{i,j}^{x,y,g,h}} = E_{\mathcal{Z}} \left[(\mathcal{D}_{x,y,g,h} \mathcal{Z})_{i,j}^2 \right] - (\mathcal{D}_{x,y,g,h} \mathbf{Z})_{i,j}^2, \qquad (7.13)$$

where $\mathcal{Z}$ corresponds to the maximum a posteriori estimate of the high-resolution image, and $E_{\mathcal{Z}}$ corresponds to the expectation operator. Therefore, the estimation problem consists of solving system:

$$\{ E_{\mathcal{Z}} \left[(\mathcal{D}_{x,y,g,h} \mathcal{Z})_{i,j}^2 \right] = (\mathcal{D}_{x,y,g,h} \mathbf{Z})_{i,j}^2) \}. \qquad (7.14)$$

It is sufficient to compute the variance of each pixel difference with respect to the prior law $E_{\mathcal{Z}} \left[(\mathcal{D}_{x,y,g,h} \mathbf{Z})_{i,j}^2 \right]$. The authors in [57] propose the simplest approximation of the local variance. The variance of the gradient $(\mathcal{D}_{x,y,g,h} \mathbf{Z})_{i,j}$ is equal to the variance of the same gradient in the homogeneous region, that is, when all the parameters are equal to $b_{i,j}^{x,y,g,h}$. Because the covariance matrix of the homogeneous prior distribution is diagonalized by a Fourier transform, this variance can be calculated and is equal to $\frac{1}{4b}$ [58]. This gives

$$\hat{b}_{i,j}^{x,y,g,h} = \frac{1}{4(\mathcal{D}_{x,y,g,h} \mathbf{Z})_{i,j}^2}. \qquad (7.15)$$

Because the true high-resolution image $\mathbf{Z}$ is not available, we use the close approximation $\mathbf{Z}_0$ obtained using the transform domain-based learning approach and obtain the parameters using

$$\hat{b}_{i,j}^{x,y,g,h} = \frac{1}{4(\mathcal{D}_{x,y,g,h} \mathbf{Z}_0)_{i,j}^2}. \qquad (7.16)$$

Because we use true LR-HR pairs for learning the initial HR estimate, we can expect that the estimated parameters are close to their true values. The refined estimates of the IGMRF prior parameters are obtained using the following equations:

$$\hat{b}_{i,j}^x \simeq \frac{1}{8[(z_0(i,j) - z_0(i-1,j))^2]},$$

$$\hat{b}_{i,j}^y \simeq \frac{1}{8[(z_0(i,j) - z_0(i,j-1))^2]},$$

$$\hat{b}_{i,j}^g \simeq \frac{1}{8[(z_0(i,j) - z_0(i-1,j+1))^2]},$$

$$\hat{b}_{i,j}^h \simeq \frac{1}{8[(z_0(i,j) - z_0(i-1,j-1))^2]},$$

$$(7.17)$$

where $z_0(i,j)$ is the pixel intensity of the initial estimate at location (i,j). Thus, we estimate four parameters at each pixel location. These parameters cannot be approximated from degraded versions of the original image. The parameters estimated from the blurred image have high values, which leads to an oversmooth solution and the parameters estimated from the noisy image are of very low values, which leads to noisy solutions. Hence, we use the already learned high-resolution estimation in order to obtain a better estimate of these parameters. In order to avoid computational difficulties, we set an upper bound $\hat{b} = \frac{1}{8}$ whenever the gradient becomes zero, that is, whenever the neighboring pixel intensities are the same. Thus, we set a minimum spatial difference of 1 for practical reasons. This avoids obtaining a high regularization parameter that would slow down the optimization. It ensures that the pixels with zero intensity difference are weighted almost the same as those with a small intensity difference (in this case with a pixel intensity difference of 1).

7.3.4.3 MAP Estimation

We now explain how a MAP estimation of the dense intensity field (super-resolved image) can be obtained. The data fitting term is derived from the forward model, which describes the image-formation process. The data fitting term contains the degradation matrix estimated using the initial HR image and the test image. In order to use maximum a posteriori estimation to super-resolve the test image, we need to obtain the estimate as

$$\hat{\mathbf{z}} = \arg \max_z P(\mathbf{z}|\mathbf{y}). \tag{7.18}$$

Using Bayes' rule we can write,

$$\hat{\mathbf{z}} = \arg \max_z \frac{P(\mathbf{y}|\mathbf{z})P(\mathbf{z})}{P(\mathbf{y})}. \tag{7.19}$$

Beacuse the denominator is not a function of **z**, Equation (7.19) can be written as

$$\hat{\mathbf{z}} = \arg \max_z P(\mathbf{y}|\mathbf{z})P(\mathbf{z}). \tag{7.20}$$

Now, taking the *log* we can write

$$\hat{\mathbf{z}} = \arg \max_z [log P(\mathbf{y}|\mathbf{z}) + log P(\mathbf{z})]. \tag{7.21}$$

Finally, using Equations (7.1) and (7.11), the final cost function to be minimized can be expressed as

$$\hat{\mathbf{z}} = \arg \min_z \left[\frac{\| \mathbf{y} - A\mathbf{z} \|^2}{2\sigma_n^2} + U(\mathbf{z}) \right]. \tag{7.22}$$

In Equation (7.22), the first term ensures the fidelity of the final solution to the observed data through the image formation model. The second term is the inhomogeneous smoothness prior. Because this cost function is convex, it can be easily minimized using a simple gradient descent optimization technique, which quickly leads to the minima. This optimization process is an iterative method, and the choice of initial solution fed to the optimization process determines the speed of convergence. Use of a close approximate to the solution as an initial estimate speeds up the optimization process. In order to provide good initial guess, we use the already learned HR estimate.

7.3.5   Applying the Algorithm to Color Images

Different image processing systems use different color models for different reasons. The RGB color space consists of three additive primary colors: red, green, and blue. The RGB model is widely used in color monitors, computer graphic systems, electronic displays, and digital storage. Although this model simplifies the design of computer graphic systems, it is not suitable for super resolution algorithms. The reason behind this is that the red, green, and blue color components are highly correlated in color space. This makes it difficult to apply the monochrome super resolution technique to each of the R, G, and B color components and still maintain the natural correspondences between the color components in the solution. The imbalanced correspondences between the color components produces certain artifacts in the super-resolved color image. In addition, applying super resolution techniques to each of these components separately increases the computational burden.

In order to avoid the drawbacks of RGB color space, we separate the luminance channel and chrominance components by applying color space transformation and represent the color image in a YC_bC_r color space. The YC_bC_r color model represents the color image using separate luminance components and chrominance (color) components. The luminance is encoded in Y, and the blueness and redness are encoded in C_b and C_r, respectively. Because the human eye is more sensitive to the details in the luminance component of an image than the details in the chrominance component, we super-resolve the luminance component Y using the proposed approach and expand the chrominance components using a simple interpolation technique. The authors in [59] propose an interpolation technique for gray-scale images by interpolating the wavelet coefficients at finer scale. We apply their approach for expanding the chrominance components C_b and C_r. The frequency domain interpolation of these components leads to enhanced edges as compared to spatial interpolation methods like bilinear interpolation or bicubic interpolation. We use the super-resolved luminance component and the interpolated chrominance components to obtain the super-resolved color image by converting YC_bC_r to RGB color space.

7.4   Transform Domain-Based Learning of the Initial HR Estimate

Having introduced the common framework for the proposed approaches for super resolution, we now discuss transform domain-based techniques for learning the initial HR estimate. Given a low-resolution observation (test image), we learn its high-frequency contents from a database consisting of a set of low-resolution images and their high-resolution versions. Here, we consider three different transform for learning. We first explain the learning using the discrete wavelet transform-based method. The transform coefficients corresponding to the high-frequency contents are learned from the database, and an initial estimate for the high-resolution version of the observation is obtained by taking the inverse DWT. This initial HR estimate is used for degradation estimation as well as for estimating the IGMRF parameters.

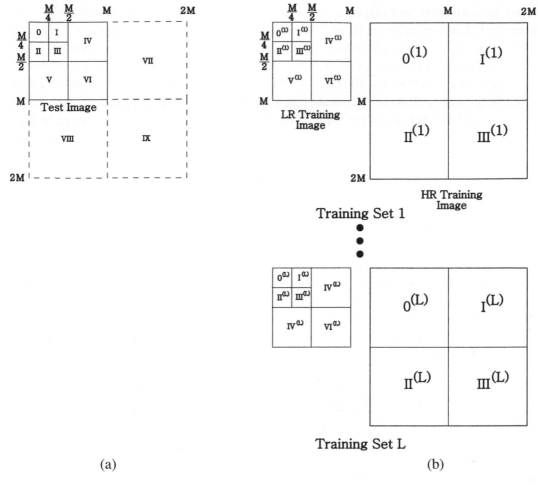

(a)
(b)

Figure 7.5 Discrete wavelet transform of test image and training images. (a) Two-level decomposition of test image (LR observation). Wavelet coefficients are to be learned for the subbands VII, VIII, and IX shown with the dashed lines, and (b) a training set of LR and HR images in the wavelet domain. (LR training images are decomposed into two levels and the HR training images into one level.)

7.4.1 Learning the Initial HR Estimate Using DWT

Here we begin to describe technique to learn the discrete wavelet transform coefficients for the initial estimate of the super-resolved image for a decimation (upsampling) factor of $q = 2$. We use two-level wavelet decomposition of the test image for learning the wavelet coefficients at the finer scale. Figure 7.5(a) shows the subbands $0 - VI$ of the low-resolution test image. The LR training images in the database are also decomposed into two levels, while their HR versions are decomposed into one level. Figure 7.5(b) displays the subbands $0^{(m)} - VI^{(m)}$, $m = 1, \ldots, L$, of the LR training images and subbands $0^{(m)} - III^{(m)}$, of the HR training images. The reason for taking one-level decomposition for HR training images is as follows. With one-level decomposition, the subband 0 represents the scaled version of the HR image, and the subbands I, II, and the III represent the detailed coefficients (vertical, horizontal, and diagonal edges) at high resolution. This means that for $q = 2$, both the LR image (subband 0) and the edge details at finer scales (subbands I, II, and III) are available in the HR transformed image. This motivates us to compare the edges in the test image with those present in the LR training set and choose the best matching wavelet coefficients from the HR images. Thus, given an LR test image, we learn its edges (high-frequency content) at finer scale using these LR-HR training images. Figure 7.5 illustrates the block schematic for learning the wavelet coefficients of the test image at finer scales using a set of L training image pairs for a decimation factor of 2. We compare the coefficients in subbands I to VI of the test image with those in subbands $I^{(m)}$ to $VI^{(m)}$ of the LR training images and obtain the best matching coefficients for subbands VII, VIII, and IX of the test image.

Let $\psi(i, j)$ be the wavelet coefficient at a location (i, j) in subband 0, where $0 \leq i, j < M/4$. For each of the subbands IV to VI in the test image, for every location (i, j) in subband 0, we learn a total $16 \times 3 = 48$ coefficients for the subbands VII to IX. We search for the LR training image that has a best match with the test image by comparing the wavelet coefficients in the subbands I to VI in the minimum absolute difference (MAD) sense. The corresponding wavelet coefficients from the subbands I to III of the HR training image are then copied into the subbands VII to IX of the test image. For a given location (i, j), the best matching LR training image in the subbands I to III is found using the following equation for MAD:

$$
\begin{aligned}
\hat{c}(i, j) \;=\; & \arg\min_{m} \Bigg[\left| \psi(i, j + \frac{M}{4}) - \psi^{(m)}(i, j + \frac{M}{4}) \right| \\
& + \left| \psi(i + \frac{M}{4}, j) - \psi^{(m)}(i + \frac{M}{4}, j) \right| \\
& + \left| \psi(i + \frac{M}{4}, j + \frac{M}{4}) - \psi^{(m)}(i + \frac{M}{4}, j + \frac{M}{4}) \right| \\
& + \sum_{k=2i}^{k=2i+1} \sum_{l=2j}^{l=2j+1} \left| \psi(k, l_1) - \psi^{(m)}(k, l_1) \right| \\
& + \sum_{k=2i}^{k=2i+1} \sum_{l=2j}^{l=2j+1} \left| \psi(k_1, l) - \psi^{(m)}(k_1, l) \right| \\
& + \sum_{k=2i}^{k=2i+1} \sum_{l=2j}^{l=2j+1} \left| \psi(k_1, l_1) - \psi^{(m)}(k_1, l_1) \right| \Bigg],
\end{aligned}
$$

(7.23)

where $k_1 = k + \frac{M}{2}$, $l_1 = l + \frac{M}{2}$, and $\hat{c}(i, j)$ is an index to the best matching LR image in the database for the location (i, j) and $1 \leq \hat{c}(i, j) \leq L$. Here, $\psi^{(m)}(i, j)$, $m = 1, 2, \ldots, L$, denotes the wavelet coefficient for the mth training image at location (i, j). For each location in subband I to III of the low-resolution observation, a best fit 4×4 block of wavelet coefficients in subbands I to III from the HR image of the training pairs given by $\hat{c}(i, j)$

are copied into subbands $VII, VIII$, and IX of the test image. Thus, we have

$$\{\psi(s, t_1)_{s=4i,t=4j}^{s=i_1,t=j_1}\} := \{\psi^{(\hat{c}(i,j))}(s, t_1)_{s=4i,t=4j}^{s=i_1,t=j_1}\}$$

$$\{\psi(s_1, t)_{s=4i,t=4j}^{s=i_1,t=j_1}\} := \{\psi^{(\hat{c}(i,j))}(s_1, t)_{s=4i,t=4j}^{s=i_1,t=j_1}\}$$

$$\{\psi(s_1, t_1)_{s=4i,t=4j}^{s=i_1,t=j_1}\} := \{\psi^{(\hat{c}(i,j))}(s_1, t_1)_{s=4i,t=4j}^{s=i_1,t=j_1}\}.$$

$$(7.24)$$

Here, $s_1 = s + M$, $t_1 = t + M$, $i_1 = 4i + 3$, and $j_1 = 4j + 3$. The inverse wavelet transform of the learned HR image then gives an initial HR estimate. The pseudocode of the DWT-based learning technique is shown in Algorithm 7.1. The learned HR estimate

---

Algorithm 7.1: Learning the HR approximation using DWT

Data: low-resolution test image and a database of LR-HR training images
Result: close approximation to the HR image
perform 2-level wavelet decomposition of the test image and LR training images;
perform 1-level wavelet decomposition of the HR training images;
foreach *pixel* (i, j) *in subband* 0 *of the test image* **do**
> search best matching LR training image $\hat{c}$ using Equation (7.23);
> copy the 4×4 block of DWT coefficients starting from location $(4i, 4j)$ in subband I in $\hat{c}$ to corresponding locations in subband VII in the test image;
> copy the 4×4 block of DWT coefficients starting from location $(4i, 4j)$ in subband II in $\hat{c}$ to corresponding locations in subband VIII in the test image;
> copy the 4×4 block of DWT coefficients starting from location $(4i, 4j)$ in subband III in $\hat{c}$ to corresponding locations in subband IX in the test image;
end
obtain the close approximation to the HR image by taking inverse DWT;

---

provides a sufficiently good approximation of the ground truth, and its properties enable robust estimation of degradation and adaptive parameters needed for regularization. For a decimation factor of $q = 4$, we first learn the initial estimate for $q = 2$ using the database consisting of LR-HR pairs with a resolution factor of 2. We then use this estimate as the test image for $q = 4$. We thus apply the single octave learning algorithm in two steps in order to obtain image super resolution for $q = 4$.

7.4.2 Initial Estimate Using Discrete Cosine Transform

In this section we explore the use of the DCT for predicting the missing high-frequency information. We obtain a close approximation to the high-resolution image using a learning technique based on DCT. The discrete cosine transform was introduced in 1974 by Ahmed and colleagues [60–62]. The DCT is used in JPEG and MPEG standards for still image compression and video compression, respectively. Considering this fact, we exploit the use of DCT in the learning approach. The proposed learning approach can readily be extended for super-resolving compressed images/video so that it does not require decoding of low-resolution images/video prior to learning the close approximation of the high-resolution video. The motivation to use DCT for learning comes from the fact that JPEG compression uses the DCT for removing the high-frequency detail assuming that the human eyes are insensitive to it. However, we use it for learning the same high-frequency detail that is very necessary for HR imaging.

We now describe a DCT-based approach to learn high-frequency detail for a decimation factor of 2 ($q = 2$). We first upsample the test image and all LR training images by a factor of 2 and create images of size $2M \times 2M$ pixels each. A standard interpolation technique can be used for the same. We divide each of the images, (i.e., the upsampled test image, upsampled LR images and their HR versions), in blocks of size 4×4. The motivation for dividing into 4×4 blocks is due to the theory of JPEG compression where an image is divided into 8×8 blocks in order to extract the redundancy in each block. However, in this case we are interested in learning the non-aliased frequency components from the HR training images using the aliased test image and the aliased LR training images. This is done by taking the DCT on each of the blocks for all the images in the database as well as the test image. Figure 7.6(a) shows the DCT blocks of the upsampled test image, whereas Figure 7.6(b) shows the DCT blocks of upsampled LR training images and HR training images. For most images, much of the signal energy lies at low frequencies (corresponding to large DCT coefficient magnitudes); these are relocated in the upper-left corner of the DCT array. Conversely, the lower-right values of the DCT array represent higher frequencies, and these turn out to be small. We compare the DCT coefficients in each block of the upsampled test image with that of the upsampled blocks of LR training images and find best matching block. We then copy the DCT coefficients from the corresponding block in the HR version to the block under consideration. Thus, for a particular block in the test image, we make use of all the blocks in every upsampled LR training image for learning HR DCT coefficients. It is reasonable to assume that when we interpolate the test image and the low-resolution training images to obtain $2M \times 2M$ pixels, the distortion is minimum in the lower frequencies. Hence, we can learn those DCT coefficients that correspond to high frequencies (already aliased) and are now distorted due to interpolation. Let $C_T(i,j)$, $1 \le i, j \le 4$, be the DCT coefficient at location (i, j) in a 4×4 block of the test image. Similarly, let $C_{LR}^{(m)}(i,j)$ and $C_{HR}^{(m)}(i,j)$, $m = 1, 2, \ldots, L$, be the DCT coefficients at location (i, j) in a block in the mth upsampled LR image and mth HR image, respectively. Here, L is the number of the training sets in the database. Now the best matching HR image block for the considered test image block (upsampled) is obtained as

$$l = \underset{m}{argmin} \sum_{i+j>Threshold} \|C_T(i,j) - C_{LR}^{(m)}(i,j)\|^2. \tag{7.25}$$

Here, l is the index for the training image, which gives the minimum for the block. Those non-aliased DCT coefficients from the corresponding block in the best matching HR image are now copied in the block of the upsampled test image. In effect, we learn non-aliased DCT coefficients for the test image blocks from the set of LR-HR images. The coefficients that correspond to low frequencies are not altered. Thus, at location (i, j) in the block, we have

$$C_T(i,j) = \begin{cases} C_{HR}^{(l)}(i,j) & \text{if } (i + j) > Threshold, \\ C_T(i,j) & \text{else.} \end{cases}$$

This is repeated for every block in the test image. After learning the DCT coefficients for the entire test image, we take the inverse DCT transform to get a high spatial resolution image

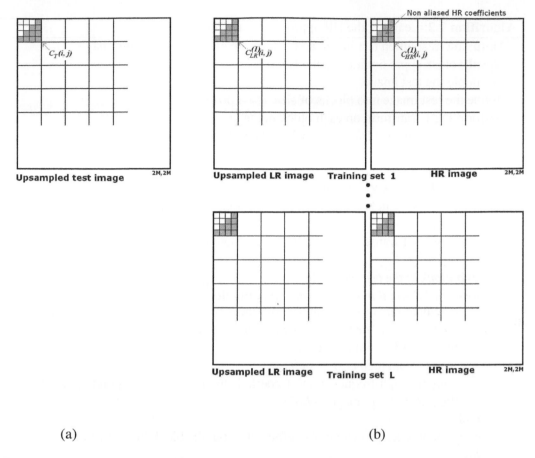

(a)   (b)

Figure 7.6 Learning DCT coefficients from a database of sets of LR-HR images for $q = 2$. (a) Upsampled test image and (b) sets of upsampled LR images and HR images for different scenes. DCT coefficients for the shaded locations in the upsampled test image are copied from corresponding locations of the best matching HR image.

and use it as the close approximation to the HR image. The pseudocode of the proposed technique is provided in Algorithm 7.2.

In order to fix the threshold, we conducted an experiment using different $Threshold$ values. We begin with $Threshold = 0$, where all sixteen coefficients in each 4×4 block are learned. The top image of the first row in Figure 7.7 shows the learned image. The image is highly degraded. We then set $Threshold$ to 2, where all the coefficients except the DC coefficient ($i = j = 1$) in each block are learned, and obtain the image shown in the second image of the top row in the same figure. There is a little improvement but still it is hard to interpret the content if the image. The subsequent increase in the $Threshold$ value (up to $Threshold = 4$) reduces the degradation. The fourth image in top row, obtained with $Threshold = 4$, is the most clear. As can be seen from the images in the bottom row, further increase in the $Threshold$ value introduces blockiness in the learned image. The reason for such behavior is that for the values of $Threshold$ beyond 5, the number of DCT coefficients learned in each block of 4×4 decreases from 5 for $Threshold = 6$ down to

---

Algorithm 7.2: Learning the HR approximation using DCT

Data: low-resolution test image and a database of LR-HR training images
Result: close approximation to the HR image
upsample the test image;
divide the test image into blocks of size 4×4 pixels;
perform DCT transform on each block in the test image;
initialize an array *cost* for each block in the test image;
initialize a matrix *hrdct* for reconstructed DCT image;
foreach *LR training image in the database* **do**
> upsample the LR training image;
> divide the upsampled LR training image and its HR version into blocks of size 4×4 pixels;
> perform DCT transform on each block in both the training images;

end
foreach *block in the test image* **do**
> **foreach** *LR training image in the database* **do**
> > calculate *sum* of the squared differences between high-frequency DCT coefficients in the block of the test image and those in the corresponding block in LR training image;
> > If minimum then *cost* = *sum*;
> > copy the high-frequency DCT coefficients from the corresponding block in the HR training image to *hrdct*;
>
> **end**
> copy the low-frequency DCT coefficients from the block in the test image to *hrdct*;

end
obtain the close approximation to the HR image by taking inverse DCT of *hrdct* ;

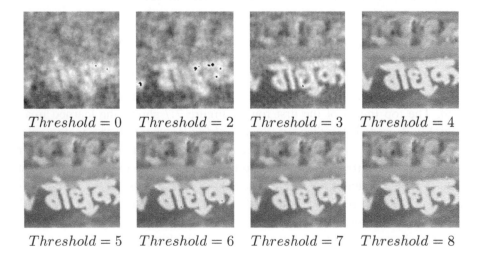

$Threshold = 0$　　　$Threshold = 2$　　　$Threshold = 3$　　　$Threshold = 4$

$Threshold = 5$　　　$Threshold = 6$　　　$Threshold = 7$　　　$Threshold = 8$

Figure 7.7 Learned images with different $Threshold$ values.

Table 7.1 MSE Comparison of Images Learned with Different *Threshold* Values

Threshold	MSE
0	0.036874
2	0.028089
3	0.017775
4	0.017483
5	0.017900
6	0.018458
7	0.019079
8	0.019866

0 for *Threshold* $= 8$. Thus, for higher values of *Threshold*, the contribution of learned DCT coefficients decreases, and hence there is little or no improvement in the interpolated image. The mean squared errors of the images learned for different *Threshold* values are given in Table 7.1. It can be seen that the best results are obtained when the *Threshold* is set to 4, which corresponds to learning a total of ten coefficients in a block.

7.4.3 Learning the Initial HR Estimate Using Contourlet Transform

The wavelet transform-based learning technique suffers from the limitations of wavelets in capturing the edges present in the horizontal, vertical, and diagonal directions only. Do and Vetterli present a new multi-resolution and directional transform called the *contourlet transform* [63]. The contourlet transform offers several advantages over the wavelet transform in terms of representing geometrical smoothness more effectively. The contourlet transform provides a high degree of directionality, anisotropy, and flexible aspect ratios. The contourlet transform employs a double filter bank structure in which at first the Laplacian pyramid (LP) [64] is used to capture the point discontinuities, followed by a directional filter bank (DFB) [65] to link point discontinuities into linear structures. DFB was designed to capture the high-frequency components present in different directions. Hence, the low-frequency components were handled poorly. So it is combined with LP where low-frequency components are removed before applying to DFB. Figure 7.8 shows the block diagram for the contourlet filter bank. First, a multiscale decomposition into octave bands by the LP is computed, and then a directional filter bank (DFB) is applied to each bandpass channel. The multiscale representation obtained from the LP structure is used as the first stage in contourlet transform. For more details on the contourlet transform, the reader is referred to [64–66].

We discuss here the learning of the initial HR estimate for a resolution factor of $q = 2$. The LR test image and LR training images are decomposed in a two-level contourlet transform as shown in Figure 7.9(a) and Figure 7.9(b), respectively. The HR training images are decomposed into a three-level contourlet transform (see Figure 7.9(c)). The contourlet coefficients in the blocks shown using dasshed lines in Figure 7.9(a) are to be learned. For each coefficient in the subbands I to IV and the corresponding 2×2 blocks in the subbands

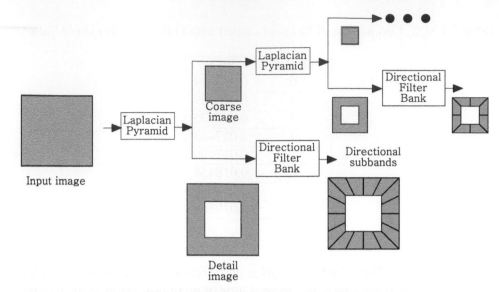

Figure 7.8 The contourlet filter bank.

V to VIII, we extrapolate a block of 2×4 contourlet coefficients in each of the subbands IX, X, XI, XII and 4×2 contourlet coefficients in each of the subbands XIII, XIV, XV, and XVI. Consider the subbands I to VIII of the low-resolution image. Let $c(i,j)$ be the contourlet coefficient at a location (i,j) in subband I, where $0 \leq i, j < M/4$. The contourlet-coefficients $c_I(i,j)$, $c_{II}(i,j+M/4)$, $c_{III}(i+M/4,j)$, $c_{IV}(i+M/4,j+M/4)$ corresponding to the subbands $I - IV$ and their corresponding 2×2 blocks in subbands $V - VIII$ in the low-resolution test image and the LR training images are considered for learning. For the test image, we learn a 2×4 contourlet block in each of the subbands IX to XII, consisting of unknown coefficients $c_{IX}(k,l)$, $c_X(k,l+M)$, $c_{XI}(k+M/2,l)$, $c_{XII}(k+M/2,l+M)$ for $k = 2i : 2i+1$ and $l = 4j : 4j+3$ and 4×2 contourlet block in each of the subbands XIII to XVI consisting of unknown coefficients $c_{XIII}(k+M,l)$, $c_{XIV}(k+M,l+M/2)$, $c_{XV}(k+M,l+M)$, $c_{XVI}(k+M,l+3M/2)$ for $k = 4i : 4i+3$ and $l = 2j : 2j + 1$. Thus, for a given set of a total of twenty contourlet coefficients in subbands I to VIII in the low-resolution image, we perform a search in the corresponding coarser levels of all the LR training images at all pixel locations for the best match in the minimum absolute difference (MAD) sense and copy the corresponding 2×4 in contourlet block in subband IX to XII and 4×2 in contourlet block in subband XIII to XVI in the test image. We search for the best matching LR image in the training database using MAD criteria as below:

$$
\begin{aligned}
\hat{m}(\hat{i},\hat{j}) &= \operatorname*{argmin}_{m,s,t}[| \, c_I(i,j) - c_I^{(m)}(s,t) \, | \\
&+ \, | \, c_{II}(i+M_I,j) - c_{II}^{(m)}(s+M_I,t) \, | \\
&+ \, | \, c_{III}(i,j+M_I) - c_{III}^{(m)}(s,t+M_I) \, | \\
&+ \, | \, c_{IV}(i+M_I,j+M_I) - c_{IV}^{(m)}(s+M_I,t+M_I) \, | \\
&+ \, S_V + S_{VI} + S_{VII} + S_{VIII}],
\end{aligned}
$$

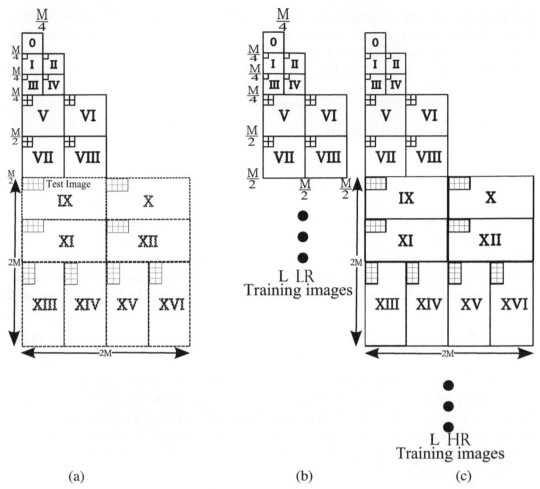

(a)   (b)   (c)

Figure 7.9 Learning contourlet transform coefficients: (a) Two-level contourlet decomposition of LR observation, and three-level contourlet decomposition of the super-resolved image is shown using dashed lines. The coefficients in the subbands shown using dashed lines are to be learned. (b) Two-level contourlet decomposition of LR training images in the database, and (c) three-level contourlet decomposition of HR training images in the database.

where $M_I = M/4$ and

$$S_V = \sum_{\substack{S=2s \\ I=2i}}^{\substack{I=2i+1 \\ S=2s+1}} \sum_{\substack{T=2t \\ J=2j}}^{\substack{J=2j+1 \\ T=2t+1}} \left| c_V(I, J) - c_V^{(m)}(S, T) \right|.$$

Similarly, S_{VI}, S_{VII}, and S_{VIII} are the corresponding sums for subbands VI, VII, and VIII, respectively, and $m = 1, 2, \ldots L$, where L is the number of image pairs in the database. Here, $c_l^{(m)}$ denotes the contourlet coefficient for mth training image at lth subband. $\hat{m}(\hat{i}, \hat{j})$ denotes the $(\hat{i}, \hat{j})$th location for the $\hat{m}$th training image that best matches the test image at the (i, j)th location in terms of contourlet coefficients. Once we get the best match, we pick the HR contourlet coefficients in subband IX to XVI of the corresponding HR training image and place them in the subband IX to XVI of the test image. We follow the same procedure while learning for a resolution factor $q = 4$. Here, the input test image is the learned image for $q = 2$, and the training database consists of pairs of $2M \times 2M$ and $4M \times 4M$ images.

7.5   Experimental Results

In this section we demonstrate the efficacy of the learning-based methods to super-resolve a low-resolution observation. All experiments are conducted on real-world images. The test images are of size 64×64, and the super resolution is shown for upsampling (decimation) factors of $q = 2$ and $q = 4$. Thus, the size of the super-resolved images are 128×128 and 256×256, respectively. We use "$DB4$" wavelets in our experiment while estimating the initial HR image using DWT. In order to compare the results using a quantitative measure, we use the peak signal-to-noise ratio (PSNR) and structural similarity (SSIM) as quantitative measures to assess the quality of the reconstructed images. The PSNR is calculated using

$$PSNR = 10 \log \frac{255^2}{\frac{1}{MN} \sum_{i,j} (f(i,j) - \hat{f}(i,j))^2}. \tag{7.26}$$

Here, $f(i,j)$ and $\hat{f}(i,j)$ are the pixel intensities at location (i,j) in the ground truth image and the super-resolved image, respectively, and MN represents the number of pixels in the images having size of $M \times N$. SSIM calculates the similarity in a linked local window by combining differences in average and variation and correlation [67]. The local similarity $S(\mathbf{x}, \mathbf{y})$ between image patches $\mathbf{x}$ and $\mathbf{y}$ taken from the same locations of the two images under comparison is calculated as

$$S(\mathbf{x}, \mathbf{y}) = \left(\frac{2\mu_x \mu_y + C_1}{\mu_x^2 + \mu_y^2 + C_1} \right) \left(\frac{2\sigma_x \sigma_y + C_2}{\sigma_x^2 + \sigma_y^2 + C_2} \right) \left(\frac{2\sigma_{xy} + C_3}{\sigma_x \sigma_y + C_3} \right), \tag{7.27}$$

where μ_x and μ_y are local sample means of image patches $\mathbf{x}$ and $\mathbf{y}$, respectively. Similarly σ_x and σ_y are the local sample standard deviations of $\mathbf{x}$ and $\mathbf{y}$, respectively. Here, σ_{xy} is the sample cross-correlation of $\mathbf{x}$ and $\mathbf{y}$ after removing their means. C_1, C_2, and C_3 are small positive constants preventing numerical instability. The SSIM score of the entire image is computed by averaging the SSIM values of the patches across the image [67]. A higher value of $S(\mathbf{x}, \mathbf{y})$ indicates better performance. For more details, see [68]. In order to compare the results, we choose an LR image from the database as a test image, and hence the true high-resolution image is available for comparison. It must be mentioned here that the HR images of the test images are removed from the database during the learning process. All the experiments were conducted on a computer with a Pentium M, 1.70-GHz processor.

7.5.1   Construction of the Training Database

In the proposed approaches, we obtain finer details of the super-resolved images using a training database of LR images and their HR versions. We construct the database of images at different resolutions that are captured by varying the zoom setting of a real camera. For each scene, there are two or three images, depending on the upsampling factor. For example, for an upsampling factor of 2 $(q = 2)$, each scene has two images, an LR image and its HR version. If $q = 4$, then there are three images, an LR image and two HR versions of the same. All the images in the database are of real-world scenes. A computer-controlled camera was used to capture the images for the database. In order to avoid motion of the camera while capturing images of a scene at different resolutions, a stable and isolated physical setup was used. The camera was triggered by a MATLAB® program at successive but three different time instances were used for capturing images at three different resolutions. The resolution setting of the camera was changed by the program before each trigger. The time duration between two successive triggers was less than a millisecond. Figure 7.10 shows randomly selected training images in the database. It may be mentioned here that we make use of images in columns (a) and (b) while learning for $q = 2$, and use images in columns (a), (b), and (c) for $q = 4$. The images of live subjects were captured under a controlled environment. We assume that the motion of subjects like human and other moving objects during one thousandths of a second is negligible. We applied mean correction in order to maintain the same average intensity among the images. Once the database is ready, it can be used for super-resolving images captured by the same camera or by a different camera having low resolution. Our database consists of images of 750 scenes. LR-HR images in the database include indoor as well as outdoor scenes captured at different times. The LR and HR images are of sizes 64×64, 128×128, and 256×256, respectively, and the test image (to be super-resolved) is of size 64×64. It may be noted here that the size of the test image needs to be an integer power of 2 and should not exceed that of the LR training images. We have a total of $3 \times 750 = 2250$ images used for learning the initial HR estimate. Because we show the results for $q = 2$ and $q = 4$, we make use of $2 \times 750 = 1500$ and $3 \times 750 = 2250$ images for $q = 2$ and $q = 4$, respectively. Here, the multiplication factors 2 and 3 correspond to the number of images of a scene.

7.5.2 Results on Gray-Scale Images

We first describe the performance of the proposed approaches for gray-scale images. We assess the performance of the SR approaches using various test images that consist of varying amounts of different textures, sharp edges, and smooth areas. We conducted the experiments on several gray-scale images. The test images used in the experiments are displayed in Figure 7.11. Figure 7.12 shows the results of the SR techniques for the upsampling factor $q = 2$. Figure 7.12(a) shows images super-resolved using the approach proposed by Kim and Kwon [45]. Figure 7.12(b) and (c) show the images upsampled using the DWT-based learning approach and DCT-based learning approach, respectively. The results obtained using contourlet-based learning approach are presented in Figure 7.12(d).

Image 1 is a picture of a person riding on a horse. This image contains sharp edges as well as smooth area. The SR image obtained using the Kim and Kwon approach (Figure 7.12(a)) exhibits inconsistent overall brightness. The image quality is improved in the images obtained using DWT-based learning and DCT-based learning. However, edges look a little blurred. In the image obtained using DCT-based learning, some chessboard effect is introduced, especially in the region near the forehead of the person. The result obtained

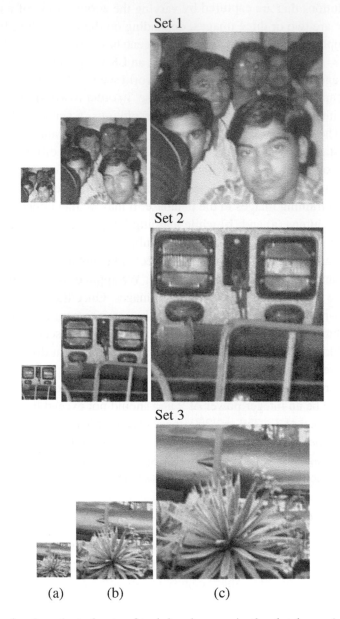

Set 1

Set 2

Set 3

(a)   (b)   (c)

Figure 7.10 Randomly selected sets of training images in the database: (a) low-resolution images; (b) high-resolution images with upsampling factor of 2 ($q = 2$); (c) high-resolution images with upsampling factor of 4 ($q = 4$).

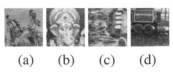

(a)   (b)   (c)   (d)

Figure 7.11 Gray-scale test images: (a) *Image 1*, (b) *Image 2*, (c) *Image 3*, (d) *Image 4*.

Image 1

Image 2

Image 3

Image 4

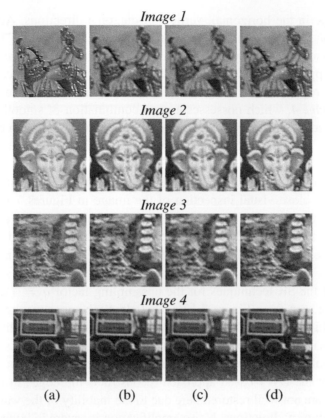

(a)   (b)   (c)   (d)

Figure 7.12 Comparison of gray-scale image super resolution for $q = 2$. Images super-resolved using (a) Kim and Kwon approach in [45]; (b) DWT-based learning approach; (c) DCT-based learning approach; (d) Contourlet-based learning approach.

using contourlet transform-based learning (Figure 7.12(d)) shows distinct boundaries between the foreground and background. The leg of the horse and the hand of the person appear sharper than those in images super-resolved using the other approaches. The overall brightness across the image area is also consistent. *Image 2* is a picture of the face of the Indian god *Lord Ganesha*. This image has lesser smooth region and more edges than those in *Image 1*. Figure 7.12(a) shows the image super-resolved using the Kim and Kwon approach. Although edges in this image, such as the contour of Ganesha's ear, trunk, and eyes are well restored, the image does not appear visually pleasant. We observe that the DWT-based algorithm does a better job of enhancing the weak edges in the textures. In the image obtained using the DCT-based approach, the artifacts due to estimation error of the DCT coefficients are visually very noticeable. Figure 7.12(d) displays the image super-resolved using contourlet transform-based learning. Notice that the facial features, ears, eyes, and crown are much clearer in this image. *Image 3* contains ample texture of sand at the seashore and has circularly shaped clay work. The periphery of these circular regions is well reconstructed by the Kim and Kwon approach but the sand texture appears artificial. There is clear improvement in the texture of images obtained using DWT-based learning. However, the boundaries of the circles are not reconstructed well. The overall perceived image quality is not much improved. We believe this is because there are no strong edges in the image in the horizontal, vertical or diagonal directions. Artifacts caused by noise amplification are visible in the image super-resolved using the DCT-based approach. The contourlet-based approach yields the output that combines the goodness of both the Kim and Kwon approach and the DWT-based approach. We can notice improved edges and better texture regions in the images displayed in Figure 7.12(d). Next we conduct the experiment on *Image 4*, which possesses a good combination of smooth regions, texture, text, and various edges of different directions and magnitudes. We can observe in Figure 7.12(a) that the finer texture details such as wheel and text in the rectangle are lost. The region near the wheels appears smeared. In Figures 7.12(b–d) sharp edges in the image, such as the contour of the wheel, the rectangular body, and the vertical pole, are well restored. In addition, close visual inspection of the image in Figures 7.12(d) confirms the superior results of the contourlet transform-based algorithm as fine details can be seen in the super-resolved image. It is clear that the contourlet transform-based approach yields a significant improvement of the level of details visible on the images and greatly reduced smearing effects.

The results of the SR techniques for the upsampling factor $q = 4$ are shown in Figure 7.13. It can be seen from Figure 7.13(a) that although super resolution could enlarge the image for the higher upsampling factor, we observe a higher possibility of the smearing phenomenon. This may be due to construction of the LR training images synthetically. Such a database does not represent true relationship between spatial features across the scales. It can be seen that the DWT-based approach reconstructs the edges in limited directions only. Some edges are not well restored here due to the inability of the wavelet transform to handle edges in arbitrary directions. Noise amplification is severe in images obtained using the DCT- based approach (See Figure 7.13(c)). Noise components in these images increase with upsampling factor. Images in Figure 7.13(d) represent the results obtained using contourlet transform based approach. It can be seen that many details are retrieved in the super-resolved image. The approach efficiently generates finer detail for the higher scaling factor without further noise amplification. We believe that the initial HR estimate learned using the contourlet transform produces better final solution when compared to using initial estimates learned using the other learning approaches. This is because the degradation matrix entries and the IGMRF parameters estimated from the initial HR estimate learned using the contourlet transform are close to the true values when compared to those estimated from the initial HR estimate learned using DWT- and DCT-based learning approaches.

The quantitative comparison in terms of the peak signal-to-noise ratio (PSNR) and the structural similarity (SSIM) of these images is shown in Table 7.2. It can bee seen that the results obtained using the contourlet-based method for both $q = 2$ and $q = 4$ show perceptual as well as quantifiable improvements over the other approaches.

7.5.3   Results on Color Images

We now show the experiments for color image super resolution. We conduct the experimentation on color image super resolution by working in YC_bC_r color space. Taking advantage of the fact that human eyes are not very sensitive to color components we apply the SR to the luminance component only. As already discussed in Section 7.3.5, we use the DWT-based interpolation technique to expand the chrominance components. This also reduces

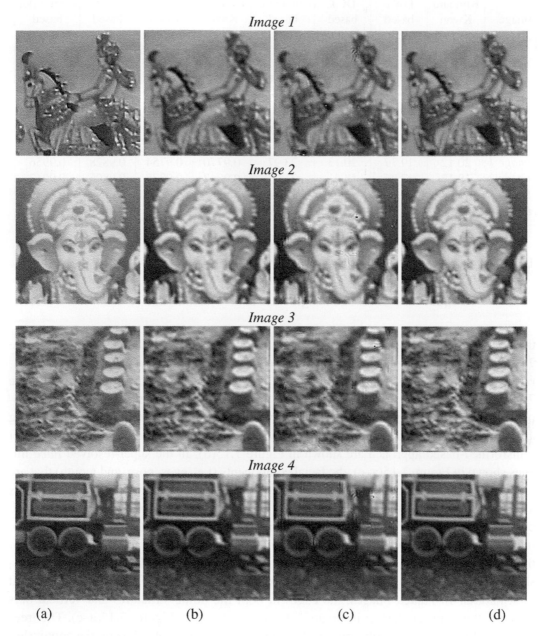

Figure 7.13 Comparison of gray-scale image super resolution for $q = 4$. Images super-resolved using (a) Kim and Kwon approach in [45]; (b) DWT-based learning approach; (c) DCT-based learning approach; (d) Contourlet-based learning approach.

Table 7.2 Performance Comparison for Gray-Scale Image Super Resolution

Image	PSNR (in dB)				SSIM			
	Kim and Kwon approach [45]	DWT based learning approach	DCT based learning approach	Contourlet based learning approach	Kim and Kwon approach [45]	DWT based learning approach	DCT based learning approach	Contourlet based learning approach
	$q = 4$							
Image 1	19.89	22.26	22.81	24.25	0.9866	0.9895	0.9931	0.9951
Image 2	19.68	19.87	20.42	20.97	0.9799	0.9804	0.9833	0.9895
Image 3	18.79	21.48	22.05	22.80	0.9708	0.9798	0.9832	0.9962
Image 4	18.51	22.08	22.40	24.45	0.9757	0.9783	0.9841	0.9907
	$q = 4$							
Image 1	20.12	19.77	20.13	23.97	0.9770	0.9754	0.9828	0.9956
Image 2	18.12	17.96	18.41	21.93	0.9847	0.9878	0.9896	0.9943
Image 3	18.64	19.56	20.00	23.44	0.9881	0.9841	0.9899	0.9939
Image 4	18.20	20.07	20.29	25.69	0.9769	0.9792	0.9852	0.9943

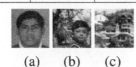

(a)   (b)   (c)

Figure 7.14 Color test images. (a) *Face*, (b) *Child*, and (c) *Palace*.

the time complexity, as one need not perform the learning for the color components. Figure 7.14 displays three color test images: *Face*, *Child*, and *Palace*. The SR is obtained separately for each of the test images, and the results for the same are shown for $q = 2$ in Figure 7.15. Images in Figure 7.15 (a) show images super-resolved using the Kim and Kwon approach [45]. Images super-resolved using wavelet-based learning and DCT-based learning are shown in Figures 7.15(b) and (c), respectively. The results obtained using the contourlet transformed-based approach are shown in Figure 7.15(d). In the super-resolved *Face* images shown in Figure 7.15(a–c), the facial components (i.e., eyes, nose, and mouth) that are crucial to determine face expressions and face identities are not sharp. These components are well reconstructed in Figure 7.15(d). The contourlet-based algorithm does a better job of enhancing these components. The *Child* image in Figure 7.15(d) shows a significant improvement in the level of details visible in the eye region and the textured region of the tree. Some fine textures of the tree are retrieved. Compared with the other approaches, sharp boundaries are achieved using the contourlet-based method, thus making the result look natural. In *Palace* images, the edges in both the dome structures and vertical bars of the gallery appear sharper in Figure 7.15(d) than those in Figures 7.15(a–c). The trees in the background are visibly improved in the image super-resolved using the contourlet-based approach. A comparison of the figures shows more clear details in the SR images obtained using the contourlet transform-based approach. It can be clearly observed that for the contourlet-based approach, preservation of edges is much better when compared to the other techniques, that is, the contourlet-based approach provides reconstruction HR images with superior visual quality, in particular the super-resolved images have less artifacts in smooth regions and sharp recovery of edges and texture regions. Figure 7.16 displays re-

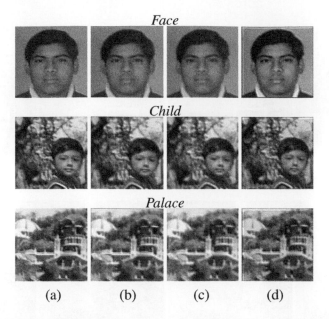

Figure 7.15 Comparison of color image super resolution for $q = 2$. Images super-resolved using (a) Kim and Kwon approach in [45]; (b) DWT-based learning approach; (c) DCT-based learning approach; (d) Contourlet-based learning approach.

sults for $q = 4$. It is clear from Figure 7.16 that even for the higher upsampling factor, the contourlet transform method performs quite well for color image as well. The quantitative comparisons of the results using PSNR and SSIM are given in Table 7.3. We can observe that in addition to perceptual betterment in all the observed images, there is also a considerable gain in PSNR and SSIM for the proposed approach. This illustrates the usefulness of edge preserving super-resolving of the images.

7.6 Conclusions and Future Research Work

7.6.1 Conclusions

In this chapter we addressed the learning-based approach for image super resolution. This technique obtains super resolution of single observations. Because it is impossible to extract additional nonredundant information from a single observation, we needed to seek other sources for deriving additional information from a training database. Further, the fact that the richness of the natural images is difficult to capture analytically has motivated us to construct a database of low-resolution images and its high-resolution versions, and use it to get nonredundant information for the super-resolved image. We have proposed learning-based techniques that attempt to derive the details of high-resolution images from the database by analyzing the spatial relationship between the image features across the scales. We have obtained the initial HR estimate using three transforms—DWT, DCT, and contourlet transform—separately and reconstructed the final solution using the regularization framework. An accurate estimation of the prior model parameters requires the high-resolution image. However, in super resolution problems, the high-resolution images are

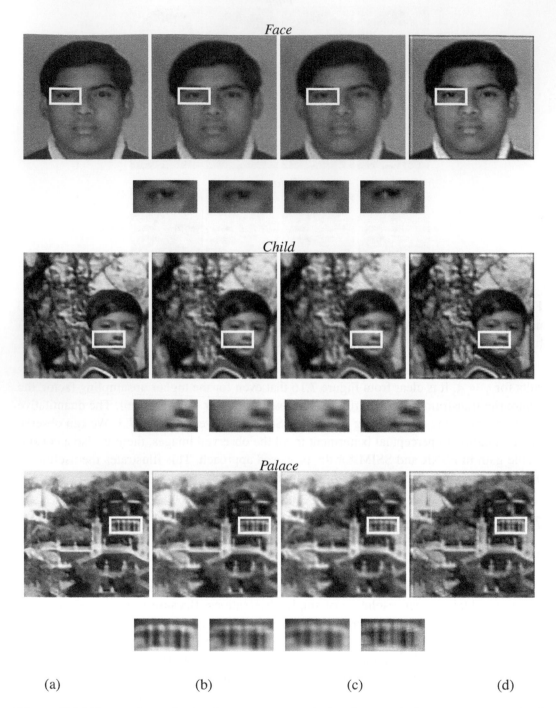

(a)   (b)   (c)   (d)

Figure 7.16 Comparison of color image super resolution for $q = 4$. Images super-resolved using (a) the Kim and Kwon approach in [45]; (b) DWT-based learning approach; (c) DCT-based learning approach; (d) Contourlet-based learning approach. Below each image, a zoomed-in version of the rectangle region is shown for easier visual inspection.

Table 7.3 Performance Comparison for Color Image Super Resolution.

Image	PSNR (in dB)				SSIM			
	Kim and Kwon approach [45]	DWT based learning approach	DCT based learning approach	Contourlet based learning approach	Kim and Kwon approach [45]	DWT based learning approach	DCT based learning approach	Contourlet based learning approach
	$q = 2$							
Face	22.04	22.24	22.45	24.65	0.8543	0.8701	0.8829	0.9922
Child	18.54	18.70	19.55	23.20	0.8975	0.8789	0.8940	0.9877
Palace	18.15	18.77	18.75	21.23	0.9681	0.9814	0.9858	0.9808
	$q = 4$							
Face	24.34	24.86	25.07	26.15	0.9871	0.9830	0.9910	0.9956
Child	14.95	15.08	15.24	15.25	0.6557	0.6445	0.6904	0.9824
Palace	18.05	18.40	18.78	19.50	0.7353	0.7457	0.7612	0.9822

not available. We have obtained better estimates of these parameters from the estimated close HR approximation in order to use them for obtaining final solution. Because, the natural images consist of edges and a wide variety of textured regions, we have proposed an inhomogeneous Gaussian Markov random field prior model that can adapt to the local structure in an image. Finally, super resolution is obtained by minimizing the cost function using a gradient descent technique. We have applied the algorithm to color images and shown the results for real-world images.

7.6.2 Future Research Work

Most super resolution techniques suffer from one or more disadvantages, such as high computational complexity, the need for a huge training database, the need for registration and poor visual quality. Therefore, several of these issues will be appropriately addressed in the future. In this section we discuss the directions for future research work to produce good, practical super resolution.

- In the image formation process model, we have assumed space-invariant blur. This enables a simple formulation for the data fitting term. However, in practice, the blur need not be space invariant and hence must be estimated. A simple approach to handle the space-varying blur is to divide the entire image into small blocks and super-resolve each of them individually.

- In our experiment we have employed a simple linear model for the image formation process. One can consider splines to model the image formation process and obtain better representation of the process by considering more HR pixels in the larger neighborhood of an LR pixel.

- One of the limitations of the super resolution approaches discussed in this chapter is the inability of the approaches to reconstruct the images for upsampling factors other than those that are powers of 2. SR approaches attempting to super-resolve the images for integer upsampling factors are also reported in the literature. However, these approaches may not be useful in applications where images need to be super-resolved to a rational scale or to continuous scales. One can investigate the issue of scale-independent representation of images and build an algorithm that can super-resolve LR images to any arbitrary scale.

- The prior for regularization is obtained by modeling the HR image as an MRF that can be expressed as a joint probability density function using the Gibbs distribution. One can consider a marginal distribution prior that can be obtained by passing the HR image or its close approximation through different filters and computing the histograms of the filtered outputs. It is possible to construct a bank of different types of filter such as Laplacian or Gaussian and Gabor filters with different parameters, and arrive at a cost function that can be optimized using simpler global optimization techniques such as particle swarm optimization or graph-cuts.

- The proposed techniques for image super resolution can be extended for video super resolution. Video super resolution requires super resolution reconstruction in both the spatial domain and the temporal domain. One can develop an algorithm for reconstructing additional temporal frames using motion fields. With the spatial super resolution of all the frames and the temporal super resolution of the video sequence, it is possible to obtain super-resolved video by concatenating the resulting SR frames.

- Most super resolution algorithms are based on iterative processes. This limits the use of algorithms for real-time applications such as video super resolution. It will be interesting to see the development toward the computationally efficient super resolution algorithms that may use a strong but differentiable prior such that the optimization converges fast and leads to a good-quality output image.

Bibliography

[1] M. S. Lee, M. Y. Shen, and C. C. J. Kuo, "Techniques for flexible image/video resolution conversion with heterogeneous terminals," *IEEE Communications Magazine*, pp. 61–67, 2007.

[2] J. E. Estes and D. S. Simonett, *Manual of Remote Sensing*. Bethesda, MD: American Society for Photogrammetry and Remote Sensing, 1975, ch. "Fundamentals of Image Interpretation," pp. 869–1076.

[3] T. Komatsu, K. Aizawa, T. Igarashi, and T. Saito, "Signal processing based method for acquiring very high resolution images with multiple cameras and its theoretical analysis," in *Proceedings of Institute of Electrical Engineers*, vol. 140, no. 1, 1993, pp. 19–25.

[4] H. Stark and P. Oskui, "High resolution image recovery from image-plane arrays using convex projections," *Journal of the Optical Society of America - A*, vol. 6, no. 11, pp. 1715–1726, 1989.

[5] K. Aizawa, T. Komatsu, and T. Saito, "A scheme for acquiring very high resolution images using multiple cameras," in *Proceedings of International Conference on Aucostics, Speech, Signal Processing*, 1992, pp. 289–292.

[6] N. K. Bose and K. J. Boo, "High-resolution image reconstruction with multisensors," *International Journal of Imaging Systems and Technology*, vol. 9, no. 4, pp. 294–304, 1998.

[7] M. Elad and A. Feuer, "Super resolution restoration of an image sequence: Adaptive filtering approach," *IEEE Transactions on Image Processing*, vol. 8, no. 3, pp. 387–395, 1999.

[8] J. C. Gillette, T. M. Stadtmiller, and R. C. Hardie, "Aliasing reduction in staring infrared imagers utilizing subpixel techniques," *Optical Engineering*, vol. 34, no. 11, pp. 3130–3137, 1995.

[9] G. Jacquemod, C. Odet, and R. Goutte, "Image resolution enhancement using subpixel camera displacement," *Signal Processing*, vol. 26, no. 1, pp. 139–146, 1992.

[10] T. Komatsu, K. Aizawa, T. Igarashi, and T. Saito, "Signal processing based method for acquiring very high resolution images with multiple cameras and its theoretical analysis," in *IEE Proceedings. I, Communications, Speech and Vision*, vol. 140, no. 1, 1993, pp. 19–24.

[11] R. Y. Tsai and T. S. Huang, "Multiframe image restoration and registration," *Advances in Computer Vision and Image Processsing*, vol. 1, pp. 317–339, 1984.

[12] S. Borman and R. Stevenson, "Spatial resolution enhancement of low-resolution image sequences: A comprehensive review with directions for future research," in *Technical Report*, University of Notre Dame, 1998.

[13] S. Farsiu, D. Robinson, M. Elad, and P. Milanfar, "Advances and challenges in super resolution," *International Journal of Imaging Systems and Technology*, vol. 14, no. 2, pp. 47–57, 2004.

[14] S. C. Park, M. K. Park, and M. G. Kang, "Super resolution image reconstruction: A technical overview," *IEEE Signal Processing Magazine*, vol. 20, pp. 21–36, 2003.

[15] S. P. Kim, N. K. Bose, and H. M. Valenzuela, "Recursive reconstruction of high resolution image from noisy undersampled multiframes," *IEEE Transactions on Acoustics, Speech, and Signal Processing*, vol. 38, no. 6, pp. 1013–1027, 1990.

[16] S. P. Kim and W. Y. Su, "Recursive high-resolution reconstruction of blurred multiframe images," *IEEE Transactions on Image Processing*, vol. 2, no. 4, pp. 534–539, 1993.

[17] S. H. Rhee and M. G. Kang, "Discrete cosine transform based regularized high-resolution image reconstruction algorithm," *Optical Engineering*, vol. 38, no. 8, pp. 1348–1356, 1999.

[18] N. K. Bose, H. C. Kim, and H. M. Valenzuela, "Recursive implementation of total least squares algorithm for image construction from noisy, undersampled multiframes," in *Proceedings of IEEE Conference on Acoustics, Speech and Signal Processing*, vol. 5, pp. 269–272, 1993.

[19] R. R. Schultz and R. L. Stevenson, "A Bayesian approach to image expansion for improved definition," *IEEE Transactions on Image Processing*, vol. 3, no. 3, pp. 233–242, 1994.

[20] S. Farsiu, M. Elad, and P. Milanfar, "Multi-frame demosaicing and super resolution of color images," *IEEE Transactions on Image Processing*, vol. 15, no. 1, pp. 141–159, 2006.

[21] M. Irani and S. Peleg, "Improving resolution by image registration," *CVGIP: Graphical Models and Image Processing*, vol. 53, pp. 231–239, 1991.

[22] S. Park, M. K. Park, and M. Kang, "Super resolution image reconstruction: A technical overview," *IEEE Signal Processing Magazine*, no. 20, pp. 21–36, 2003.

[23] P. Vandewalle, S. Susstrunk, and M. Vetterli, "A frequency domain approach to registration of aliased images with application to super resolution," *EURASIP Journal of Appllied Signal Processing*, vol. 2006, Article ID 71459, 2006.

[24] A. N. Rajagopalan and V. P. Kiran, "Motion-free super resolution and the role of relative blur," *Journal of the Optical Society of America A*, vol. 20, no. 11, pp. 2022–2032, 2003.

[25] M. V. Joshi, S. Chaudhuri, and P. Rajkiran, "Super resolution imaging: Use of zoom as a cue," *Image and Vision Computing*, vol. 14, no. 22, pp. 1185–1196, 2004.

[26] M. Ng, H. Shen, S. Chaudhuri, and A. Yau, "A zoom based super resolution reconstruction approach using total variation prior," *Optical Engineering*, vol. 46, 127003, 2007.

[27] D. Rajan and S. Chaudhuri, "Generation of super resolution images from blurred observations using an MRF model," *Journal of Mathematical Imaging and Vision*, vol. 16, pp. 5–15, 2002.

[28] ——, "Simultaneous estimation of super-resolved intensity and depth maps from low resolution defocussed observations of a scene," in *Proceedings of IEEE Conference on Computer Vision*, 2001, pp. 113–118.

[29] M. V. Joshi and S. Chaudhuri, "Simultaneous estimation of super-resolved depth map and intensity field using photometric cue," in *Computer Vision and Image Understanding*, vol. 101, 2006, pp. 31–44.

[30] S. Sharma and M. V. Joshi, "A practical approach for super resolution using photometric stereo and graph cuts," in *Proceedings of British Machine Vision Conference*, 2007.

[31] M. V. Joshi, S. Chaudhuri, and R. Panuganti, "A learning based method for image super resolution from zoomed observations," *IEEE Transactions Systems, Man and Cybernetics, Part B, Special Issue on Learning in Computer Vision and Pattern Recognition*, vol. 35, no. 3, pp. 527–537, 2005.

[32] R. R. Sahay and A. N. Rajagopalan, "Extension of the shape from focus method for reconstruction of high-resolution images," *Journal of the Optical Society of America A*, vol. 24, no. 11, pp. 3649–3657, 2007.

[33] W. T. Freeman, T. R. Jones, and E. C. Pasztor, "Example-based super resolution," *IEEE Computer Graphics and Applications*, vol. 22, no. 2, pp. 56–65, 2002.

[34] C. V. Jiji and S. Chaudhuri, "Single-frame image super resolution through contourlet learning," *EURASIP Journal on Applied Signal Processing*, vol. 2006, Article ID 73767, 2006.

[35] S. Rajaram, M. D. Gupta, N. Petrovic, and T. S. Huang, "Learning based nonparametric image super resolution," *EURASIP Journal on Applied Signal Processing*, vol. 2006, no. 2, pp. 1 – 11, 2006.

[36] Q. Wang, X. Tang, and H. Shum, "Patch based blind image superresolution," in *Proceedings of IEEE Conference on Computer Vision*, vol. 1, pp. 709 – 716, 2005.

[37] T. A. Stephenson and T. Chen, "Adaptive markov random fields for example-based super resolution of faces," *EURASIP Journal on Applied Signal Processing*, vol. 2006, 2006.

[38] L. C. Pickup, S. J. Roberts, and A. Zisserman, "A sampled texture prior for image super resolution," in *Proceedings of Neural Information Processing Systems*, pp. 1587 – 1594, 2004.

[39] W. T. Freeman, E. C. Pasztor, and O. T. Carmichael, "Learning low-level vision," *International Journal of Computer Vision*, vol. 40, no. 1, pp. 25–47, 2000.

[40] D. Capel and A. Zisserman, "Super resolution from multiple views using learnt image models," *Proceedings of IEEE International Conference on Computer Vision and Pattern Recognition*, pp. 627–634, 2001.

[41] S. Baker and T. Kanade, "Limits on super resolution and how to break them," *IEEE Transactions on Pattern Analysis and Machine Intelligence*, vol. 24, no. 9, pp. 1167–1183, 2002.

[42] C. V. Jiji, M. V. Joshi, and S. Chaudhuri, "Single-frame image superresolution using learned wavelet coefficients," *International Journal of Imaging Systems and Technology*, vol. 14, no. 3, pp. 105–112, 2004.

[43] J. Sun, N. Zheng, H. Tao, and H. Shum, "Image hallucination with primal sketch priors," in *Proceedings of IEEE International Conference on Computer Vision and Pattern Recognition*, vol. II, pp. 729–736, 2003.

[44] A. Chakrabarti, A. N. Rajagopalan, and R. Chellappa, "Super resolution of face images using kernel PCA-based prior," *IEEE Transactions on Multimedia*, vol. 9, no. 4, pp. 888–892, 2007.

[45] K. I. Kim and Y. Kwon, "Example-based learning for single-image super resolution," *Thirtieth Annual Symposium of the Deutsche Arbeitsgemeinschaft für Mustererkennung*, pp. 456–465, 2008.

[46] H. Chang, D. Y. Yeung, and Y. Xiong, "Super resolution through neighbor embedding," in *Proceedings of the IEEE Conference on Computer Vision and Pattern Recognition*, vol. 1, pp. 275 – 282, 2004.

[47] K. S. Ni and T. Q. Nguyen, "Image superresolution using support vector regression," *IEEE Transactions on Image Processing*, vol. 16, no. 6, pp. 1596–1610, 2007.

[48] B. S. Morse and D. Schwartzwald, "Image magnification using level set reconstruction," in *Proceedings of Conference on Computer Vision and Pattern Recognition*, 2001.

[49] Y. W. Tai, W. S. Tong, and C. K. Tang, "Perceptually-inspired and edge-directed color image super resolution," in *Proceedings of Conference on Computer Vision and Pattern Recognition*, 2006.

[50] V. Rabaud and S. Belongie, "Big little icons," in *1st IEEE Workshop on Computer Vision Applications for the Visually Impaired*, 2005.

[51] T. M. Chan, J. P. Zhang, J. Pu, and H. Huang, "Neighbor embedding based super resolution algorithm through edge detection and feature selection," *Pattern Recognition Letters*, vol. 30, no. 5, pp. 494–502, 2009.

[52] C. Liu, H. Shum, and C. Zhang, "A two-step approach to hallucinating faces: Global parametric model and local nonparametric model," in *Proceedings of IEEE International Conference on Computer Vision and Pattern Recognition*, pp. 192–198, 2001.

[53] D. Glasner, S. Bagon, and M. Irani, "Super resolution from a single image," in *ICCV*, 2009. [Online]. Available: http://www.wisdom.weizmann.ac.il/ vision/SingleImageSR.html

[54] G. Freeman and R. Fattal, "Image and video upscaling from local self-examples," *ACM Transactions on Graphics*, vol. 28, no. 3, pp. 1–10, 2010.

[55] J. Yang, J. Wright, T. Huang, and Y. Ma, "Image super resolution via sparse representation," *IEEE Transactions on Image Processing*, vol. 19, no. 11, pp. 2861–2873, 2010.

[56] R. G. Aykroyd, "Bayesian estimation for homogeneous and inhomogeneous Gaussian random fields," *IEEE Transactions on Pattern Analysis and Machine Intelligence*, vol. 20, no. 5, pp. 533–539, 1998.

[57] A. Jalobeanu, L. Blanc-Féruad, and J. Zerubia, "An adaptive Gaussian model for satellite image blurring," *IEEE Transactions on Image Processing*, vol. 4, no. 13, pp. 613–621, 2004.

[58] A. Jalobeanu, L. Blanc-Fraud, and J. Zerubia, "Adaptive parameter estimation for satellite image deconvolution," in *Rep. 3956*, 2000.

[59] N. Kaulgud and U. B. Desia, *Super resolution imaging*. Dordrecht: Kluwer, 2001, ch. "Image Zooming: Use of Wavelets," pp. 21–44.

[60] N. Ahmed, T. Natarajan, and K. R. Rao, "Discrete cosine transform," *IEEE Transactions on Computers*, vol. C-23, pp. 90–93, 1974.

[61] N. Ahmed and K. R. Rao, *Orthogonal Transforms for Digital Signal Processing*. Berlin: Springer Verlag, 1975.

[62] K. R. Rao and P. Yip, *Discrete Cosine Transform: Algorithms, Advantages, Applications*. New York: Academic Press, 1990.

[63] M. N. Do and M. Vetterli, "The contourlet transform: An efficient directional multiresolution image representation," *IEEE Transactions on Image Processing*, vol. 14, no. 12, pp. 2091–2106, 2005.

[64] M. Do and M. Vetterli, "Framing pyramids," *IEEE Transaction on Signal Processing*, vol. 51, no. 9, pp. 2329–2342, 2003.

[65] H. B. Roberto and J. T. S. Mark, "A filter bank for the directional decomposition of images: Theory and design," *IEEE Transactions on Signal Processing*, vol. 40, no. 4, 882–893, 1992.

[66] P. J. Burt and E. H. Adelson, "The Laplacian pyramid as a compact image code," *IEEE Transactions on Communication*, vol. 31, no. 4, pp. 532–540, 1983.

[67] Z. Wang, A. Bovik, H. Sheikh, and E. Simoncelli, "Image quality assessment: From error visibility to structural similarity," *IEEE Transactions on Image Processing*, vol. 13, no. 4, pp. 600–612, Apr 2004.

[68] Z. Wang and A. C. Bovik, "Mean squared error: Love it or leave it?" *IEEE Signal Processing Magazine*, pp. 98–117, 2009.

Chapter 8

Super Resolution for Multispectral Image Classification

FENG LI
Chinese Academy of Sciences

XIUPING JIA
University of New South Wales

DONALD FRASER
University of New South Wales

ANDREW LAMBERT
University of New South Wales

8.1   Introduction

Super resolution (SR) is the reconstruction of a high-resolution image with more detail from a series of low-resolution (LR) images that can either be acquired from different viewpoints or from the same view point at different times. This is possible because every LR image may contribute some information that is not in other LR images. Example applications include satellite HR image reconstruction both for panchromatic images and multispectral images, long-range camera images affected by atmospheric turbulence, multispectral classification [1], to name a few. Although many SR algorithms have been proposed, most of them suffer from several impractical assumptions: for example, that any shift or rotation between LR images is global, that the motion occurring between LR images is known exactly, or that the LR images are noise-free. However, the imaging procedure is generally much more complicated, and may include local warping, blurring, decimation, and noise-contamination. These difficulties have led us to develop a novel SR algorithm in order to avoid them and to cover all real applications.

For remote sensing imagery, both spatial resolution and spectral resolution are crucial in evaluating the quality of multispectral images. High spatial resolution and spectral resolution for multispectral images are required to provide detailed information for land use and land cover mapping. There are some factors that limit the spatial resolution of multispectral sensors, such as instantaneous field of view (IFOV), density of CCD sensors, the optical system, the speed of the platform and revisit cycle, etc. Since the late 1960s, when the first multispectral sensor was used on Apollo-9, improvement in spatial resolution has been achieved progressively from the 79-m resolution of Landsat-1 MSS to the 20-m resolution of EO-1 Hyperion. There are also some factors that affect the spectral resolution, including the spectral splitter lens and calibration accuracy. In general, high spatial resolution is associated with low spectral resolution, and vice versa. The conflict comes from the property of CCD cells; to keep a reasonable signal-to-noise ratio, a certain number of photons is necessary to be accumulated against noise with a fixed exposure time. The narrower the spectral band (high spectral resolution), the smaller the number of photons that will be collected during a fixed exposure time. To keep high spectral resolution imagery and collect more photons for each CCD cell, a larger IFOV must be adopted, which sacrifices the spatial resolution for the spectral resolution. That is the reason why hyper-spectral data, whose spectral resolution can be as high as 10-nm wavelength, has relatively low spatial resolution. Similarly, a panchromatic band covering a wide spectral range of over 300-nm, has higher spatial resolution than individual multispectral bands [2]. Hence, the study of super resolution for remote sensing imagery has been conducted actively as an alternative means of overcoming this conflict to reconstruct images with both high spectral resolution and high spatial resolution.

Super resolution reconstruction has been tested for the multispectral sensor CHRIS-Proba (Compact High Resolution Imaging Spectrometer onboard the Project for On-board Autonomy) [3], MODIS (Moderate Resolution Imaging Spectroradiometer) [4], and SPOT4 onboard sensors [5]. However, the impact of the reconstructed data toward image classification has not been addressed previously. In this chapter, SR is applied to several groups of multispectral data captured at different times to reconstruct a group of HR multispectral images. Then, the performance of classification is investigated based on the super-resolved multispectral images. Classification is a procedure for quantitative analysis of multispectral images. It includes two different strategies: separating the pixels into several groups without knowing the ground cover types (unsupervised classification or clustering) and labeling the pixels of a region as representing particular ground cover types (supervised classification).

Satellites usually capture multispectral images across the visual range and out to the longer wavelength nonvisual range. A multispectral image is a collection of several monochromatic (as opposed to panchromatic) images of the same scene, each of them taken with a different sensor in a particular spectral range. Each image is referred to as a band. Landsat7, for example, with the onboard sensor Enhanced Thematic Mapper Plus (ETM+), produces seven bands covering the wavelength range from 450 to 1,250 nm. Multispectral images with more channels and higher spectral resolution are called "hyperspectral images." The reason why multispectral sensors or even hyperspectral sensors are adopted as payloads on satellite platforms is that different objects display different spectral reflection "signatures." Therefore, we can well recognize ground objects (clusters or classes) based on the multispectral images. Here, only multispectral images are discussed. However, the research covering multispectral images can be applied to hyperspectral images directly, but at greater cost and greater computational effort.

Because an image consists of pixels, a pixel size is the minimum unit to interpret or recognize the ground object (clusters or classes) based on multispectral images. There are several ways to improve the performance for distinguishing different clusters and classes. First of all, developing a higher spatial resolution sensor is an obvious way to achieve better performance. However, there are many factors that limit the spatial resolution of multispectral sensors, such as the signal-to-noise ratio of the sensors, IFOV, the detector technology used, the space available on the platform, the conflict between spatial resolution and spectral resolution, etc. In particular, once a satellite is launched, it is difficult to update its imaging devices. Second, developing efficient classification algorithms is another option for better classification, and has been investigated for decades. Third, super resolution mapping (SRM) [6–8] is another way to achieve better classification results. SRM is built on the top of spectral unmixing, and it can be achieved based on the Markov random field theorem [6], or via pixel-swapping optimization algorithms [8] in order to generate a realistic spatial structure on a finer thematic map. Alternatively, a geostatistical indicator (Cokriging) can be introduced for a noniterative SRM procedure [7].

All the methods mentioned above are based on a single data group. In papers [3–5, 9], however, super resolution is tested for a series of multispectral sensors CHRIS/Proba and MODIS, HRVIR, and VEGETATION on SPOT4, respectively. Unfortunately, the authors evaluate the SR methods by the comparison of channel by channel without the testing on classification.

In this chapter we propose a different approach to improve the classification result by using multitemporal data groups within the same region captured by the same sensor. A super resolution (SR) algorithm is applied to reconstruct higher spatial resolution multispectral images before classification is conducted. This can increase the number of informative pixels by making use of the (slight) ground footprint mismatch from image to image. Classification can then be carried out on these reconstructed higher spatial resolution images without any limitation of classification techniques adopted. Super resolution reconstruction is a preprocessing step for clustering or classification. The super-resolved data can then be ready for segmentation using standard clustering techniques, such as ISODATA, or thematic mapping using classification methods. In this chapter we only focus on the preprocess step, rather than the clustering or classification methods.

This chapter is organized as follows: In Section 8.2 we propose a new super resolution method, Maximum a Posteriori, based on the universal Hidden Markov Tree model (MAP-uHMT), and improved classification is targeted using MAP-uHMT on several data groups of multispectral images captured on different dates; unsupervised classification and supervised classification tests are carried out on the reconstructed higher resolution multispectral images in Section 8.3; both visual results and numerical results of the comparison are also provided in Section 8.3; the final conclusions are given in Section 8.4.

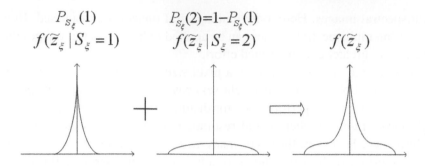

Figure 8.1 The wavelet coefficient $\widetilde{z}_\xi$ is conditionally Gaussian, given its state variable S_ξ, and has an overall $Y = 2$ state Gaussian mixture model.

8.2 Methodology

8.2.1 Background

The method proposed in this chapter is based on the hidden Markov tree (HMT) theory in the wavelet domain [10, 11]. The reason that we choose to work in the wavelet domain, rather than in the spatial or frequency domains is that the probability density function (pdf) of the wavelet coefficient can be well characterized by a mixed Gaussian pdf, and the dependencies between the coarse scale wavelet coefficients, "parent," and the finer scale wavelet coefficients, "children," can be well described by HMT in the wavelet domain.

The ξth wavelet coefficient $\widetilde{z}_\xi$ is conditionally Gaussian given its state variable S_ξ, and the wavelet coefficient has an overall Y-state Gaussian mixture model:

$$f(\widetilde{z}_\xi) = \sum_{y=1}^{Y} P_{S_\xi}(y) f(\widetilde{z}_\xi | S_\xi = y), \tag{8.1}$$

where $f(\widetilde{z}_\xi | S_\xi = y)$, which follows a Gaussian density $N(0, \sigma_\xi(y))$, denotes the probability of $\widetilde{z}_\xi$ under the condition of the hidden state equal to y. $P_{S_\xi}(y)$ represents the probability of the wavelet coefficient $\widetilde{z}_\xi$ belonging to the hidden state y. The more states that are used, the closer the approximation is to the distribution of the wavelet coefficients, but the computation becomes exponentially large. As in [10], we also solve the problem for two hidden states, that is, $Y = 2$, signifying whether the coefficient is small or large. As shown in Figure 8.1, hidden state 1 means that the magnitude of wavelet coefficients is small; on the contrary, 2 denotes that the magnitude of wavelet coefficient is large. Therefore, $\sigma_\xi(1) < \sigma_\xi(2)$. This is also similar to the property of energy compaction of the wavelet transform, as the wavelet transforms of real-world images tend to be sparse. Most of the wavelet coefficients are small, because real-world images have many smooth regions separated by edges.

The wavelet domain HMT theory was first proposed by Crouse et al. [10]. By connecting state variables vertically across scales, a graph with tree-structured dependencies between state variables is obtained. A state transition matrix R_ξ for each link quantifies statistically the degree of persistence of the states between "parent" and "child," as shown by the arrows in Figure 8.2. The format of R_ξ can be written as

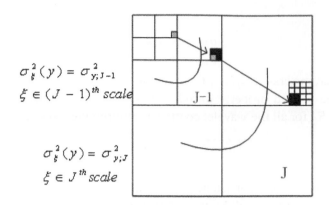

$$\sigma_\xi^2(y) = \sigma_{y;J-1}^2$$
$$\xi \in (J-1)^{th} \text{ scale}$$

$$\sigma_\xi^2(y) = \sigma_{y;J}^2$$
$$\xi \in J^{th} \text{ scale}$$

Figure 8.2 The tree structure of the state dependencies between the parent and the children is indicated by arrows; tying together all the wavelet coefficients within the same scale j, the variance of the ξth wavelet coefficient given a hidden state y within the same scale will be the same value $\sigma_{y;j}^2$.

$$R_\xi = \begin{bmatrix} P^{1\to1} & 1 - P^{1\to1} \\ P^{2\to1} & 1 - P^{2\to1} \end{bmatrix}, \tag{8.2}$$

where $P^{1\to1}$ denotes the probability that a wavelet coefficient is small, given that its parent is small; on the other hand, $P^{2\to1}$ denotes the probability that a wavelet coefficient is small, given that its parent is large.

Because $P_{S_\xi}(1) + P_{S_\xi}(2) = 1$, four parameters $P_{S_\xi}(1)$, R_ξ, $\sigma_\xi(1)$, and $\sigma_\xi(2)$ must be specified for every wavelet coefficient position. Although the authors in [10] reduce the number of parameters to four parameters only for all wavelet coefficient positions in a single scale by tying together all wavelet coefficients and hidden states of that scale, there are $4J$ parameters that must be specified (J denotes the number of scales of the wavelet decomposition) in every iteration during the training procedure. Therefore, the variances of the wavelet coefficients, given a hidden state within the same scale of the wavelet transform, will be regarded as the same. As shown in Figure 8.2, all wavelet coefficients are tied together within the same scale $j = 1 \cdots J$, $\sigma_\xi(1)^2 = \sigma_{1;j}^2$, and $\sigma_\xi(2)^2 = \sigma_{2;j}^2$ for all ξ locating in the jth scale. Here, $\sigma_{1;j}^2$ denotes the variance of the wavelet coefficients within the jth scale, given the hidden state of 1; $\sigma_{2;j}^2$ denotes the variance of the wavelet coefficients within the jth scale given the hidden state of 2.

Using the two tertiary properties of the wavelet coefficients, the values of the wavelet coefficients of real-world images decrease exponentially across scales; the persistence of the magnitudes of the (large/small) wavelet coefficients tend to be exponentially stronger at finer scales, a universal HMT (uHMT) model was proposed in [11]. Applying the property: The values of the wavelet coefficients of real-world images decrease exponentially across scale, and the simplified expression of the standard deviations of the wavelet coefficients within the jth scale given the hidden sates can be written as

$$\sigma_{1;j} = C_{\sigma_1} 2^{-j\eta_1};$$
$$\sigma_{2;j} = C_{\sigma_2} 2^{-j\eta_2}. \tag{8.3}$$

These four parameters—C_{σ_1}, η_1, C_{σ_2} and η_2—characterize the exponential decrease across scales. The second property introduced above brings the expression of the state transition matrix R_ξ for all the wavelet coefficients within the jth scale:

$$R_\xi = \begin{bmatrix} 1 - C_{11}2^{-r_1 j} & C_{11}2^{-r_1 j} \\ \frac{1}{2} - C_{22}2^{-r_2 j} & \frac{1}{2} + C_{22}2^{-r_2 j} \end{bmatrix}, \xi \in j^{th} scale. \tag{8.4}$$

So the exponentially stronger persistence of the hidden states of the wavelet coefficients at finer scales can be described well by these four parameters: C_{11}, r_1, C_{22} and r_2. One extra parameter is the probability of the hidden state value of the root coefficients, $P_{S_{\xi=1}}(1)$ (only one number in the coarsest level, because the wavelet decomposition was carried to completion). Therefore, just nine parameters that are independent of the size of the image and the number of wavelet scales are needed, in contrast to $4J$ of the original HMT. The nine-parameter set is denoted by θ in this chapter. Given a set of observations $\tilde{z}$ and the uHMT model θ, the Upward-Downward algorithm [10] is adopted to determine the probability of that location under a given state such as $P_{S_\xi}(1)$ and $P_{S_\xi}(2)$. The high efficiency and the good results of uHMT in denoising [11] motivate us to develop a super resolution method based on the uHMT.

8.2.2   Super Resolution Based on a Universal Hidden Markov Tree Model

When a satellite captures an image, there is, of course, a limit to the spatial resolution at the ground due to the pixel spacing along a scan-line and the scan-line separation and the overall optical magnification. In addition, there are optical distortions (defocus, diffraction effects, etc.), motion blur, warping caused by variations in the satellite track, atmospheric turbulence, and additional detector noise. For these reasons, the image obtained is likely to suffer from geometric distortion, blurring, and added noise. To simulate these effects for algorithm development and testing, observed low-resolution images are modeled as [12,13]

$$g_i = DH_i M_i z + n_i, \quad i = 1, \cdots, K, \tag{8.5}$$

where z can be regarded as an ideal data set obtained by sampling a continuous scene at high resolution; g_i is the ith observed low-resolution image, K is the number of low-resolution images, M_i is a warping matrix, H_i is a blur matrix, D is a sub-sampling matrix, and n_i represents additive noise of Gaussian distribution with zero mean. The aim of a super resolution algorithm, then, is to estimate z from the set of LR images, g_i.

Choose the oth LR image, as a reference image, then $g_o = DH_o z$ or $g_o = DH_o M_o z$, where M_o is an identity matrix in this case. In practice, the warping matrix M_i must be estimated initially from the LR images between g_i and the reference LR image g_o, so it is impossible to calculate the exact warping matrix M_i. Thus, only an approximation M'_i is available. Similarly, only an approximation, H'_i, to the blur matrix H_i is available. Due to the difference between M'_i and M_i and the difference between H'_i and H_i, Equation (8.5) can be rewritten as:

$$g_i = DH'_i M'_i z + v_i + n_i, \quad i = 1, \cdots, K, \tag{8.6}$$

where v_i is an additive "noise" term representing the error caused by the incorrect blur matrix and warping matrix. If we assume that v_i follows an independent and identically distributed (i.i.d.) Gaussian distribution with zero mean as also assumed in [14], and although it may not represent the closest distribution in some circumstances, then we have

$$g_i = DH'_i M'_i z + n'_i, \quad i = 1, \cdots, K \tag{8.7}$$

where $n'_i = v_i + n_i$, and n'_i follows a zero mean Gaussian distribution also. Set $N \times N$ as the number of pixels in the original SR image z, and c as the decimating (sub-sampling) rate. The terms g_i, z and n'_i are considered in the form of lexicographically ordered vectors, with lengths N^2/c^2, N^2, and N^2/c^2, respectively. So, H'_i and M'_i are $N^2 \times N^2$ matrices. The size of D is $N^2/c^2 \times N^2$.

Because the imaging procedure is not invertible, Equations. (8.5) and (8.7 represent a typically ill-conditioned problem. This implies that an exact solution does not exist. A well posed problem is preferred, and an approximation to the original super resolution image z is desired.

8.2.2.1 Derivation

In this section we construct a new prior model for the Maximum a Posteriori (MAP) estimate in the wavelet domain. By using Steepest Descent (SD), a stable and approximate estimate z^* is available for the ill-conditioned problem.

An orthogonal wavelet transform is performed on both sides of Equation (8.7), which can be written as

$$\widetilde{g}_i = \widetilde{D}\widetilde{H'}_i\widetilde{M'}_i\widetilde{z} + \widetilde{n'}_i \quad i = 1, \cdots, K, \tag{8.8}$$

where $\widetilde{g}_i$, $\widetilde{z}$, and $\widetilde{n'}_i$ are the lexicographically ordered vector expressions of g_i, z, and n'_i in the wavelet domain, respectively; $\widetilde{D}$, $\widetilde{H'}_i$, and $\widetilde{M'}_i$ are the wavelet domain expressions of D, H'_i, and M'_i, respectively (note, these are still matrices). Similarly, z^* is denoted as $\widetilde{z}^*$ in the wavelet domain.

Using the MAP estimator and assuming $P(\widetilde{z}|\theta)$ can be expressed as $exp(-\phi(\widetilde{z}, \theta))$, where θ is the learnt uHMT parameter set, we have

$$\widetilde{z}^* = argmin\, L(\widetilde{z}), \tag{8.9}$$

where

$$L(\widetilde{z}) = \sum_{i=1}^{K} \frac{1}{2\lambda_i^2} \parallel \widetilde{g}_i - \widetilde{D}\widetilde{H'}_i\widetilde{M'}_i\widetilde{z} \parallel^2 + \phi(\widetilde{z}, \theta). \tag{8.10}$$

λ_i^2 is the noise variance of n'_i in the ith image, which keeps the same value in the wavelet domain. The function $\phi(\widetilde{z}, \theta)$ is called the energy function in [15, 16], and is regarded as the energy attributed to $\widetilde{z}$. This term comes from the prior probability.

The problem now is to construct the energy function (or the prior model) $\phi(\widetilde{z}, \theta)$. We introduce the wavelet domain uHMT model originally developed for signal denoising and signal estimation [11] to this application. For simplicity, let us assume that the marginal probability density functions of the wavelet coefficients are conditionally independent, although the state values between the parent wavelet coefficients and the child wavelet coefficients are dependent. Then $P(\widetilde{z}|\theta)$ can be denoted as

$$P(\widetilde{z}|\theta) = \prod_{\xi=1}^{N^2} \{P_{S_\xi}(1)f(\widetilde{z}_\xi|S_\xi = 1) + P_{S_\xi}(2)f(\widetilde{z}_\xi|S_\xi = 2)\}. \tag{8.11}$$

Because we regard the marginal pdf of the ξth wavelet coefficient $\widetilde{z}_\xi$ as a Gaussian mixture as shown in Equation (8.1), we have

$$logP(\widetilde{z}|\theta) = \sum_{\xi=1}^{N^2} log\{P_{S_\xi}(1)\frac{e^{-\frac{\widetilde{z}_\xi^2}{2\sigma_\xi(1)^2}}}{\sqrt{2\pi}\sigma_\xi(1)} + P_{S_\xi}(2)\frac{e^{-\frac{\widetilde{z}_\xi^2}{2\sigma_\xi(2)^2}}}{\sqrt{2\pi}\sigma_\xi(2)}\}. \tag{8.12}$$

where $\sigma_\xi(1)^2$ and $\sigma_\xi(2)^2$ are the variances of the pdf of the ξth wavelet coefficient, given the hidden Markov state as 1 and 2, respectively.

We have the following definition:

$$logf(\widetilde{z}_\xi) = log\{P_{S_\xi}(1)\frac{e^{-\frac{\widetilde{z}_\xi^2}{2\sigma_\xi(1)^2}}}{\sqrt{2\pi}\sigma_\xi(1)} + P_{S_\xi}(2)\frac{e^{-\frac{\widetilde{z}_\xi^2}{2\sigma_\xi(2)^2}}}{\sqrt{2\pi}\sigma_\xi(2)}\}. \tag{8.13}$$

The derivative of $logf(\widetilde{z}_\xi)$ will be required in searching for the minimum value of Equation (8.10). It is difficult to get an analytical expression; therefore we adopt a treatment similar to that proposed in [17–19].

Let us consider two extreme circumstances, if $P_{S_\xi}(1) = 1$, which means the state of the ξth wavelet coefficient is small, so $P_{S_\xi}(2) = 0$; then

$$logf(\widetilde{z}_\xi) = -\frac{\widetilde{z}_\xi^2}{2\sigma_\xi(1)^2} - c1, \tag{8.14}$$

where $c1$ is a constant. The mixed Gaussian distribution will degenerate into a single Gaussian with a variance of $\sigma_\xi(1)^2$. On the other hand, if $P_{S_\xi}(2) = 1$, then

$$logf(\widetilde{z}_\xi) = -\frac{\widetilde{z}_\xi^2}{2\sigma_\xi(2)^2} - c2, \tag{8.15}$$

where $c2$ is a constant. Then the mixed Gaussian distribution will degenerate into a single Gaussian with the variance of $\sigma_\xi(2)^2$.

Using a single Gaussian distribution to approximate the mixed Gaussian distribution as shown in Equation (8.13) will simplify this problem; understandably, the variance of the Gaussian will be decided by the probability of the hidden states for the wavelet coefficient. The larger the probability of a hidden state, the closer will be the approximation of the variance of the Gaussian to the variance of the hidden state. We approximate the variance of the Gaussian as $P_{S_\xi}(1)\sigma_\xi(1)^2 + P_{S_\xi}(2)\sigma_\xi(2)^2$. Therefore, we arrive at the following approximation:

$$\log f(\widetilde{z}_\xi) \approx \log \frac{e^{-\frac{\widetilde{z}_\xi^2}{2(P_{S_\xi}(1)\sigma_\xi(1)^2+P_{S_\xi}(2)\sigma_\xi(2)^2)}}}{\sqrt{2\pi(P_{S_\xi}(1)\sigma_\xi(1)^2 + P_{S_\xi}(2)\sigma_\xi(2)^2)}}$$

$$\approx -\frac{\widetilde{z}_\xi^2}{2(P_{S_\xi}(1)\sigma_\xi(1)^2 + P_{S_\xi}(2)\sigma_\xi(2)^2)} - c3, \tag{8.16}$$

where $c3$ is also a constant. Equation (8.16) is identical to Equation (8.13) for these two above extreme cases. Based on our observations, the probability of the hidden states is distributed mainly in the two extremes (for example, $P_{S_\xi}(1)$ is either larger than 80% or smaller than 20%) for most images, which means the expression in Equation (8.16) can approximate well to Equation (8.13). While it is not an ideal simplification, it is a compromise to make it easy to be managed further, borne out by successful results in Section 8.3.

Based on the simplification, $L(\widetilde{z})$ in Equation (8.9) can be rewritten as

$$L(\widetilde{z}) = \sum_{i=1}^{K} \frac{1}{2\lambda_i^2} \parallel \widetilde{g}_i - \widetilde{D}\widetilde{H'}_i\widetilde{M'}_i\widetilde{z} \parallel^2$$

$$+ \sum_{\xi=1}^{N^2} \frac{\widetilde{z}_\xi^2}{2(P_{S_\xi}(1)\sigma_\xi(1)^2 + P_{S_\xi}(2)\sigma_\xi(2)^2)}. \tag{8.17}$$

The minimized formulation can be rewritten further as

$$L(\widetilde{z}) = \frac{\alpha}{K} \sum_{i=1}^{K} \parallel \widetilde{g}_i - \widetilde{D}\widetilde{H'}_i\widetilde{M'}_i\widetilde{z} \parallel^2 + \widetilde{z}^T Q\widetilde{z}, \tag{8.18}$$

where Q is a diagonal matrix ($N^2 \times N^2$) consisting of $q_{(\xi,\xi)}$:

$$q_{(\xi,\xi)} = \frac{1}{2(P_{S_\xi}(1)\sigma_\xi(1)^2 + P_{S_\xi}(2)\sigma_\xi(2)^2)}. \tag{8.19}$$

As introduced in Figure 8.2, $\sigma_\xi(1)^2 = \sigma_{1;j}^2$ and $\sigma_\xi(2)^2 = \sigma_{2;j}^2$, $\xi \in j$th scale. Therefore, if $j = 1, 2, \cdots J$, there are J pairs of variances only. Using the uHMT parameter set and Equation (8.3), $\sigma_\xi(1)$ and $\sigma_\xi(2)$ can be easily assigned for all locations. Moreover, using the Upward-Downward algorithm [10] based on the uHMT parameter set, observations $\widetilde{z}$ and Equation (8.4), $P_{S_\xi}(1)$ and $P_{S_\xi}(2)$ for all locations can be calculated in a straightforward manner. Therefore, the diagonal components $q_{(\xi,\xi)}$ of the matrix Q can be assigned. Then Q is normalized between $[0, 1]$.

In Equation (8.18), α is a parameter to describe the variance of the noise as well as the error caused by the incorrect warping and blur matrices (for simplicity, λ_i is assumed to be the same value for all LR images). In order to eliminate the effect of the different number of LR images in different circumstances, the denominator K is brought in. Therefore, α can actually balance the contribution from the prior model against all LR images, no matter how many LR images are used. This makes it easy to select α experimentally without considering how many LR images are used in this algorithm.

The estimation of the wavelet coefficients of the original image is found using the SD method to minimize Equation (8.18) as

$$\widetilde{z}^{r+1} = \widetilde{z}^r + ad, \tag{8.20}$$

where d is the descent gradient, r is the rth internal iteration, and a is the step size. So, d is written as

$$d = -[\frac{\alpha}{K} \sum_{i=1}^{K} \widetilde{M'_i}^T \widetilde{H'_i}^T \widetilde{D}^T (\widetilde{D}\widetilde{H'_i}\widetilde{M'_i}\widetilde{z} - \widetilde{g}_i) + Q\widetilde{z}], \tag{8.21}$$

and the step size a is given by

$$a = \frac{d^T d}{d^T(\frac{\alpha}{K} \sum_{i=1}^{K} \widetilde{M'_i}^T \widetilde{H'_i}^T \widetilde{D}^T \widetilde{D}\widetilde{H'_i}\widetilde{M'_i})d + d^T Q d}. \tag{8.22}$$

Because the prior model is achieved using the previous pseudo-super-resolved result, Q must to be updated $Q^l \rightarrow Q^{l+1}$ using Equation (8.19), where l denotes the lth outer iteration.

8.2.2.2   Implementation Details

A critical step in this SR method is to calculate M'_i (which is image registration). If we were to adopt IBP for SR in remote sensing, the image registration method used in [20] limits the overall efficiency of the SR algorithm, although Irani and Peleg modified their original registration method in order to consider a more general motion model [21]. In [21], the number of parameters in the motion model was increased from 3 to 8 to describe the 2-D transformation. However, it is still not suitable to accurately describe 2-D warps of satellite images because of their particular method of capture. Based on Ardy Goshtasby's projective transformation analysis [22], affine is a suitable transform function when registering images taken from a distant platform of a flat scene. The capture of the remote sensing images accords closely with this rule, as the distance between the sensors on satellites and the scene is more than 500 km, in general. Second, there are different affine, translation, and intensity changes caused by cloud, atmospheric turbulence, etc., not only within the different LR remote sensing images, but also within different regions within each LR remote sensing image. Therefore, we have adopted an efficient elastic image registration algorithm [23] in lieu of other methods to calculate M'_i and $M'_i{}^T$. Within this image registration method, a pyramid method is adopted to deal with possible large spatial variations within images. Also, a smoothness constraint is used to deal with local affine transform and intensity variation in dealing with local warps within the remote sensing images. For the $\widetilde{M'_i}$ and $\widetilde{M'_i}^T$ matrix, we can achieve this registration by warping the image with matrix M'_i and $M'_i{}^T$ before transforming back to the wavelet domain, and thus avoid the need to set up the exact matrixes.

In the case of H'_i, we consider the point spread function as a Gaussian blur kernel, and its size can be approximately estimated from the blurring of small points or edges in the original images. The calculation of $\widetilde{H'}_i$ requires a huge memory capacity and greatly reduces the efficiency. However, the blurring can be carried out by multiplying the Fourier transform matrix of the blur kernel and the Fourier transform matrix of the image in the frequency domain and then transforming back to the wavelet domain. $\widetilde{H'}_i^T$ can be solved by multiplying the conjugate Fourier transform matrix of the blur kernel and the Fourier transform matrix of the image in the frequency domain and then transforming back to the wavelet domain.

Similarly, D is a decimating operator in the spatial domain, and D^T can be resolved by upsampling the image (e.g., by bilinear interpolation) with an upsampling factor c in the spatial domain. $\widetilde{D}$ and $\widetilde{D^T}$ can be achieved in the spatial domain with downsampling and upsampling, and then transforming back to the wavelet domain.

Our proposed SR method — Maximum a Posteriori based on the universal Hidden Markov Tree model—is abbreviated as MAP-uHMT. The SR method of MAP-uHMT can be briefly described by the flowchart in Figure 8.3 and by the following steps:

1. Initialize all parameters and matrices, including the selection of a reference image g_o; calculate the warp matrix M'_i and estimate the blur kernel size for H'_i; apply the wavelet transform to an enlarged g_o, interpolated by the bilinear method as $\widetilde{z}^0$; and calculate an initial prior Q^0. Begin the outer iterations, loop index $l = 0$.

2. Begin the inner iterations, loop index $r = 0$. Apply the SD method to calculate $\widetilde{z}^{r+1}$ based on Equations. (8.21) and (8.22), $r = r + 1$. This is the inner iteration loop. Iterate until $\|z^{r+1} - z^r\|/\|z^r\| \leq \epsilon$ or the iteration count exceeds a maximum; then skip to Step 3. ϵ is a threshold that is chosen to be 10^{-5}, which has been found to provide a good result with a reasonable computation time.

3. Update the prior Q^l to Q^{l+1} by recalculating the $q_{(\xi,\xi)}$ for every location. Set $l = l + 1$, and go back to step 2. When the outer iteration index, l, exceeds a maximum count, then skip to step 4. This is the outer iteration loop. In updating Q^l, $P_{S_\xi}(1)$ and $P_{S_\xi}(2)$ for every wavelet coefficient $\widetilde{z}_\xi$ in $\widetilde{z}^r$ can be recalculated using the Upward-Downward algorithm [10]; $\sigma_\xi(1)$ and $\sigma_\xi(2)$ can be easily assigned for all locations based on the uHMT parameter set and Equation (8.3), because $\sigma_\xi(1)^2 = \sigma_{1;j}^2$ and $\sigma_\xi(2)^2 = \sigma_{2;j}^2, \xi \in j$th scale, by tying all the wavelet coefficients in the same scale j, as shown in Figure 8.2. Then, $q_{(\xi,\xi)}$ can be worked out by Equation (8.19).

4. Quit and apply the 2-D inverse discrete wavelet transform to $\widetilde{z}^r$ to reconstruct the SR image z^*.

As introduced previously, α acts as a trade-off parameter to balance the influence of the prior model (the enlarged reference LR image) and the other LR images. If α is too large, the reconstructed image will become a mean of the registered and enlarged LR images; however, if it is selected with a very small value, the SR method will degrade into an interpolation method. We use the Daubechies-8 wavelet in the orthogonal wavelet transform and fully decompose the images. That is to say, there is only one scale coefficient in the $\widetilde{z}$. We assume that there is no Markovian relationship between the scale wavelet and the wavelet coefficients; the prior model does not provide any information for the scale coefficient. So

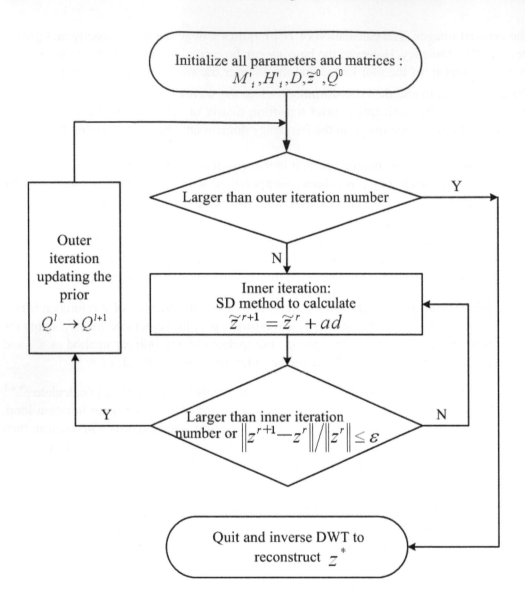

Figure 8.3 Flowchart for MAP-uHMT computation.

$q_{(0,0)}$ is set to 0. Note that Q is normalized between $[0, 1]$, and all intensities of the images are normalized between $[0, 1]$.

8.2.3 MAP-uHMT on Multispectral Images

The electromagnetic radiation reflection from the objects of the same nature is similar overall, and these objects will thus have similar spectral signatures. This is the most important principle in classification. Using suitable algorithms and some prior knowledge, it is possible to classify pixels into different ground covers or types. Classification methods can be divided into supervised and unsupervised classifications.

This chapter does not try to cover research on classification methods, which is outside its scope. What is emphasized is that SR methods can be used to improve the performance of clustering and classification, provided that several groups of multispectral images are captured within a short period and of the same scene. For example, satellites such as TERRA and AQUA cover the entire Earth's surface every 1 to 2 days. Because the revisit cycle is so frequent, we can assume that the ground truth has not changed during that short period. Even for some long cycle period satellites such as LANDSAT7 (16 days), if the ROI (regions of interest) are urban areas, concrete buildings, etc., then SR methods can still be used on multispectral classification because these areas will not change much during that period.

Let us assume that the spatial resolution of each of the bands of a multispectral image is the same. Therefore, the imaging procedure for the multispectral bands can be described as

$$g_{bi} = DH_i M_i z_b + n_{bi}, \quad i = 1, \cdots, K \ ; b = 1, \cdots, B, \tag{8.23}$$

where z_b can be regarded as the bth band ideal high spatial resolution image, g_{bi} is the ith observed low-resolution image of band b, K is the number of low-resolution images captured on different dates, B is the number of multispectral bands, M_i is a warping matrix, H_i is a blur matrix, D is a sub-sampling matrix, and n_{bi} represents additive noise of Gaussian distribution with zero mean. Because the incoming light passes through a beam splitter to arrive at different spectral range sensors on a satellite, all the multispectral bands should be aligned. This will be so, no matter whether whiskbroom (across-track) or pushbroom (along-track) sensors are used. Therefore, all the observed low-resolution multispectral images are assumed to suffer the same warping effect; the warping matrix M_i is adopted for all the bands. Moreover, we can also assume that all the bands suffer the same blur matrix H_i.

For this ill-conditioned problem, the super resolution method MAP-uHMT proposed above will be adopted to reconstruct z_b from the set of LR images g_{bi} channel by channel. Misaligned multispectral images will make the classification worse. Therefore, M_i is calculated from one-channel data and applied to all other channels. In this way, all reconstructed multispectral channels with higher resolution are still aligned, which is a precondition for classification.

Before the commencement of experiments, some approaches to evaluate the classification results are discussed. There are a number of popular ways of quantitatively evaluating the agreement between the ground truth and the results of unsupervised classification and supervised classification. For example, in an error matrix table, each row of the table is reserved for one of the classes used by the classification algorithm, and each column displays the corresponding ground truth classes in the same order. Based on the error matrix, user's accuracy, producer's accuracy, overall accuracy, and the Kappa coefficient are calculated to assess the classification accuracy. User's accuracy, which evaluates the percentage of the correctly labeled pixels of that class in the thematic map, is defined as

$$User's\ accuracy\ of\ the\ k th\ class = \frac{x_{kk}}{x_{k+}}, \tag{8.24}$$

where x_{kk} means the kth diagonal element of the error matrix, and x_{k+} means the sum over all columns for row k of the error matrix. Producer's accuracy, which is a measure of the percentage of the land in each category classified correctly, is defined as

$$Producer's\ accuracy\ of\ the\ k^{th}\ class = \frac{x_{kk}}{x_{+k}}, \qquad (8.25)$$

where x_{kk} means the kth diagonal element of the error matrix, and x_{+k} means the sum over all rows for column k of the error matrix. A popular accuracy assessment as a whole is overall accuracy, which is defined as

$$Overall\ accuracy = \frac{\sum\limits_{k} x_{kk}}{N^2}, \qquad (8.26)$$

where x_{kk} means the kth diagonal element of the error matrix and N^2 is the number of all the pixels for the classification.

Another whole measure of map accuracy is the *Kappa* (or called KHAT in [24]) coefficient, which is a measure of the proportional (or percentage) improvement by the classifier over a purely random assignment to classes:

$$Kappa = \frac{N^2 \sum\limits_{k} x_{kk} - \sum\limits_{k} x_{k+} x_{+k}}{N^4 - \sum\limits_{k} x_{k+} x_{+k}} \qquad (8.27)$$

where x_{kk} is the kth diagonal element of the error matrix, x_{k+} is the sum over all columns for row k of the error matrix, N^2 is the number of all the pixels in the error matrix, and x_{+k} is the sum over all rows for column k of the error matrix.

Differences in overall accuracy and *Kappa* are to be expected, in that each incorporates different forms of information from the error matrix. The overall accuracy only includes the diagonal data and excludes the errors of omission and commission [24]. In contrast, *Kappa* incorporates the nondiagonal elements of the error matrix as a product of the row and column marginal. Accordingly, it is not possible to give definitive advice as to when each measure should be adopted in any given application. Therefore, both overall accuracy and *Kappa* are selected as classification accuracy assessment in this chapter.

8.3   Experimental Results

8.3.1   Testing with MODIS data

Based on the new SR method MAP-uHMT introduced in the previous section, high-resolution multispectral images are reconstructed from a sequence of LR multispectral images channel by channel. The performance of the clustering and classification are tested based on these reconstructed super resolution multispectral MODIS images from the satellite TERRA. TERRA's orbit around the Earth is timed so that it passes from north to south across the equator in the morning. Because TERRA is a sun-synchronized satellite, all images are captured at the same local time and under similar illumination conditions; the changes in the track of TERRA allow the sensor to capture the same ground area from slightly different view-points during each pass, although the orbit is periodically adjusted after a specified amount of drift has occurred in order to bring the satellite back to an orbit that is nearly coincident with the initial orbit. Such variations will benefit the SR method.

Five different TERRA Level 1B multispectral data groups (within the Canberra area of Australia), captured on August 24 and September 2, 9, 11, 27, 2008, respectively, are selected as the test images. Changes of objects on the ground over time will be ignored, so data captured on the closest dates are best. However, the repeating cycle of TERRA is 2 days, and clouds appeared on other dates between August and September in 2008 within Canberra region, so that the selection of those five data groups is a compromise under the circumstances. These MODIS data from the Australian Centre for Remote Sensing (ACRES) in Australia, which includes the same scene in several of the same spectral ranges with different spatial resolutions, is particularly suitable for super resolution testing and classification validation. ACRES provides three resolutions for bands 1 and 2 (250 m, 500 m, and 1 km), and two resolutions for bands 3, 4, 5, 6, and 7 (500 m and 1 km) by post-processing methods. Classification and clustering results of the reconstructed super-resolved images using low-resolution data can be compared with that of the actual high-resolution data to give both visual and numerical comparisons.

Data from the ACT (Australian Capital Territory) area in Australia are extracted using the center longitude $149.20°$and latitude $-35.40°$, and two-dimensional correlations are applied to allow the source images to be coarsely aligned (only integer pixel shifts are applied). This procedure helps to extract the same area in the series of LR images without damaging any detail information. Blocks of size 64×64 pixels from the coarsest resolution (1 km) multispectral images of the first seven bands (620–670 nm, 841–876 nm, 459–479 nm, 545–565 nm, 1230–1250 nm, 1628–1652 nm, 2105–2155 nm) of the five data groups were extracted. To cover the same region, blocks of size 128×128 from the 500 m resolution seven bands of the five data groups are also extracted. In the same way, blocks of size 256×256 from the 250 m resolution two bands (620–670 nm and 841–876 nm) of the five data group are extracted, because there are only two bands available on that resolution level in MODIS. The blocks of 256×256 pixels and 128×128 pixels of Band 02 from resolution 250 m and 500 m can be seen in Figures 8.4(a) and (b), which were captured on September 2, 2008. This data group is regarded as the reference image set for the super resolution method, which is regarded as ground truth for cluster and classification experiments.

Two different tests with these blocks of the five MODIS data groups are conducted. In the first experiment, the spatial resolution of the 250-m Band 02 is regarded as the real high-resolution image, and the SR method MAP-uHMT is applied to 500 m resolution Band 01 and Band 02 of these five MODIS data groups, respectively. Then clustering performance is tested on super-resolved Band 01 and Band 02. The super-resolved SR images are based on an expansion factor of 2×2, the Gaussian blur kernel size 5×5 with standard deviation of 0.8, and 256×256 pixels. The balancing parameter $\alpha = 0.5$, and all intensities of the images, are normalized between $[0, 1]$. The Band 02 image of spatial resolution 500 m is regarded as the reference image for the image registration step in the super resolution method. But selecting Band 02 as an example, the bilinear interpolation expansion of

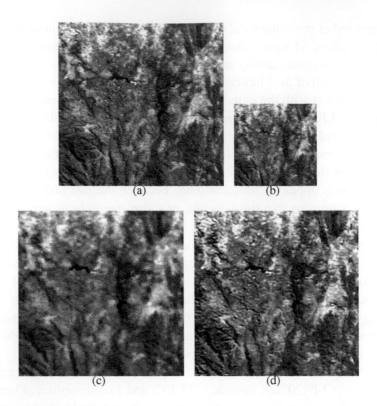

Figure 8.4 One Band 02 MODIS data and its relevant results within Canberra, Australia: (a) Band 02 with spatial resolution of 250 m (High Resolution, HR) captured on September 2, 2008; (b) Band 02 captured at the same time by MODIS but coarser spatial resolution of 500 m (Low Resolution, LR); (c) coarse resolution image of (b) enlarged 2×2 by bilinear interpolation of (b); (d) super resolution enlargement resolved by MAP-uHMT.

Figure 8.4(b) is shown in Figure 8.4(c). Figure 8.4(d) is the reconstructed image of spatial resolution of 250-m of Band 02 applying MAP-uHMT to the 500-m resolution images of all five MODIS data groups. Again, it can be seen that the super-resolved image includes more detail than Figure 8.4(c).

In a second experiment, the original seven bands multi-spectral images with spatial resolution of 500-m are regarded as the reference high-resolution data. Then classification performance is tested on these seven reconstructed high-resolution bands. The SR method MAP-uHMT is then applied to the relevant resolution 1-km seven bands of these five MODIS data groups band by band. The recovered SR images have the following parameters: an expansion factor 2, a Gaussian blur kernel size 5×5 with a standard deviation of 0.8, balancing parameter $\alpha = 0.5$, and all intensities of the images are normalized between $[0, 1]$. Because the LR 1-km resolution multispectral images are 64×64 pixels, the recovered seven bands SR images are 128×128 pixels with a resolution of 500-m. The Band 02 image of spatial resolution 1-km is regarded as the reference for the image registration step in the super resolution method.

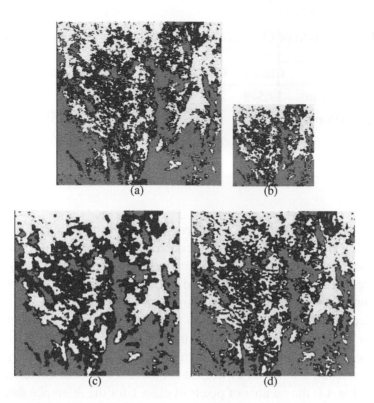

Figure 8.5 Unsupervised classification results of MODIS Band 01 and Band 02: (a) MODIS 250-m data clustering; (b) MODIS 500-m data clustering; (c) clustering of bilinear interpolated images; (d) clustering of MAP-uHMT super-resolved images.

8.3.1.1 SR for MODIS Data Clustering

Here, the following data sets are tested: (a) the actual SR set of 250-m resolution captured on September 2, 2008 (the clustering result using this data set is treated as "ground truth"), (b) the data set of 500-m resolution, (c) a 2×2 expansion by bilinear interpolation of the 500-m resolution set, and (d) a 2×2 super resolution expansion of the 500-m resolution set by MAP-uHMT. Thus, three of the multispectral images (a), (c), and (d) have a resolution of 250-m and a size of 256×256 pixels before classification, while (b) remains at 128×128 pixels.

The Iterative Self-Organizing Data Analysis (ISODATA) unsupervised method is adopted to classify two multispectral bands—Band 01 and Band 02; application on the other bands is straightforward. Again, research on unsupervised classification methods is beyond the scope of this chapter. ISODATA uses the minimum spectral distance formula to form clusters. It begins with arbitrary cluster means, and each time the clustering repeats, the means of these clusters are shifted. The new cluster means are used for the next iteration. It repeats the clustering of the multispectral data cube until a maximum percentage (98% is adopted here) of unchanged pixels has been reached between two iterations. All images are classified into three clusters, and the minimum cluster size is set to 30 pixels.

Table 8.1 Comparison of Classification Results in Number of Pixels and Accuracy

	Class 1	Class 2	Class 3	Overall accuracy	Kappa
Reference (250 m)	19,549	23,511	22,476		
Source (500 m)	15,017	15,911	18,188	74.95%	0.62
Bilinear	15,377	16,409	18,085	76.10%	0.64
MAP-uHMT	17,077	18,904	19,634	84.86%	0.77

The data group of spatial resolution 500-m captured on September 2, 2008, is adopted as the reference, and the unsupervised classification result is shown in Figure 8.5(b). Clustering results using the data sets (a) to (d) are given in Figure 8.5. Visually, the clustering result of Figure 8.5(d) provides more detailed information, similar to that of Figure 8.5(a), showing the superior performance provided by the SR data set.

Furthermore, unsupervised classification results of Figure 8.5 are compared in a numerical way as listed in Table 8.1. As far as the low-resolution multispectral data are concerned, because the size of the LR multispectral data is 128×128, after the classification, the classification result is enlarged by nearest neighbor for a better numerical comparison. As shown in Table 8.1, the number of pixels of class 1 for the reference data classification result is 19,549, while the correctly classified pixels in class 1 for LR multispectral data, bilinear data, and MAP-uHMT data are 15,017, 15,377 and 17,077, respectively. So, for class 1, unsupervised classification on super-resolved images by MAP-uHMT provides the closest approximation to the ground truth. For a better and easier comparison, the overall accuracy is also adopted. The overall accuracy of 84.86% is achieved for the unsupervised classification results on the super-resolved images by MAP-uHMT compared to 76.10% and 74.95% for those bilinear interpolated images and LR images, respectively. Moreover, the *Kappa* coefficient is improved to 0.77 from 0.62 (using the bilinear interpolated data) and 0.64 (using the LR data).

All MODIS data were downloaded from the Australian Centre for Remote Sensing (ACRES), courtesy of Geoscience Australia. Classification and clustering were performed by MultiSpec, courtesy of Purdue University.

8.3.1.2 SR for MODIS Data Classification

In this subsection the comparisons using seven bands (Band 1 to Band 7) of MODIS multispectral data are conducted. The following data sets are tested: (a) the original set of 500-m resolution captured on September 2, 2008 (the classification results using this data set are treated as "ground truth"), (b) the data set of 1-km resolution, (c) a 2×2 expansion by bilinear interpolation of the 1-km resolution set, and (d) a 2×2 super resolution expansion of the 1-km resolution set by MAP-uHMT. Thus, three of the multispectral images—(a), (c), and (d)—have a resolution of 500-m and a size of 128×128 pixels before classification, while (b) remains at 64×64 pixels.

A maximum likelihood supervised method [2] is used to classify the data into 6 classes:

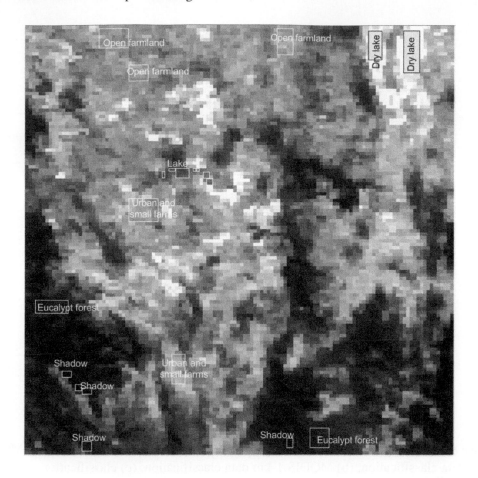

Figure 8.6 Training fields in data set (a) of MODIS.

(Dry lake, Open farmland, Eucalypt forest, Lake, Shadow, and Urban and small farms) from the 7 bands of these test images. The supervised classification method maximum likelihood is the most common method used with remote sensing data. It is based on statistics such as mean, variance/covariance, *etc.*. Probability functions or multivariate normal class models are calculated from the inputs for classes from the training fields. Then each pixel is judged as to the class to which it most probably belongs. In each multivariate normal class model, the method can work on spreads of data in some spectral directions. Because covariance data is adopted, maximum likelihood provides greater reliability and better accuracy in contrast with the minimum distance method, as long as it has enough training data.

Training fields are selected from data set (a) as shown in Figure 8.6, in which Band 03 is displayed as blue; Band 04 is displayed as green; and Band 01 is displayed as red. Because all the images are normalized between 0 to 1, it does not look like a true color image. These training fields are then transferred to other data sets. For each image classification, the statistics are updated based on the new training data from the corresponding training fields. The classification results for the four data sets are given in Figures 8.7(a) to (d), respectively. We can see that Figure 8.7(d) provides the closest classification result to (a) compared to the other results. In particular, the shape of Lake Burley Griffin in Canberra in

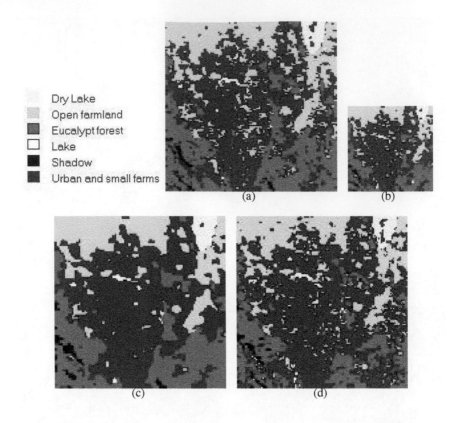

Figure 8.7 Supervised classification results of the first seven bands of MODIS. (a) MODIS 500m data classification; (b) MODIS 1-km data classification; (c) classification of bilinear interpolated images; (d) classification of MAP-uHMT super-resolved images.

the upper central areas (described as class "Lake" here) more closely matches the shape in the ground truth result than do the other classification results.

Furthermore, numerical comparisons also support the visual comparisons. The error matrices for the LR images, the bilinear interpolated images, and the reconstructed SR images by MAP-uHMT are listed in Tables 8.2, 8.3, and 8.4. Note that the size of the seven bands LR multispectral images are 64×64 pixels; before supervised classification, the classification result is enlarged by nearest neighbor for a better numerical comparison. From these tables we can also see that the overall accuracy of the classification for the super resolved images by MAP-uHMT provides the best performance, 69.23%, compared to the bilinear interpolated images, 57.65%, and the LR images, 51.64%. Similar results are achieved for the *Kappa* coefficient in evaluating the mapping accuracy, which can be seen in the last row in each table.

Based on the above tests, we can see that super-resolved multispectral MODIS data can achieve a better classification, both visually and numerically.

Table 8.2 Error Matrix on LR Seven Bands Multispectral Images

	Dry Lake	Open Farmland	Eucalypt Forest	Lake	Shadow	Urban and Small Farms	User's Accuracy
Dry lake	298	344	806	0	4	1,156	11.43%
Open farmland	11	2,735	86	0	0	1,064	70.20%
Eucalypt forest	0	18	1,674	2	54	216	85.23%
Lake	12	183	728	61	14	1,346	2.60%
Shadow	0	11	878	0	190	181	15.08%
Urban and small farms	40	630	122	16	1	3,503	81.24%
Producer's accuracy	82.55%	69.75%	38.98%	77.22%	72.24%	46.92%	
Overall accuracy			51.64%				
Kappa			0.61				

Table 8.3 Error Matrix on Interpolated Seven Bands Multispectral Images by Bilinear

	Dry Lake	Open Farmland	Eucalypt Forest	Lake	Shadow	Urban and Small Farms	User's Accuracy
Dry lake	301	323	896	0	2	1,215	11.00%
Open farmland	8	2,645	51	0	0	828	74.89%
Eucalypt forest	0	38	1,854	2	29	401	79.78%
Lake	4	36	277	58	2	556	6.22%
Shadow	0	5	1,018	0	227	106	16.74%
Urban and small farms	48	874	198	19	3	4,360	79.24%
Producer's accuracy	83.38%	67.46%	43.18%	73.42%	86.31%	58.40%	
Overall accuracy			57.65%				
Kappa			0.63				

Table 8.4 Error Matrix on Super-Resolved Seven Bands Multispectral Images by MAP-uHMT

	Dry Lake	Open Farmland	Eucalypt Forest	Lake	Shadow	Urban and Small Farms	User's Accuracy
Dry lake	269	5	0	0	0	23	90.57%
Open farmland	1	2,719	65	0	0	677	78.54%
Eucalypt forest	0	5	1,899	0	24	52	95.91%
Lake	2	13	800	68	43	503	4.76%
Shadow	0	0	603	1	189	12	23.48%
Urban and small farms	89	1,179	927	10	7	6,199	73.70%
Producer's accuracy	74.51%	69.35%	44.23%	86.08%	71.86%	83.03%	
Overall accuracy				69.23%			
Kappa				0.68			

8.3.2   Testing with ETM+ data

Next, this method will be tested again using the ETM+ data from LANDSAT7, which is is quite different from MODIS data. The ETM+ consists of eight spectral bands, with a spatial resolution of 30 m for Bands 1 to 5 and Band 7. Resolution for Band 6 (thermal infrared) is 60, m and resolution for Band 8 (panchromatic) is 15 m. The spatial resolution of the ETM+ data is much higher than that of the MODIS data; however, the spectral resolution of MODIS is better than that of ETM+ data. LANDSAT7 is a polar, sun-synchronous satellite, meaning it scans across the entire Earth's surface. The main instrument on board is ETM+. Although the two satellites (TERRA and LANDSAT7) orbit at an altitude that is almost the same, the revisit cycle is different. For LANDSAT7, the revisit cycle is 16 days compared with every 1 to 2 days for TERRA. This long revisit cycle of ETM+ data will bring more challenge for super resolution's application on both clustering and classification, because the ground object might change during that long time period.

Four different L1R (Level-1 data product after radiometric correction) LANDSAT7 multispectral data groups (within the Canberra region of Australia) are selected as the testing data. These four data groups were captured on September 13, 29 and December 2, 18, 2000, respectively. Changes in the objects on the ground over time will be ignored, so data captured on the closest dates are best. Although a single location on the Earth's surface can be imaged every 16 days, clouds appeared between October and November in 2000 within the Canberra region, so that the selection of those four images is a compromise under the circumstances.

All four data groups are extracted using the center longitude 149.03°and latitude −35.25°and two-dimensional correlations are applied to allow all multispectral images to be coarsely aligned between each data group. Bands 1 to 5 are tested with both unsupervised classification and supervised classification. The wavelength of Band 1 is 450–515 nm, Band 2 is 525–605 nm, Band 3 is 630–690 nm, Band 4 is 775–900 nm, and Band 5 is 1550–1750 nm. In order to evaluate cluster and classification benefits from super resolution and compare with the ground truth, all the original four data groups are first aggregated from 30-m resolution down to 60-m. Then Band 1 to 5 with higher spatial resolution (30 meters) images are reconstructed from these sub-sampled low resolution multispectral images using MAP-uHMT.

The data captured on September 13, 2000, is selected as the ground truth. Then Band 3 of this data group is regarded as the reference image for all three Band 3s in the other three data groups. The SR method MAP-uHMT is applied to the relevant resolution 60 m five bands of these four LANDSAT7 data groups band by band. The recovered SR images are under the parameter set: an expansion factor 2×2, the Gaussian blur kernel size 5×5 with standard deviation of 0.8, balancing parameter $\alpha = 0.4$, and all intensities of the images are normalized between $[0, 1]$.

The relative multi-spectral data cube can be seen in Figure 8.8. The original five bands of 30-m resolution multispectral images of size 256×256 pixels captured on September 13, 2000, are shown in Figure 8.8(a); the sub-sampled low-resolution five bands of 60-m resolution multispectral images of size 128×128 pixels captured on September 13, 2000, are shown in Figure 8.8(b); enlarged 2×2 by bilinear interpolation images from the sub-sampled data group are shown in Figure 8.8(c), and the super-resolved images by MAP-uHMT from all four low-resolution data groups are shown in Figure 8.8(d). The comparison for both unsupervised classification and supervised classification is conducted between these data cubes.

All LANDSAT7 ETM+ data were purchased from the Australian Centre for Remote Sensing (ACRES). Classification and clustering were performed by MultiSpec, courtesy of Purdue University.

8.3.2.1 SR for ETM+ Data Clustering

The ISODATA unsupervised method is adopted to cluster five multispectral bands: Band 1 to 5. As far as ISODATA is concerned, all images are clustered into three clusters, and the minimum cluster size is set to 30 pixels. The data group of spatial resolution 30 m captured on September 13, 2000, is adopted as the reference, and the classification result is shown in Figure 8.9(a). The classification of this scene is regarded as the ground truth. Figure 8.9(b) is the unsupervised classification of the multispectral data as shown in Figure 8.8(b). The unsupervised classification result of the bilinear interpolation enlarged multispectral data (Figure 8.8(c)) is shown in Figure 8.9(c). The classification result on the five bands, which are super resolution reconstructed by MAP-uHMT, is shown in Figure 8.9(d).

Furthermore, unsupervised classification results of Figure 8.9 are compared in a numerical way as listed in Table 8.5. As far as the LR multispectral data are concerned, because the size of the LR multispectral data is 128×128, after the classification, the classification result by nearest neighbor is enlarged by nearest neighbor interpolation for a better numerical comparison. As shown in Table 8.5, the number of pixels of class 2 for the real classification result is 40,946, while the number of correctly classified pixels in class 2 for LR multispectral data, bilinear data, and MAP-uHMT data is 35,634, 35,528 and 38,195, respectively. So, for class 2, unsupervised classification on super-resolved images by MAP-uHMT provides the closest approximation to the ground truth. However, the classification results of MAP-uHMT do not always provide good classification results; for example, the

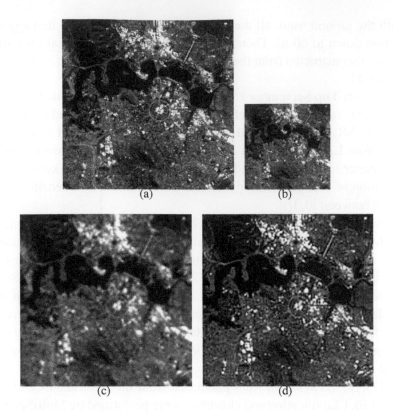

Figure 8.8 Five bands of ETM+ data displayed in color composite with Bands 1 (blue), 2 (green), 3 (red): (a) original high-resolution (HR) images; (b) sub-sampled LR images; (c) interpolated images; (d) MAP-uHMT super-resolved images.

number of pixels of the class 1 for real classification result is 10,912, while the number of correctly classified pixels in class 2 for LR multispectral data, bilinear data, and MAP-uHMT data is 8,913, 9,018 and 7,279, respectively. The reason is that, the spectral reflection signatures possibly changed during the period from September to December.

As shown in Table 8.5, the overall accuracy 85.99% is achieved for the unsupervised classification results on the super-resolved images compared to 84.47% and 84.70% for those bilinear interpolated images and LR images, respectively. Kappa coefficient of the classification results of MAP-uHMT is 0.73 which is slightly improved compared to the other classification results.

8.3.2.2   SR for ETM+ Data Classification

In this section, the comparisons using Five bands of ETM+ multispectral data are conducted. The following data sets are tested: (a) the original set of 30-m resolution captured on September 13, 2000 (the classification results of this data set are regarded as "ground truth" as shown in Figure 8.8(a)); (b) the data set of the sub-sampled data of September 13, 2000, as shown in Figure 8.8(b); (c) a 2×2 expansion by bilinear interpolation of the data set (b) as shown in Figure 8.8(c); and (d) a 2×2 super resolution expansion by the

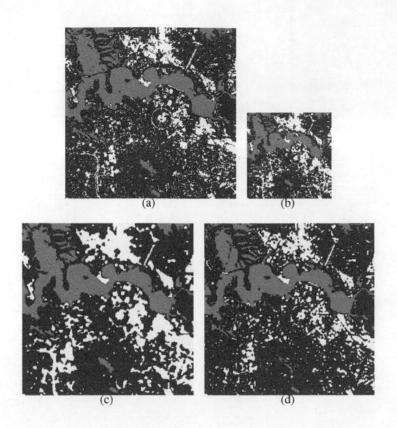

Figure 8.9 Unsupervised classification results of ETM+ Band 1 to 5: (a) clustering of original high-resolution data; (b) clustering of LR data; (c) clustering of bilinear interpolated data; (d) clustering of MAP-uHMT super-resolved data.

Table 8.5 Comparison of Unsupervised Classification Results of ETM+ Data in Number of Pixels and Accuracy

	Class 1	Class 2	Class 3	Overall accuracy	Kappa
Real	10912	40946	13678		
LR	8913	35634	10963	84.70%	0.72
Bilinear	9018	35528	10815	84.47%	0.71
MAP-uHMT	7279	38195	10882	85.99%	0.73

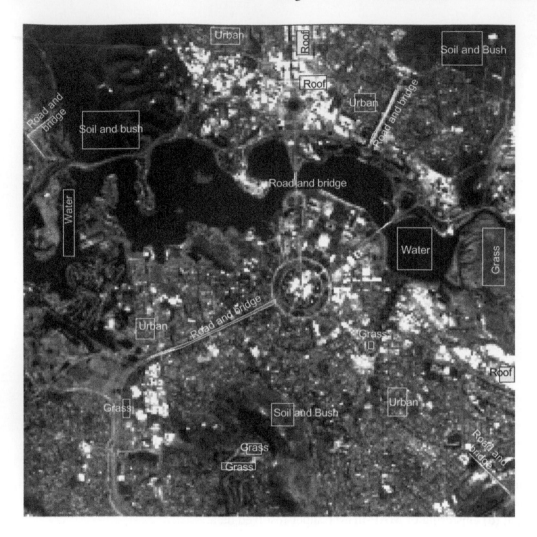

Figure 8.10 Training fields in data set (a) of ETM+.

MAP-uHMT method as shown in Figure 8.8(d). Thus, three of the multispectral images (a), (c), and (d) have a resolution of 30 m for all five bands and a size of 256×256 pixels in each band before classification, while (b) remains at 128×128 pixels in each band.

A maximum likelihood supervised method is also used to classify the data into six classes (water, soil and bush, grass, roof, road and bridge, urban) from the five bands of these test images. Training data are selected from data set (a) and then transferred to other data sets. The training fields of data set (a) is shown in Figure 8.10, where Band 1 is displayed as blue, Band 2 is displayed as green, and Band 3 is displayed as red. However, this is not a true color image, as the bands are normalized between 0 and 1. For each image classification, the statistics are updated based on the new training data from the corresponding training fields. The classification results for the four data sets are given in Figures 8.11(a) to (d), respectively. We can see that Figure 8.11(d) provides the closest classification result to (a) compared to the other results.

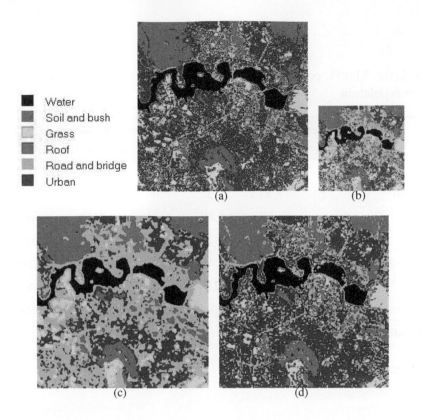

Water
Soil and bush
Grass
Roof
Road and bridge
Urban

(a)

(b)

(c)

(d)

Figure 8.11 Supervised classification results of the five bands of Landsat7: (a) classification results of the five bands of the original HR images; (b) classification results of the five bands of the LR images; (c) classification of the bilinear interpolated images; (d) classification of the MAP-uHMT super-resolved images.

Moreover, numerical comparisons also support the visual comparisons. Note that although the size of the five-band LR multispectral image is 128×128 pixels, before supervised classification, its classification result is enlarged by 2×2 using nearest neighbor for a better numerical comparison. Compared with the ground truth, all the error matrixes of data sets (b), (c), and (d) are listed in Tables 8.6, 8.7, and 8.8, respectively. The diagonal numbers in each error matrix are the correctly classified pixels. As far as the class of "Water" is concerned, the classification result of MAP-uHMT indicates that 5,392 pixels are correctly classified as "Water", which is much more accurate than with other methods. From these tables, we can also see that the overall accuracy of the classification for the super-resolved images by MAP-uHMT provides the best performance (71.24%) compared to the bilinear interpolated images (63.34%) and the LR images (66.06%). Similar results are archived for *Kappa* coefficient in evaluating the mapping accuracy, which can be seen in the last row in each table. *Kappa* coefficient is improved to 0.62 by the MAP-uHMT method compared to 0.53 for bilinear interpolation and 0.57 for nearest-neighbor interpolation.

Super-resolved multispectral bands can achieve improved classification results as shown in this section for LANDSAT7 ETM+ data. The selection of testing fields is critical for the combination of both super resolution and supervised classification in which we

Table 8.6 Error Matrix on Super-Resolved Five Bands of LANDSAT7 by Nearest-Neighbor Interpolation

	Water	Soil and Bush	Grass	Roof	Road and Bridge	Urban	User's Accuracy
Water	5,214	40	0	0	139	3	96.63%
Soil and bush	5	8,375	3	0	24	1,081	88.27%
Grass	0	18	4,897	1	100	880	83.06%
Roof	0	1	2	2,410	664	63	76.75%
Road and bridge	358	1,122	2,426	1,336	9,559	10,863	37.25%
Urban	0	1,000	213	20	1,880	12,839	80.49%
Producer's accuracy	93.49%	79.34%	64.94%	63.98%	77.30%	49.90%	
Overall accuracy				66.06%			
Kappa				0.57			

Table 8.7 Error Matrix on Super-Resolved Five Bands of LANDSAT7 by Bilinear Interpolation

	Water	Soil and Bush	Grass	Roof	Road and Bridge	Urban	User's Accuracy
Water	4,953	2	0	0	9	1	99.76%
Soil and bush	0	8,046	0	0	5	939	89.50%
Grass	0	20	4,867	0	59	727	85.79%
Roof	0	2	0	2,410	549	40	80.31%
Road and bridge	624	1,523	2,528	1,344	10,272	13,060	35.00%
Urban	0	963	146	13	1,472	10,962	80.87%
Producer's accuracy	88.81%	76.22%	64.54%	63.98%	83.07%	42.61%	
Overall accuracy				63.34%			
Kappa				0.54			

Table 8.8 Error Matrix on Super-Resolved Five Bands of LANDSAT7 by MAP-uHMT

	Water	Soil and Bush	Grass	Roof	Road and Bridge	Urban	User's Accuracy
Water	5,392	33	0	0	129	0	97.08%
Soil and bush	2	8,259	6	0	137	1,381	84.41%
Grass	0	14	4,348	1	94	643	85.26%
Roof	0	0	12	2,729	674	66	78.40%
Road and bridge	183	1,110	365	993	7,475	5,155	48.92%
Urban	0	1,140	2,810	44	3,857	18,484	70.19%
Producer's accuracy	96.68%	78.24%	57.66%	72.45%	60.45%	71.84%	
Overall accuracy				71.24%			
Kappa				0.62			

should bear in mind that the spectral reflection signatures probably changed during the long period for ETM+ data.

8.4 Conclusion

In this chapter we proposed our new SR method: MAP-uHMT. The application of super resolution (SR) techniques to multispectral image clustering and classification is investigated and tested using both MODIS from TERRA and ETM+ from LANDSAT7.

The basic concepts involved in MAP-uHMT are these: select one of a set of LR images as a reference image and set up a prior model using the uHMT method with the enlarged reference image in the wavelet domain; the SD optimization method is adopted to calculate the SR image for a particular prior model. The prior model is updated during each one of the outer iterations. Because the elastic image registration method [23] is used to deal with local warps in remote sensing images, and uHMT is adopted to characterize the prior model very well, a good SR remote sensing image can be reconstructed.

Generally speaking, classifications are carried out on multispectral bands captured at a particular time. However, in this chapter, by applying the SR method MAP-uHMT on several groups of multispectral bands captured on different dates over a study area, a set of multispectral images with better spatial resolution is obtained. Applying cluster and classification on these reconstructed high-resolution multispectral images, improved clustering and classification performance are achieved. The experimental results are demonstrated visually and quantitatively for the reconstructed high-resolution images of both MODIS data and ETM+ data.

In applying SR techniques to multispectral images, one critical step is the calculation of the warping matrix. One band captured on a certain date is selected as a representative in calculating the warping matrix, and the warping matrix is applied to all the other bands. The reason is that all bands are aligned originally.

This method will bring a new technique for classification improvement. Rather than developing higher-density sensors, this first departure from traditional approaches brings an alternative way to improve classification results with a given classification method. This chapter illustrated the possibility and feasibility of the use of super resolution reconstruction for finer scale classification of remote sensing data, which is encouraging as a means of breaking through current satellite detectors' resolution limits.

Bibliography

[1] F. Li, X. Jia, and D. Fraser, "Super resolution reconstruction of multispectral data for improved image classification," *IEEE Geoscience and Remote Sensing Letters*, vol. 6, no. 4, pp. 689–693, 2009.

[2] J. Richards and X. Jia, *Remote Sensing Digital Image Analysis*. Berlin: Springer-Verlag, 2006.

[3] J. Chan, J. Ma, and F. Canters, "A comparison of superresolution reconstruction methods for multi-angle CHRIS/Proba images," *Proceedings of the SPIE, Image and Signal Processing for Remote Sensing XIV*, vol. 7109, p. 710904, 2008.

[4] H. Shen, M. Ng, P. Li, and L.Zhang, "Super resolution reconstruction algorithm to MODIS remote sensing images," *The Computer Journal*, vol. 52, pp. 90–100, 2009.

[5] F. Laporterie-Dejean, G. Flouzat, and E. Lopez-Ornelas, "Multitemporal and multiresolution fusion of wide field of view and high spatial resolution images through morphological pyramid," in *Proceedings of SPIE*, vol. 5573, p. 52, 2004.

[6] T. Kasetkasem, M. Arora, and P. Varshney, "Super resolution land cover mapping using a Markov random field based approach," *Remote Sensing of Environment*, vol. 96, no. 3-4, pp. 302–314, 2005.

[7] A. Boucher and P. Kyriakidis, "Super resolution land cover mapping with indicator geostatistics," *Remote Sensing of Environment*, vol. 104, no. 3, pp. 264–282, 2006.

[8] P. Atkinson, "Super resolution target mapping from soft classified remotely sensed imagery," *Photogrammetric Engineering and Remote Sensing*, vol. 71, no. 7, pp. 839–846, 2005.

[9] J. Chan, J. Ma, P. Kempeneers, and F. Canters, "Superresolution enhancement of hyperspectral CHRIS/Proba images with a thin-plate spline nonrigid transform model," *IEEE Transactions on Geoscience and Remote Sensing*, vol. 48, no. 6, pp. 2569–2579, 2010.

[10] M. Crouse, R. Nowak, and R. Baraniuk, "Wavelet-based statistical signal processing using hidden Markov models," *IEEE Transactions on Signal Processing*, vol. 46, no. 4, pp. 886–902, 1998.

[11] J. Romberg, H. Choi, and R. Baraniuk, "Bayesian tree-structured image modeling using wavelet-domain hidden Markov models," *IEEE Transactions on Image Processing*, vol. 10, no. 7, pp. 1056–1068, 2001.

[12] M. Elad and A. Feuer, "Restoration of a single superresolution image from several blurred, noisy, and undersampled measured images," *IEEE Transactions on Image Processing*, vol. 6, no. 12, pp. 1646–1658, 1997.

[13] S. Park, M. Park, and M. Kang, "Super resolution image reconstruction: A technical overview," *IEEE Transaction on Signal Processing,*, vol. 20, no. 3, pp. 21–36, 2003.

[14] R. Schultz and R. Stevenson, "Extraction of high-resolution frames from video sequences," *IEEE Transactions on Image Processing*, vol. 5, pp. 996–1011, June 1996.

[15] M. Belge, M. Kilmer, and E. Miller, "Wavelet domain image restoration with adaptive edge-preserving regularization," *IEEE Transactions on Image Processing*, vol. 9, pp. 597–608, April 2000.

[16] M. Belge and E. Miller, "Wavelet domain image restoration using edge preserving prior models," *IEEE International Conference on Image Processing*, pp. 103–107, 1998.

[17] S. Zhao, H. Han, and S. Peng, "Wavelet-domain HMT-based image super resolution," *IEEE International Conference on Image Processing*, vol. 2, pp. 953–956, Sept. 2003.

[18] F. Li, X. Jia, and D. Fraser, "Universal HMT based super resolution for remote sensing images," *IEEE International Conference on Image Processing*, pp. 333–336, Oct. 2008.

[19] F. Li, D. Fraser, X. Jia, and A. Lambert, "Super resolution for remote sensing images based on a universal hidden Markov tree model," *IEEE Transactions on Geoscience and Remote Sensing*, vol. 48, no. 3, pp. 1270–1278, 2010.

[20] M. Irani and S. Peleg, "Improving resolution by image registration," *CVGIP: Graphical Models and Image Processing*, vol. 53, no. 3, pp. 231–239, 1991.

[21] M. Irani and S. Peleg, "Motion analysis for image enhancement: Resolution, occlusion, and transparency," *Journal of Visual Communication and Image Representation*, vol. 4, no. 4, pp. 324–335, 1993.

[22] A. Goshtasby, *2-D and 3-D Image Registration for Medical, Remote Sensing, and Industrial Applications*. New York: Wiley-Interscience, 2005.

[23] F. Li, D. Fraser, X. Jia, and A. Lambert, "Improved elastic image registration method for SR in remote sensing images," presented at the *Signal Recovery and Synthesis*, p. SMA5, 2007.

[24] T. Lillesand, R. Kiefer, and J. Chipman, *Remote Sensing and Image Interpretation*. New York, John Wiley & Sons Ltd. Chichester, UK, 2008.

Chapter 9

Color Image Restoration Using Vector Filtering Operators

RASTISLAV LUKAC
Foveon, Inc. / Sigma Corp.

9.1 Introduction

Over the past two decades, color image processing has become a daily necessity in practical life. Color is essential in human perception and provides important information for both human observers and data processing machines in various digital imaging, multimedia, computer vision, graphics, and biomedical applications. Unfortunately, images are often corrupted by noise, which can significantly degrade the value of the conveyed visual information, decrease the perceptual fidelity, and complicate various image processing and analysis tasks. Therefore, noise filtering — the process of rectifying signal disturbances in order to produce an image that corresponds as closely as possible to the output of an ideal imaging system — is needed.

This chapter focuses on vector operators suitable for restoring desired color information from the corresponding noisy measurements. More specifically, it discusses popular noise reduction filters that operate inside the localized image area based on the supporting window sliding over the entire image, and utilizes robust estimation, data-adaptive filtering, and color-correlation driven principles. Such design characteristics make these filters effective in removing image noise while preserving desired structural and color features.

To facilitate the discussions on color image filtering and restoration, Section 9.2 presents color fundamentals. This section starts with describing briefly color vision basics and a typical representation of a digital color image. Then, the attention shifts to the problem of image formation and noise modeling, which are crucial for simulating the noise observed in real-life imaging applications and studying the effect of the filters on both the noise and the desired image features. This section further discusses various measures for evaluating the distances and similarities among color vectors in order to perform color discrimination, which has an important role in color image filtering.

Section 9.3 presents several typical color conversions used in the field of color image filtering. Such conversions are often needed because different color spaces are more suitable for different applications or better correspond to human visual perception. Thus, to utilize color as a cue in image restoration, an appropriate representation of the color signal is needed, for example, in order to effectively discriminate between colors, distinguish noise from the desired signal, specify object boundaries, and manipulate the image data during processing.

Section 9.4 surveys popular noise filtering methods that benefits from the aforementioned multichannel principles of robust estimation and data-adaptive image processing. These methods are formally taxonomized into two main groups: those based on the order statistics and those that combine the input samples to determine the filter output. Example order-statistic methods presented in this chapter include vector median filters, vector directional filters, selection weighted vector filters, and switching vector filters. Example combination methods include fuzzy vector filters, digital path-based filters, anisotropic diffusion-based filters, and bilateral filters. Both filter classes are able to preserve important color and structural elements, and efficiently eliminate degradations that can be modeled as impulsive, Gaussian, and mixed type of noise.

Qualitative improvements have been a significant boosting factor in digital image processing research and development. Section 9.5 describes methods that are commonly used for color image quality evaluation and quantitative manipulation.

Finally, this chapter concludes with Section 9.6 by summarizing main color image filtering ideas.

9.2   Color Imaging Basics

Color is a psycho-physiological sensation [1] used by observers to sense the environment and understand its visual semantics. The human visual system is based on two types of photoreceptors localized on the retina of the eye: the rods sensitive to light and the cones sensitive to color [2]. The perception of color depends on the response of three types of cones, commonly called S, M, and L cones for their respective sensitivity to short, middle, and long wavelengths. Color is interpreted as a perceptual result of light interacting with the spectral sensitivities of the photoreceptors. Because three signals are generated based on the extent to which each type of cones is stimulated, any visible color can be numerically represented using three numbers called tristimulus values as a three-component vector within a three-dimensional coordinate system with color primaries lying on its axes [3]. The set of all such vectors constitutes a color space.

The well-known Red-Green-Blue (RGB) space was derived based on color matching experiments, aimed at finding a match between a color obtained through an additive mixture of color primaries and a color sensation [4]. Its standardized version [5] is used in most of today's image acquisition and display devices. It provides a reasonable resolution, range, depth, and stability of color reproduction while being efficiently implementable on various hardware platforms.

(a) (b)

Figure 9.1 Additive color mixing: (a) image obtained by shifting the three gray-scale images on the left, top, and bottom of the figure, which correspond to the red, green, and blue color channels, respectively; (b) original RGB image.

The RGB space is additive (Figure 9.1); any color can be obtained by combining the three primaries through their weighted contributions [4]. Equal contributions of all three primaries give a shadow of gray. The two extremes of gray are black and white, which correspond, respectively, to no contribution and the maximum contributions of the primaries. When one primary contributes greatly, the resulting color is a shade of that dominant primary. Any pure secondary color is formed by maximum contributions of two primary colors: cyan is obtained using green and blue, magenta using red and blue, and yellow using red and green. When two primaries contribute greatly, the result is a shade of the corresponding secondary color.

9.2.1 Numeral Representation

An RGB color image $\mathbf{x}$ with K_1 rows and K_2 columns represents a two-dimensional matrix of three-component samples $\mathbf{x}_{(r,s)} = [x_{(r,s)1}, x_{(r,s)2}, x_{(r,s)3}]$ occupying the spatial location (r, s), with $r = 1, 2, ..., K_1$ denoting the image row and $s = 1, 2, ..., K_2$ denoting the image column. In the color vector $\mathbf{x}_{(r,s)}$, the terms $x_{(r,s)1}$, $x_{(r,s)2}$, and $x_{(r,s)3}$ denote the R, G, and B components, respectively. In standard eight-bit RGB representation, color components can range from 0 to 255. The large value of $x_{(r,s)k}$, for $k = 1, 2, 3$, denotes high contribution of the kth primary in the color vector $\mathbf{x}_{(r,s)}$. The process of displaying an image creates a graphical representation of the image matrix where the pixel values represent particular colors (Figure 9.1).

Each color vector $\mathbf{x}_{(r,s)}$ is uniquely defined by its magnitude

$$\sqrt{x_{(r,s)1}^2 + x_{(r,s)2}^2 + x_{(r,s)3}^2} \tag{9.1}$$

and direction

$$\mathbf{x}_{(r,s)} / \sqrt{x_{(r,s)1}^2 + x_{(r,s)2}^2 + x_{(r,s)3}^2}, \tag{9.2}$$

which indirectly indicate luminance and chrominance properties of RGB colors, and thus are important for human perception [6]. Note that the magnitude represents a scalar value,

(a)   (b)

Figure 9.2 Luminance representation of the image shown in Figure 9.1(b) using (a) Equation (9.1) and (b) Equation (9.3).

whereas the direction, as defined above, is a vector. Because the components of this vector are normalized, such vectors form the unit sphere in the vector space.

In practice, the luminance value $L_{(r,s)}$ of the color vector $\mathbf{x}_{(r,s)}$ is usually obtained as follows:

$$L_{(r,s)} = 0.299x_{(r,s)1} + 0.587x_{(r,s)2} + 0.114x_{(r,s)3}. \tag{9.3}$$

The weights assigned to individual color channels reflects the perceptual contributions of each color band to the luminance response of the human visual system [2]. Figure 9.2 shows the two luminance versions of the input image.

The chrominance properties of color pixels are often expressed as the point on the triangle that intersects the RGB color primaries in their maximum value. The coordinates of this point are obtained as follows [7]:

$$\frac{x_{(r,s)k}}{x_{(r,s)1} + x_{(r,s)2} + x_{(r,s)3}}, \quad \text{for } k = 1, 2, 3, \tag{9.4}$$

and their sum is equal to unity. Both above vector formulations of the chrominance represent the parametrization of the chromaticity space, where each chrominance line is entirely determined by its intersection point with the chromaticity sphere or triangle in the vector space. Figure 9.3 shows the two chrominance versions of the input image.

9.2.2   Image Formation

In practice, image acquisition is usually affected by the aliasing, blurring, and noise processes, which cause information loss [8]. Aliasing effects result from sampling the visual information by the image sensor. Blurring effects may have various origins, including atmospheric; optical due to out-of-focus, diffraction, and lens aberrations; sensor due to spatial averaging built in the imager; and motion caused by the long exposure time relative to the motion in the scene. Noise present in digital images can also have various origins, as discussed in detail below. In addition to the acquisition process, images can also degrade during their transmission and/or due to faulty storage. Any of these undesired effects can significantly degrade the value of captured visual information, decrease the perceptual fidelity of an image, and complicate many image processing and analysis tasks. Therefore,

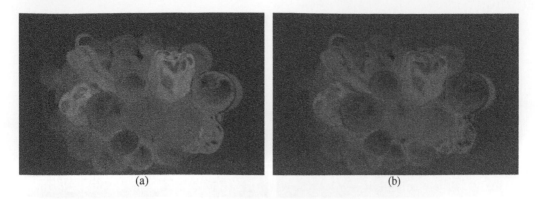

(a) (b)

Figure 9.3 Chrominance representation of the image shown in Figure 9.1(b) using (a) Equation (9.2) and (b) Equation (9.4).

the goal of image restoration is to digitally remove or reduce the effects of degradation and produce an image with desired characteristics.

Various mathematical models are used to approximate complex relations among sources of image corruption processes. For example, spatial sampling and blurring effects are usually combined and implemented through convolution, whereas image noise is typically modeled as an additive term. Such an imaging model can be expressed as follows:

$$\mathbf{x} = \mathbf{H} * \mathbf{o} + \mathbf{v}, \tag{9.5}$$

where $\mathbf{x}$ denotes the corrupted image, $\mathbf{H}$ approximates blurring and aliasing effects, $*$ is the convolution operator, $\mathbf{o}$ denotes the original image, and $\mathbf{v}$ denotes the noise term.

9.2.3 Noise Modeling

Image sensors are the usual source of noise in digital imaging [9, 10]. Noise is caused by random processes associated with quantum signal detection, signal independent fluctuations, and inhomogeneity of the responsiveness of the sensor elements. Noise increases with the temperature and sensitivity setting of the camera, as well as the reduced length of the exposure. It can vary within an individual image; darker regions usually suffer more from noise than brighter regions. The level of noise also depends on characteristics of the camera electronics and the physical size of photosites in the sensor; larger photosites usually have better light-gathering abilities, thus producing a stronger signal and higher signal-to-noise ratio. Noise can appear as random speckles in otherwise smooth regions, altering both tone and color of the original pixels. Figure 9.4 shows examples of noise present in various real-life images.

Noise affects both the magnitude and direction of original color vectors, thus significantly influencing color perception [6]. Characterizing the noise observed in digital images using various faithful approximations (Figure 9.5) is crucial in filter design, as it allows for evaluating and further optimizing the filtering performance both objectively and subjectively. Focusing just on the noise processes, which is the goal of this chapter, the above imaging model can be simplified on the pixel level as follows [6, 11, 12]:

$$\mathbf{x}_{(r,s)} = \mathbf{o}_{(r,s)} + \mathbf{v}_{(r,s)}, \tag{9.6}$$

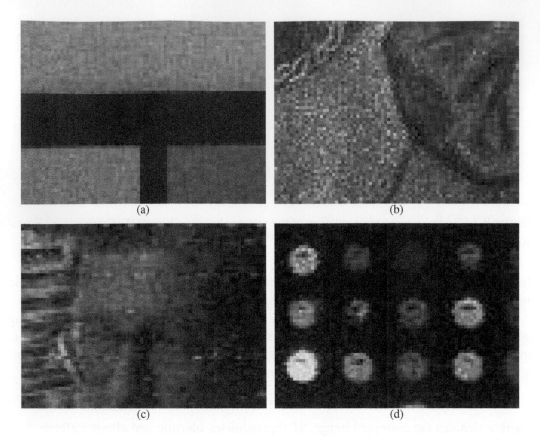

Figure 9.4 Noise present in real-life images: (a) digital camera image; (b) digitized artwork image; (c) television image; (d) microarray image.

where (r, s) characterizes the spatial position of the samples in the image, $\mathbf{x}_{(r,s)} = [x_{(r,s)1}, x_{(r,s)2}, x_{(r,s)3}]$ represents the observation (noisy) sample, $\mathbf{o}_{(r,s)} = [o_{(r,s)1}, o_{(r,s)2}, o_{(r,s)3}]$ is the desired (noise-free) sample, and $\mathbf{v}_{(r,s)} = [v_{(r,s)1}, v_{(r,s)2}, v_{(r,s)3}]$ is the vector describing the noise process. Note that $\mathbf{v}_{(r,s)}$ can be used to describe both signal-dependent and signal-independent additive noise.

Noise can often be approximated using the Gaussian noise model. Such approximations are useful, for example, when the noise is caused by thermal degeneration of materials in image sensors. Noise can also appear as abrupt local changes in the image, caused, for example, by malfunctioning of the sensor elements, electronic interference, and flaws in the data transmission process. Such noise characteristics can be modeled as impulsive sequences that occur in the form of short-time duration, high-energy spikes attaining large amplitudes with probability higher than predicted by a Gaussian density model [13, 14].

9.2.3.1 Additive Gaussian Noise

Assuming an additive Gaussian noise model with zero mean that affects each color component and spatial image pixel position independently and that the noise variance σ is the same for all three color components in the RGB color space representation, the

Figure 9.5 Color noise modeling: (a) a noise-free gray-field image used for better visualization of the noise effect; (b) additive Gaussian noise; (c) impulsive noise; (d) mixed noise comprised of additive Gaussian noise followed by impulsive noise.

noise corruption process can be reduced to a scalar perturbation. Namely, considering the mathematical model in Equation (9.6), the distribution of the noise vector magnitudes $\|\mathbf{v}\| = (v_{(r,s)1}^2 + v_{(r,s)2}^2 + v_{(r,s)3}^2)^{1/2}$ can be expressed as follows [6, 15]:

$$Pr(\|\mathbf{v}\|) = \left(\frac{1}{\sqrt{2\pi\sigma^2}}\right)^3 4\pi\|\mathbf{v}\|^2 \exp\left(-\frac{\|\mathbf{v}\|^2}{2\sigma^2}\right) \tag{9.7}$$

The magnitude perturbation can be mapped to an angular perturbation A described by

$$Pr(A) \approx A\left(\frac{\|\mathbf{o}\|^2}{\sigma^2}\right) \exp\left(-\frac{\|\mathbf{o}\|^2 A^2}{2\sigma^2}\right) \tag{9.8}$$

with the mean angular perturbation $\overline{A}$ defined as

$$\overline{A} \approx \sqrt{\frac{\sigma^2\pi}{2\|\mathbf{o}\|^2}}, \tag{9.9}$$

where $\|\mathbf{o}\| = (o_{(r,s)1}^2 + o_{(r,s)2}^2 + o_{(r,s)3}^2)^{1/2}$ denotes the magnitude of the noise-free color vector. The above derivation is in line with the observation that the perturbation due to the noise results in the creation of a cone in the RGB space.

9.2.3.2 Impulsive Noise

In the case of impulsive noise described using the additive model in Equation (9.6), the image degradation process can be formulated as follows [16]:

$$\mathbf{v}_{(r,s)} = \begin{cases} \mathbf{v}_{(r,s)} & \text{with probability } p_\mathbf{v} \\ 0 & \text{with probability } 1 - p_\mathbf{v}, \end{cases} \qquad (9.10)$$

where $p_\mathbf{v}$ is a corruption probability. The noise contribution is usually assumed to be independent from pixel to pixel, resulting in generally much larger or smaller amplitude of the corrupted pixel compared to that of neighboring pixels, at least in one of the color components. Considering the three-dimensional nature of the color signal, the noise vector can be expressed as follows [17]:

$$\mathbf{v}_{(r,s)} = \begin{cases} [v_{(r,s)1}, 0, 0] & \text{with probability } p_\mathbf{v} p_{\mathbf{v}_1} \\ [0, v_{(r,s)2}, 0] & \text{with probability } p_\mathbf{v} p_{\mathbf{v}_2} \\ [0, 0, v_{(r,s)3}] & \text{with probability } p_\mathbf{v} p_{\mathbf{v}_3} \\ [v_{(r,s)1}, v_{(r,s)2}, 0] & \text{with probability } p_\mathbf{v} p_{\mathbf{v}_1} p_{\mathbf{v}_2} \\ [v_{(r,s)1}, 0, v_{(r,s)3}] & \text{with probability } p_\mathbf{v} p_{\mathbf{v}_1} p_{\mathbf{v}_3} \\ [0, v_{(r,s)2}, v_{(r,s)3}] & \text{with probability } p_\mathbf{v} p_{\mathbf{v}_2} p_{\mathbf{v}_3} \\ [v_{(r,s)1}, v_{(r,s)2}, v_{(r,s)3}] & \text{with probability } p_\mathbf{v} p_{\mathbf{v}_1} p_{\mathbf{v}_2} p_{\mathbf{v}_3} \\ [0, 0, 0] & \text{with probability } 1 - p_\mathbf{v}, \end{cases} \qquad (9.11)$$

where $p_\mathbf{v}$ is the probability that the original color image vector $\mathbf{o}_{(r,s)}$ is being corrupted by noise. The term $p_{\mathbf{v}_k}$ denotes the probability of corruption of a particular color component of the vector $\mathbf{o}_{(r,s)}$. Note that the probability values are constrained as follows:

$$p_{\mathbf{v}_1} + p_{\mathbf{v}_2} + p_{\mathbf{v}_3} + p_{\mathbf{v}_1} p_{\mathbf{v}_2} + p_{\mathbf{v}_1} p_{\mathbf{v}_3} + p_{\mathbf{v}_2} p_{\mathbf{v}_3} + p_{\mathbf{v}_1} p_{\mathbf{v}_2} p_{\mathbf{v}_3} = 1 \qquad (9.12)$$

9.2.3.3 Mixed Noise

In practice, digital images suffer from both Gaussian and impulsive noise contributions. Such a corruption process is typically modeled using the mixed noise model [18]. Considering the additive model in Equation (9.6), the mixed noise can be defined as follows:

$$\mathbf{v}_{(r,s)} = \begin{cases} \mathbf{v}_{(r,s)}^A + \mathbf{v}_{(r,s)}^I & \text{with probability } p_\mathbf{v} \\ \mathbf{v}_{(r,s)}^A & \text{with probability } 1 - p_\mathbf{v}, \end{cases} \qquad (9.13)$$

where $\mathbf{v}_{(r,s)}^A$ denotes the Gaussian noise and $\mathbf{v}_{(r,s)}^I$ denotes the impulsive noise.

9.2.4 Distance and Similarity Measures

As discussed above, color pixels affected by noise deviate from their neighbors. Noise can be seen as fluctuations in intensity and color, and therefore the evaluation of magnitude and directional differences between vectors in a local image area constitutes the basis in numerous filtering techniques.

The difference or similarity between two color vectors $\mathbf{x}_{(i,j)} = [x_{(i,j)1}, x_{(i,j)2}, x_{(i,j)3}]$ and $\mathbf{x}_{(g,h)} = [x_{(g,h)1}, x_{(g,h)2}, x_{(g,h)3}]$ can be quantified using various distance and similarity measures. A common method is the Minkowski metric [6, 19]:

$$d(\mathbf{x}_{(i,j)}, \mathbf{x}_{(g,h)}) = \|\mathbf{x}_{(i,j)} - \mathbf{x}_{(g,h)}\|_L = c \left(\sum_{k=1}^{3} \xi_k |x_{(i,j)k} - x_{(g,h)k}|^L \right)^{1/L}, \qquad (9.14)$$

where L defines the nature of the metric, c is the nonnegative scaling parameter, and ξ_k, for $\sum_k \xi_k = 1$, is the weighting coefficient denoting the proportion of attention allocated to the kth component. This metric generalizes the Euclidean distance

$$\|\mathbf{x}_{(i,j)} - \mathbf{x}_{(g,h)}\|_2 = \sqrt{\sum_{k=1}^{3} \left(x_{(i,j)k} - x_{(g,h)k} \right)^2}, \qquad (9.15)$$

the city-block distance

$$\|\mathbf{x}_{(i,j)} - \mathbf{x}_{(g,h)}\|_1 = \sum_{k=1}^{3} |x_{(i,j)k} - x_{(g,h)k}|, \qquad (9.16)$$

and the chess-board distance defined as $\max\{|x_{(i,j)k} - x_{(g,h)k}|; \text{ for } 1 \leq k \leq 3\}$.

Another suitable measure is the normalized inner product [6, 20]:

$$s(\mathbf{x}_{(i,j)}, \mathbf{x}_{(g,h)}) = \frac{\mathbf{x}_{(i,j)} \mathbf{x}_{(g,h)}^T}{|\mathbf{x}_{(i,j)}||\mathbf{x}_{(g,h)}|}, \qquad (9.17)$$

which produces a large value when its inputs are similar and converges to zero if its inputs are dissimilar. Because this similarity measure corresponds to the cosine of the angle between $\mathbf{x}_{(i,j)}$ and $\mathbf{x}_{(g,h)}$, the difference in orientation of the two vectors can be correspondingly quantified using the angular distance as follows:

$$A(\mathbf{x}_{(i,j)}, \mathbf{x}_{(g,h)}) = \arccos \left(\frac{\mathbf{x}_{(i,j)} \mathbf{x}_{(g,h)}^T}{|\mathbf{x}_{(i,j)}||\mathbf{x}_{(g,h)}|} \right). \qquad (9.18)$$

The Minkowski metric-driven measures quantify differences between the two vectors based on their magnitude, as opposed to the measures based on the normalized inner product, which refers to the vectors' orientation. Because both magnitude and directional characteristics of the color vectors are important for human perception, combined similarity measures, such as

$$s(\mathbf{x}_{(i,j)}, \mathbf{x}_{(g,h)}) = w_1 \left(\frac{\mathbf{x}_{(i,j)} \mathbf{x}_{(g,h)}^T}{|\mathbf{x}_{(i,j)}||\mathbf{x}_{(g,h)}|} \right) w_2 \left(1 - \frac{||\mathbf{x}_{(i,j)}| - |\mathbf{x}_{(g,h)}||}{\max\left(|\mathbf{x}_{(i,j)}|, |\mathbf{x}_{(g,h)}|\right)} \right) \qquad (9.19)$$

can provide a robust solution to the problem of similarity quantification between two color vectors. The terms w_1 and w_2 denote tunable weights used to control the contribution of magnitude and directional differences. Other examples of content-based similarity measures can be found in [17] and [20].

A different concept of similarity measures is obtained when the difference between the two color vectors is modulated by a nonlinear function, such as [21]:

$$\mu(\mathbf{x}_{(i,j)}, \mathbf{x}_{(g,h)}) = \exp\left\{-\left(\|\mathbf{x}_{(i,j)} - \mathbf{x}_{(g,h)}\|/h\right)^2\right\} \tag{9.20}$$

$$\mu(\mathbf{x}_{(i,j)}, \mathbf{x}_{(g,h)}) = \exp\left\{-\|\mathbf{x}_{(i,j)} - \mathbf{x}_{(g,h)}\|/h\right\} \tag{9.21}$$

$$\mu(\mathbf{x}_{(i,j)}, \mathbf{x}_{(g,h)}) = \frac{1}{1 + \|\mathbf{x}_{(i,j)} - \mathbf{x}_{(g,h)}\|/h}, \ h \in (0;\infty) \tag{9.22}$$

$$\mu(\mathbf{x}_{(i,j)}, \mathbf{x}_{(g,h)}) = \frac{1}{(1 + \|\mathbf{x}_{(i,j)} - \mathbf{x}_{(g,h)}\|)^h} \tag{9.23}$$

$$\mu(\mathbf{x}_{(i,j)}, \mathbf{x}_{(g,h)}) = 1 - \frac{2}{\pi} \arctan\left(\|\mathbf{x}_{(i,j)} - \mathbf{x}_{(g,h)}\|/h\right) \tag{9.24}$$

$$\mu(\mathbf{x}_{(i,j)}, \mathbf{x}_{(g,h)}) = \frac{2}{1 + \exp\left\{\|\mathbf{x}_{(i,j)} - \mathbf{x}_{(g,h)}\|/h\right\}}, \ h \in (0;\infty) \tag{9.25}$$

$$\mu(\mathbf{x}_{(i,j)}, \mathbf{x}_{(g,h)}) = \frac{1}{1 + \|\mathbf{x}_{(i,j)} - \mathbf{x}_{(g,h)}\|^h}, \tag{9.26}$$

where h is a design parameter and $\mu(\mathbf{x}_{(i,j)}, \mathbf{x}_{(g,h)}) = \mu(\|\mathbf{x}_{(i,j)} - \mathbf{x}_{(g,h)}\|)$ denotes the similarity function, which is nonascending and convex in $[0;\infty)$. Note that $\mu(0) = 1$ corresponds to two identical vectors, whereas $\mu(\infty) = 0$ indicates that the similarity between the two very different color vectors should be zero or very close to zero.

9.3 Color Space Conversions

Depending on the nature of a restoration method, an image can be converted from its native color space to some other space that is more suitable for completing a given task. Such conversions are typically motivated by various design, performance, implementation, and application aspects of a given color image processing task. The following discusses several typical color representations.

9.3.1 Standardized Representations

The Commission Internationale de l'Éclairage (CIE) introduced the standardized CIE-RGB and CIE-XYZ color spaces. A relationship between these two spaces can be expressed as follows [3, 22]:

$$\begin{bmatrix} X \\ Y \\ Z \end{bmatrix} = \begin{bmatrix} 0.49 & 0.31 & 0.20 \\ 0.17697 & 0.81240 & 0.01063 \\ 0 & 0.01 & 0.99 \end{bmatrix} \begin{bmatrix} R \\ G \\ B \end{bmatrix}, \tag{9.27}$$

where the Y component corresponds to the luminance, whereas X and Z do not correspond to any perceptual attributes. The XYZ color space is device independent and thus very useful in situations where consistent color representation across devices with different characteristics is required. Although the CIE-XYZ color space is rarely used in image processing applications, other color spaces can be derived from it through mathematical transforms.

The RGB space models the output of physical devices, and therefore it is considered device dependent. To consistently detect or reproduce the same RGB color vector, some form of color management is usually required. This relates to the specification of the white point, gamma correction curve, dynamic range, and viewing conditions [23]. Given an XYZ color vector with components ranging from zero to one, and the reference white being the same as that of the RGB system, the conversion to sRGB values starts as follows:

$$
\begin{bmatrix} R \\ G \\ B \end{bmatrix} = \begin{bmatrix} 3.2410 & -1.5374 & -0.4986 \\ -0.9692 & 1.8760 & 0.0416 \\ 0.0556 & -0.2040 & 1.0570 \end{bmatrix} \begin{bmatrix} X \\ Y \\ Z \end{bmatrix}. \tag{9.28}
$$

To avoid values outside the nominal range that are usually not supported in RGB encoding, both negative values and values exceeding one are clipped to zero and one, respectively. This step is followed by gamma correction:

$$
f(\tau) = \begin{cases} 1.055\tau^{1/2.4} - 0.055 & \text{if } \tau > 0.00304 \\ 12.92\tau & \text{otherwise,} \end{cases} \tag{9.29}
$$

where τ denotes the uncorrected color component. Finally, gamma-corrected components $f(R)$, $f(G)$, and $f(B)$ are multiplied by 255 to obtain their corresponding values in standard eight-bits-per-channel encoding.

9.3.2 Luminance–Chrominance Representations

It is well-known that the human visual system is most sensitive to green and less sensitive to red and blue light. By exploiting further the characteristics of human perception, various luminance–chrominance representations of image signals have been introduced and successfully used in many applications, such as television broadcasting, image and video encoding, and digital camera imaging. Luminance refers to the perceived brightness, whereas chrominance is specified by hue, which characterizes the color tone, and saturation, which denotes the color pureness. Luminance–chrominance representations are analogous to the receptive field encoding at the ganglion cells in the human retina [24]. Because the human visual system is less sensitive to high frequencies in chrominance than in luminance [24, 25], a coarser processing can often be applied to chrominance compared to that of luminance without observing any visual degradation in the output image.

The YUV color space, used in several composite color television systems, is a popular luminance–chrominance representation (Figure 9.6). The conversion formula from RGB to YUV values is given by

$$
\begin{bmatrix} Y \\ U \\ V \end{bmatrix} = \begin{bmatrix} 0.299 & 0.587 & 0.114 \\ -0.147 & -0.289 & 0.436 \\ 0.615 & -0.515 & -0.100 \end{bmatrix} \begin{bmatrix} R \\ G \\ B \end{bmatrix}. \tag{9.30}
$$

Alternatively, U and V can be expressed using Y as $U = 0.492(B - Y)$ and $V = 0.877(R - Y)$.

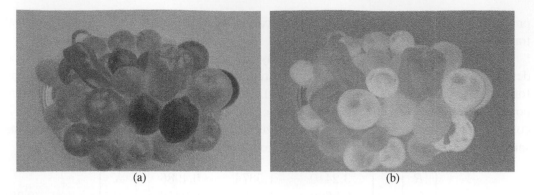

Figure 9.6 YUV representation of the image shown in Figure 9.1(b): (a) U channel and (b) V channel. The Y channel is shown in Figure 9.2(b).

The YC_bC_r color space, employed in digital image and video coding, denotes another luminance–chrominance representation. A luminance component Y and two chrominance components C_b and C_r can be obtained from an RGB triplet as follows:

$$\begin{bmatrix} Y \\ C_b \\ C_r \end{bmatrix} = \begin{bmatrix} 0.299 & 0.587 & 0.114 \\ -0.169 & -0.331 & 0.500 \\ 0.500 & -0.419 & -0.081 \end{bmatrix} \begin{bmatrix} R \\ G \\ B \end{bmatrix}. \tag{9.31}$$

The chrominance components C_b and C_r can be equivalently expressed using Y as $C_b = 0.564(B - Y)$ and $C_r = 0.713(R - Y)$.

As can be seen from the above definitions, both U and C_b represent a scaled difference between the blue signal and the luminance signal, thus indicating the extent to which the color deviates from gray toward blue. Similarly, V and C_r represent a scaled difference between the red signal and the luminance signal, thus indicating the extent to which the color deviates from gray toward red. Using the luminance–chrominance representation of a color image can be very beneficial in color image denoising, as such an approach allows for handling noise separately in the luminance and chrominance domain (Figure 9.7).

9.3.3 Cylindrical Representations

Cylindrical color spaces [20, 26], such as the hue-saturation-value (HSV), hue-saturation-lightness (HSL), hue-saturation-intensity (HSI), and hue-chroma-intensity (HCI) spaces, aim at rearranging the geometry of the RGB color space in an attempt to be more intuitive and perceptually relevant. The rationale behind cylindrical-coordinate representations of color is explained in the example of the HSV color space. The conversion formula from

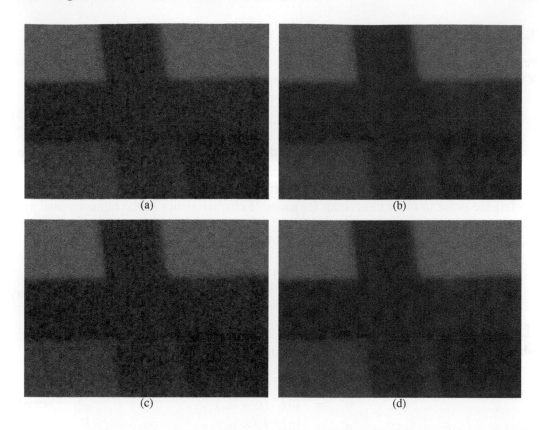

Figure 9.7 Noise reduction in a color image: (a) captured noisy image; (b) luminance noise suppression; (c) chrominance noise suppression; (d) both luminance and chrominance noise suppression.

RGB to HSV values is defined as follows [26]:

$$
\begin{aligned}
H &= \left\{ \begin{array}{ll} \eta & \text{if } B \leq G \\ 360 - \eta & \text{otherwise} \end{array} \right. \; ; \; \eta = \arccos\left(\frac{(2R - G - B)/2}{\sqrt{(R - G)^2 + (R - B)(G - B)}} \right) \\
S &= \frac{\max(R, G, B) - \min(R, G, B)}{\max(R, G, B)} \\
V &= \frac{\max(R, G, B)}{255},
\end{aligned}
\tag{9.32}
$$

where H is undefined and S is zero whenever the input RGB vector is a pure shade of gray, which happens for $R = G = B$.

The hue component H gives an indication of the spectral composition of a color. It is measured as the angle, ranging from 0 to 360 degrees, with red corresponding to 0 degrees. The saturation component S ranges from zero to one. It indicates how far a color is from a gray by referring to the proportion of pure light of the dominant wavelength. The value component V defines the relative brightness of the color. It ranges from zero to one, which corresponds to black and white, respectively.

9.3.4 Perceptual Representations

Perceptually uniform representations CIE u, v, CIELuv, and CIELab, are derived from XYZ values. In the case of CIE u and v values, the conversion formulas

$$u = \frac{4X}{X + 15Y + 3Z}, \quad v = \frac{9Y}{X + 15Y + 3Z} \tag{9.33}$$

can be used to form a chromaticity diagram that corresponds reasonably well to the characteristics of human visual perception [28].

Converting the data from CIE-XYZ to CIELuv and CIELab color spaces [22] requires a reference point in order to account for adaptive characteristics of the human visual system [4]. Denoting this point as $[X_n, Y_n, Z_n]$ of the reference white under the reference illumination, the CIELab values are calculated as

$$
\begin{aligned}
L^* &= 116 f(Y/Y_n) - 16 \\
a^* &= 500 \left(f(X/X_n) - f(Y/Y_n) \right) \\
b^* &= 200 \left(f(Y/Y_n) - f(Z/Z_n) \right),
\end{aligned}
\tag{9.34}
$$

whereas the CIELuv values are obtained as follows:

$$
\begin{aligned}
L^* &= 116 f(Y/Y_n) - 16 \\
u^* &= 13 L^* \left(u - u_n \right) \\
v^* &= 13 L^* \left(v - v_n \right),
\end{aligned}
\tag{9.35}
$$

where $u_n = 4X_n/(X_n + 15Y_n + 3Z_n)$ and $v_n = 9Y_n/(X_n + 15Y_n + 3Z_n)$ correspond to the reference point $[X_n, Y_n, Z_n]$. The terms u and v, calculated using Equation (9.33), correspond to the color vector $[X, Y, Z]$ under consideration. Because the human visual system exhibits different characteristics in normal illumination and low light levels, $f(\cdot)$ is defined as follows [22]:

$$
f(\gamma) = \begin{cases} \gamma^{1/3} & \text{if } \gamma > 0.008856 \\ 7.787\gamma + 16/116 & \text{otherwise.} \end{cases}
\tag{9.36}
$$

The component L^*, ranging from zero (black) to 100 (white), represents the lightness of a color vector. All other components describe color; neutral or near-neutral colors correspond to zero or close to zero values of u^* and v^*, or a^* and b^*. Following the characteristics of opponent color spaces [29], u^* and a^* coordinates represent the difference between red and green, whereas v^* and b^* coordinates represent the difference between yellow and blue.

9.4 Color Image Filtering

Regardless of the choice of the color space for processing the image, many restoration methods operate on the premise that an image can be subdivided into small regions, each of which can be treated as stationary [6, 30]. Namely, a processing window $\Psi_{(r,s)} = \{\mathbf{x}_{(i,j)}; (i,j) \in \zeta\}$ is used to determine a localized area of the input image $\mathbf{x}$. This window slides over the entire image, placing successively every pixel $\mathbf{x}_{(r,s)}$, for

$r = 1, 2, ..., K_1$ and $s = 1, 2, ..., K_2$, at the center of a local neighborhood denoted by ζ and replacing $\mathbf{x}_{(r,s)}$ with the output $\mathbf{y}_{(r,s)} = f(\Psi_{(r,s)})$ of a filter function $f(\cdot)$ operating over samples located inside the filter window.

The performance of a filtering method is generally influenced by the size and shape of the processing window. Different techniques may require a different processing window, in terms of its size and shape, to achieve optimal performance. For instance, unidirectional and bidirectional windows are often chosen to preserve specifically oriented image edges. However, the most commonly used window is a rectangular shape, such as a 3×3 window described by $\zeta = \{(r - 1, s - 1), (r - 1, s), ..., (r + 1, s + 1)\}$, due to its versatility and demonstrated good performance.

The following presents the two vector filtering frameworks that employ the sliding window concept. The first one comprises the methods that benefit from the theory of order statistics. The other one comprises the methods that combine the input samples to produce the filter output. Relevant filtering operators designed within these two frameworks follow the multichannel nature of color images and process the color data as vectors in order to avoid producing color artifacts in the output.

9.4.1 Order-Statistic Methods

Each vector $\mathbf{x}_{(i,j)}$ from the set of color vectors in $\Psi_{(r,s)} = \{\mathbf{x}_{(i,j)}; (i,j) \in \zeta\}$ can be represented by a scalar value

$$D_{(i,j)} = \sum_{(g,h) \in \zeta} d(\mathbf{x}_{(i,j)}, \mathbf{x}_{(g,h)}) \text{ or } D_{(i,j)} = \sum_{(g,h) \in \zeta} s(\mathbf{x}_{(i,j)}, \mathbf{x}_{(g,h)}), \quad (9.37)$$

which corresponds to the aggregated distances or the aggregated similarities. These aggregated values can be used to discriminate vectors in the population $\Psi_{(r,s)}$ by ordering scalars $D_{(i,j)}$, for $(i,j) \in \zeta$, to produce the ordered sequence $D_{(1)} \leq D_{(2)} \leq ... \leq D_{(\tau)} \leq ... \leq D_{(|\zeta|)}$ and then implying the same ordering of the corresponding vectors $\mathbf{x}_{(i,j)} \in \Psi_{(r,s)}$ as follows [6, 20], [31]:

$$\mathbf{x}_{(1)} \leq \mathbf{x}_{(2)} \leq ... \leq \mathbf{x}_{(\tau)} \leq ... \leq \mathbf{x}_{(|\zeta|)}, \quad (9.38)$$

where $\mathbf{x}_{(\tau)}$, for $\tau = 1, 2, ..., |\zeta|$, denotes the so-called τ-th vector order statistics.

The above ordering scheme, called reduced ordering, relies on overall ranking of the set $\Psi_{(r,s)}$ of input samples. Both the lowest and the highest vector order statistics are very useful in edge detection and image segmentation to determine object boundaries and localize transitions among color regions [17, 20]. For the aggregated distances, vectors that diverge greatly from the data population usually appear in higher ranks in the ordered set, whereas the most representative vectors of the local image area are associated with lower ranks. Therefore, various noise filtering solutions select one of the lower ranked vectors as the filter output. For the aggregated similarities, lower and higher ranks have an opposite meaning compared to the ordering based on the aggregated distances. Because the ordering can be used to determine the positions of the different input vectors without any a priori information regarding the signal distributions, vector order-statistics filters are considered robust estimators (Figure 9.8).

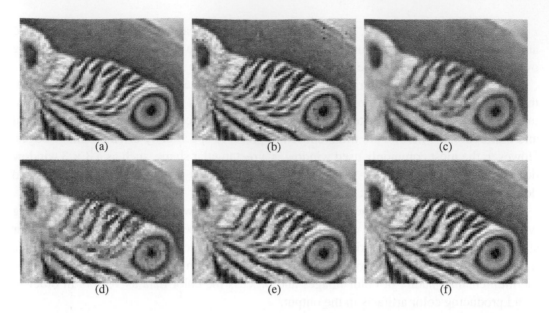

Figure 9.8 Impulsive noise suppression using vector order-statistics filters: (a) noise-free image; (b) noisy image; (c) vector median filter; (d) basic vector directional filter; (e) optimal weighted vector directional filter; (f) switching vector median filter. The filtered images are produced in a single iteration using a 3×3 supporting window.

9.4.1.1 Vector Median Filters

The vector median [11] of the population $\Psi_{(r,s)}$ corresponds to the lowest ranked vector $\mathbf{x}_{(1)}$ in $\Psi_{(r,s)}$. Equivalently, the output of the vector median filter can be defined using the minimization concept as follows:

$$\mathbf{y}_{(r,s)} = \arg \min_{\mathbf{x}_{(g,h)}} \sum_{(i,j) \in \zeta} \|\mathbf{x}_{(g,h)} - \mathbf{x}_{(i,j)}\|_L, \qquad (9.39)$$

where $\mathbf{y}_{(r,s)} = \mathbf{x}_{(g,h)} \in \Psi_{(r,s)}$ denotes the filter output. The calculation of aggregated distances is rather expensive, particularly for large supporting windows, and can be sped up using the linear approximation of the Euclidean norm [32]. Due to its minimization concept and outputting one of the input samples, the vector median is suitable for suppressing impulsive noise.

In order to take into account the importance of the specific samples in the filter window or structural contents of the image, the contribution of the input vectors $\mathbf{x}_{(i,j)}$ to the aggregated distances can be controlled using the associated weights $w_{(i,j)}$. This concept is followed in the weighted vector median filter [33] defined as

$$\mathbf{y}_{(r,s)} = \arg \min_{\mathbf{x}_{(g,h)}} \sum_{(i,j) \in \zeta} w_{(i,j)} \|\mathbf{x}_{(g,h)} - \mathbf{x}_{(i,j)}\|_L. \qquad (9.40)$$

The performance of this filter can be tuned using the optimization frameworks from [16, 34, 35]. If all the weights are set to the same value, the filter reduces to the vector median filter.

In the presence of additive Gaussian noise, the noise reduction characteristics of the vector median filter can be improved by combining it with the linear filter as follows:

$$
\mathbf{y}_{(r,s)} = \begin{cases} \mathbf{y}_{(r,s)}^{AMF} & \text{if } \sum_{(i,j)\in\varsigma} \left\| \mathbf{y}_{(r,s)}^{AMF} - \mathbf{x}_{(i,j)} \right\|_L \leq \sum_{(i,j)\in\varsigma} \left\| \mathbf{y}_{(r,s)}^{VMF} - \mathbf{x}_{(i,j)} \right\|_L \\ \mathbf{y}_{(r,s)}^{VMF} & \text{otherwise,} \end{cases} \tag{9.41}
$$

which forms the extended vector median filter. The term $\mathbf{y}_{(r,s)}^{VMF}$ denotes the vector median obtained in Equation (9.39), whereas $\mathbf{y}_{(r,s)}^{AMF}$ corresponds to an arithmetic mean filter defined over the vectors inside the neighborhood ς:

$$
\mathbf{y}_{(r,s)} = \frac{1}{|\varsigma|} \sum_{(i,j)\in\varsigma} \mathbf{x}_{(i,j)} \tag{9.42}
$$

Because the extended vector median filter in Equation (9.41) tends to apply the vector median filter near a signal edge and the arithmetic mean filter in the smooth areas, it preserves the structural information of the image while improving noise attenuation in the smooth areas. The weighted version of Equation (9.41) can be found in [33].

Because the averaging operation tends to smooth fine details and it is prone to outliers, improved design characteristics can be obtained by replacing it with its alpha-trimmed version [33]:

$$
\mathbf{y}_{(r,s)} = \frac{1}{|\varsigma_\alpha|} \sum_{(i,j)\in\varsigma_\alpha} \mathbf{x}_{(i,j)}, \tag{9.43}
$$

where α is a design parameter that can have values $\alpha = 0, 1, ..., |\varsigma|-1$. The set $\varsigma_\alpha = \{(i,j),$ for $D_{(i,j)} \leq D_{(|\varsigma|-\alpha)}\} \subset \varsigma$ consists of the spatial locations of the vectors $\mathbf{x}_{(r,s)} \in \Psi_{(r,s)},$ which have the aggregated distances smaller or equal to the $(|\varsigma| - \alpha)$-th largest aggregated distances $D_{(|\varsigma|-\alpha)} \in \{D_{(i,j)}; (i,j) \in \varsigma\}$.

The linear combination of the ordered input samples is also used in the design of multichannel L-filters [36, 37]:

$$
\mathbf{y}_{(r,s)} = \sum_{\tau=1}^{|\varsigma|} w_\tau \mathbf{x}_{(\tau)}, \tag{9.44}
$$

where w_τ is the weight associated with the τth ordered vector $\mathbf{x}_{(\tau)} \in \Psi_{(r,s)}$. Assuming the weight vector $\mathbf{w} = [w_1, w_2, ..., w_{|\varsigma|}]$ and the unity vector $\mathbf{e} = [1, 1, ..., 1]$ of the dimension identical to that of $\mathbf{w}$, the filter weights can be optimally determined using the mean-square error criterion as follows:

$$
\mathbf{w} = \frac{\mathbf{R}^{-1}\mathbf{e}}{\mathbf{e}^T\mathbf{R}^{-1}\mathbf{e}}, \tag{9.45}
$$

where $\mathbf{w}^T\mathbf{e} = 1$ is the constraint imposed on the solution and $\mathbf{R}$ is a $|\varsigma| \times |\varsigma|$ correlation matrix of the ordered noise variables. The optimization process can be sped up using the least mean square formulation $\mathbf{w} = \mathbf{w} + 2\mu e_{(r,s)} \Psi_{(r,s)}^r$ based on the ordered input set $\Psi_{(r,s)}^r$ instead.

9.4.1.2 Vector Directional Filters

The filters discussed previously operate in the magnitude domain of color images. A departure from this concept can be observed in the design of vector directional filters [38, 39], which aim at evaluating the angle between the image vectors in order to eliminate ones with atypical directions in the vector space. The computational cost of these filters associated with the inverse cosine calculation can be significantly reduced using minimax approximations [40].

The basic vector directional filter [39] minimizes the aggregated angular distance, based on Equation (9.18), to other samples inside the filter window $\Psi_{(r,s)}$:

$$\mathbf{y}_{(r,s)} = \arg \min_{\mathbf{x}_{(g,h)}} \sum_{(i,j) \in \zeta} \mathrm{A}(\mathbf{x}_{(g,h)}, \mathbf{x}_{(i,j)}). \qquad (9.46)$$

The closest variant of this method, called a spherical median [39], minimizes the same angular criterion without the constraint that the filter output is one of the samples inside $\Psi_{(r,s)}$.

The weighted vector directional filter [34, 42] determines its output using the weights $w_{(i,j)}$ associated with the input vectors $\mathbf{x}_{(i,j)}$, for $(i, j) \in \zeta$ as follows:

$$\mathbf{y}_{(r,s)} = \arg \min_{\mathbf{x}_{(g,h)}} \sum_{(i,j) \in \zeta} w_{(i,j)} \mathrm{A}(\mathbf{x}_{(g,h)}, \mathbf{x}_{(i,j)}). \qquad (9.47)$$

Depending on the weight settings, improved detail-preserving characteristics can be obtained. The weights can be optimized using devoted multichannel adaptation algorithms [16, 34].

To utilize both magnitude and directional characteristics during processing, the generalized vector directional filters defined as

$$\mathbf{y}_{(r,s)} = f(\mathbf{x}_{(1)}, \mathbf{x}_{(2)}, ..., \mathbf{x}_{(\tau)}) \qquad (9.48)$$

use the aggregated angular distance criterion to first eliminate the color vectors with atypical directions in the vector space. This operation produces a subset $\{\mathbf{x}_{(1)}, \mathbf{x}_{(2)}, ..., \mathbf{x}_{(\tau)}\}$ that consists of the τ lowest vector order statistics. Because these remaining vectors have approximately the same direction in the vector space, they can be processed in a subsequent step through the function $f(\cdot)$ according to their magnitude without creating chrominance artifacts.

The combined magnitude-directional processing strategy is also followed by the directional-distance filter [43]:

$$D_{(i,j)} = \left(\sum_{(g,h) \in \zeta} \|\mathbf{x}_{(i,j)} - \mathbf{x}_{(g,h)}\|_L \right) \left(\sum_{(g,h) \in \zeta} \mathrm{A}(\mathbf{x}_{(i,j)}, \mathbf{x}_{(g,h)}) \right). \qquad (9.49)$$

In this case, the ordering or minimization criterion is expressed through a product of the aggregated Euclidean distances and the aggregated angular distances.

The hybrid vector filters [44] operate on the direction and the magnitude of the color vectors independently to obtain the intermediate outputs, which are then combined to produce the final filter output. Because these filters use two independent ordering schemes, they are computationally demanding. In one example, the outputs of the vector median filter $\mathbf{y}_{(r,s)}^{VMF}$ defined in Equation (9.39) and the basic vector directional filter $\mathbf{y}_{(r,s)}^{BVDF}$ defined in Equation (9.46) are combined as follows:

$$\mathbf{y}_{(r,s)} = \begin{cases} \mathbf{y}_{(r,s)}^{VMF} & \text{if } \mathbf{y}_{(r,s)}^{VMF} = \mathbf{y}_{(r,s)}^{BVDF} \\ \bar{\mathbf{y}}_{(r,s)} & \text{otherwise,} \end{cases} \tag{9.50}$$

where $|\cdot|$ denotes the magnitude operator and

$$\bar{\mathbf{y}}_{(r,s)} = \left(\frac{|\mathbf{y}_{(r,s)}^{VMF}|}{|\mathbf{y}_{(r,s)}^{BVDF}|} \right) \mathbf{y}_{(r,s)}^{BVDF}. \tag{9.51}$$

In another example, the arithmetic mean filter $\mathbf{y}_{(r,s)}^{AMF}$ defined in Equation (9.42) is added to the mix to operate as follows:

$$\mathbf{y}_{(r,s)} = \begin{cases} \mathbf{y}_{(r,s)}^{VMF} & \text{if } \mathbf{y}_{(r,s)}^{VMF} = \mathbf{y}_{(r,s)}^{BVDF} \\ \bar{\mathbf{y}}_{(r,s)} & \text{if } \sum_{(i,j)\in\zeta} |\mathbf{x}_{(i,j)} - \bar{\mathbf{y}}_{(r,s)}| < \sum_{(i,j)\in\zeta} |\mathbf{x}_{(i,j)} - \hat{\mathbf{y}}_{(r,s)}| \\ \hat{\mathbf{y}}_{(r,s)} & \text{otherwise,} \end{cases} \tag{9.52}$$

where $\bar{\mathbf{y}}_{(r,s)}$ is obtained in Equation (9.51) and

$$\hat{\mathbf{y}}_{(r,s)} = \left(\frac{|\mathbf{y}_{(r,s)}^{AMF}|}{|\mathbf{y}_{(r,s)}^{BVDF}|} \right) \mathbf{y}_{(r,s)}^{BVDF}. \tag{9.53}$$

9.4.1.3 Selection Weighted Vector Filters

The selection weighted vector filters [16,41] constitute a generalized class of vector filtering operators:

$$\mathbf{y}_{(r,s)} = \arg\min_{\mathbf{x}_{(g,h)}} \left[\left(\sum_{(g,h)\in\zeta} w_{(i,j)} \|\mathbf{x}_{(g,h)} - \mathbf{x}_{(i,j)}\|_L \right)^{1-\xi} \left(\sum_{(g,h)\in\zeta} w_{(i,j)} A(\mathbf{x}_{(g,h)}, \mathbf{x}_{(i,j)}) \right)^{\xi} \right], \tag{9.54}$$

where $\mathbf{w} = \{w_{(i,j)}; (i,j) \in \zeta\}$ are nonnegative weights, which signify the importance of individual color vectors $\mathbf{x}_{(i,j)}$ in $\Psi_{(r,s)}$. The term ξ is a design parameter ranging from zero to one; any deviation from 0.5 to a lower or larger value of ξ places more emphasis during the processing on the magnitude or directional characteristics, respectively. Each setting of the above parameters represents a unique filter that can be used for a specific task. It is straightforward to show that Equation (9.54) generalizes various popular filters discussed previously, including the vector median (for $\mathbf{w} = 1$ with $\xi = 0$), basic vector directional (for $\mathbf{w} = 1$ with $\xi = 1$), directional-distance (for $\xi = 0.5$), weighted vector median (for $\xi = 0$), and weighted vector directional (for $\xi = 1$) filters.

To obtain the desired filtering performance, the weights can be determined by the user or optimized as follows [16,41]:

$$w_{(i,j)} = P\left[w_{(i,j)} + 2\mu R(\mathbf{y}_{(r,s)}^{*}, \mathbf{y}_{(r,s)}) \text{sgn}(R(\mathbf{x}_{(i,j)}, \mathbf{y}_{(r,s)})) \right], \tag{9.55}$$

where μ is a regulation factor and $\mathbf{y}^*_{(r,s)} = [y^*_{(r,s)1}, y^*_{(r,s)2}, y^*_{(r,s)3}]$ is the feature signal used to guide the adaptation process. Each weight $w_{(i,j)}$ is adjusted by adding the contributions of the corresponding input vector $\mathbf{x}_{(i,j)}$ and the filter output $\mathbf{y}_{(r,s)}$, which are expressed as the distances to the feature signal $\mathbf{y}^*_{(r,s)}$. The recommended initial setting of the adaptation process is $\mu \ll 0.5$ and $w_{(i,j)} = 1$, for $(i,j) \in \zeta$, which corresponds to robust smoothing. In order to keep the aggregated distances in Equation (9.54) positive and thus ensure low-pass filtering characteristics, the updated weights $w_{(i,j)}$ are constrained through a projection function

$$P(w_{(i,j)}) = \begin{cases} 0 & \text{if } w_{(i,j)} < 0 \\ w_{(i,j)} & \text{otherwise.} \end{cases} \qquad (9.56)$$

The adjustment of $w_{(i,j)}$ is allowed for both positive and negative contributions, using the sign sigmoidal function

$$\text{sgn}(a) = \frac{2}{1 + \exp(-a)} - 1 \qquad (9.57)$$

and the vectorial sign function, which utilizes both the magnitude and directional image characteristics, as follows:

$$R(\mathbf{x}_{(i,j)}, \mathbf{x}_{(g,h)}) = S(\mathbf{x}_{(i,j)}, \mathbf{x}_{(g,h)}) \left(\|\mathbf{x}_{(i,j)} - \mathbf{x}_{(g,h)}\|_L \right)^{1-\xi} \left(A(\mathbf{x}_{(i,j)}, \mathbf{x}_{(g,h)}) \right)^\xi, \quad (9.58)$$

where $S(\cdot, \cdot)$ is the polarity function defined as

$$S(\mathbf{x}_{(i,j)}, \mathbf{x}_{(g,h)}) = \begin{cases} +1 \text{ for } \|\mathbf{x}_{(i,j)}\| - \|\mathbf{x}_{(g,h)}\| \geq 0 \\ -1 \text{ for } \|\mathbf{x}_{(i,j)}\| - \|\mathbf{x}_{(g,h)}\| < 0. \end{cases} \qquad (9.59)$$

The adaptation algorithm described above is not the only one suitable for optimizing the filter weights. Using the uppermost $\mathbf{x}_{(|\zeta|)}$ and the lowest $\mathbf{x}_{(1)}$ ranked vectors, the weight adaptation can be performed as follows:

$$w_{(i,j)} = P[w_{(i,j)} + 2\mu\{R(\mathbf{x}_{(|\zeta|)}, \mathbf{x}_{(1)}) - 2|R(\mathbf{y}^*_{(r,s)}, \mathbf{x}_{(i,j)})| \\ - \sum_{(g,h)\in\zeta} w_{(g,h)}(R(\mathbf{x}_{(|\zeta|)}, \mathbf{x}_{(1)}) - 2|R(\mathbf{x}_{(i,j)}, \mathbf{x}_{(g,h)})|)\}], \qquad (9.60)$$

where μ is a positive stepsize.

In many situations, the feature signal $\mathbf{y}^*_{(r,s)}$ can have the form of the original, noise-free image $\mathbf{o}_{(r,s)}$. If the original signal $\mathbf{o}_{(r,s)}$ is unavailable, the adaptation process can be guided by some other feature signal instead. For example, the adaptation process can follow the acquired signal characteristics when the corrupting noise power is low and strong detail-preserving characteristics are desired. Alternatively, robust noise suppression can be obtained using a smoothed version of the original image. The trade-off between noise suppression and detail preservation can be ensured through the combination of the input signal and the obtained estimate. The presented weight adaptation framework allows optimizing the filter weights in the magnitude domain ($\xi = 0$), the directional domain ($\xi = 1$), or using both magnitude and directional characteristics in various amounts.

9.4.1.4 Switching Vector Filters

In imaging applications corrupted by impulsive noise, various switching filters can be used to achieve the trade-off between noise suppression and image-detail preservation. Typically, a switching function is designed by altering between a robust nonlinear smoothing filter outputting $\mathbf{y}^*_{(r,s)}$ and an identity operation denoted by $\mathbf{x}_{(r,s)}$ as follows [17,42]:

$$\mathbf{y}_{(r,s)} = \begin{cases} \mathbf{y}^*_{(r,s)} & \text{if } \lambda \geq \xi \\ \mathbf{x}_{(r,s)} & \text{otherwise.} \end{cases} \tag{9.61}$$

The switching mechanism is controlled by comparing the adaptive parameter $\lambda(\Psi_{(r,s)})$ and the nonnegative threshold ξ. The smoothing filter is active only when the noise is detected, that is, for $\lambda \geq \xi$. Otherwise, the input color vector $\mathbf{x}_{(r,s)}$ is considered noise-free and remains unchanged.

The switching mechanism can use a set of robust order statistics $\{\mathbf{x}_{(c)} \in \Psi_{(r,s)}, \text{ for } c = 1, 2, ..., \tau\}$, as follows [45]:

$$\lambda = d\left(\mathbf{x}_{(r,s)}, \frac{1}{\tau}\sum_{c=1}^{\tau}\mathbf{x}_{(c)}\right). \tag{9.62}$$

This data-adaptive parameter λ is compared with the predetermined value of ξ. The ordering criterion needed for ranking the input color vectors in order to perform noise detection and subsequent filtering using the lowest ranked vector can be based on the angular distances. However, a more robust solution can be obtained when the Euclidean distance is employed in the filter design.

Another switching filter concept is based on the approximation of the multivariate dispersion [46,47]. Namely, the filter calculates $\lambda = D_{(r,s)}$ defined as the aggregated distance between the window center and the other vectors inside the supporting window $\Psi_{(r,s)}$. The value of λ is compared with

$$\xi = D_{(1)} + \tau\psi = \frac{|\zeta| - 1 + \tau}{|\zeta| - 1}D_{(1)}, \tag{9.63}$$

where $\psi = D_{(1)}/(|\zeta| - 1)$ is the variance approximated using the smallest aggregated distance $D_{(1)}$ and τ is the tuning parameter used to adjust the smoothing characteristics. The variance can also be approximated as $\psi_{\bar{\mathbf{x}}} = D_{\bar{\mathbf{x}}}/|\zeta|$, where $D_{\bar{\mathbf{x}}} = \sum_{(i,j)\in\zeta} d(\bar{\mathbf{x}}_{(r,s)}, \mathbf{x}_{(i,j)})$ is the aggregated distance between input vectors $\mathbf{x}_{(i,j)} \in \Psi_{(r,s)}$ and the sample mean $\bar{\mathbf{x}}_{(r,s)}$, resulting in

$$\xi = D_{\bar{\mathbf{x}}} + \tau\psi_{\bar{\mathbf{x}}} = \frac{|\zeta| + \tau}{|\zeta|}D_{\bar{\mathbf{x}}}. \tag{9.64}$$

The selection weighted filters can also be used in the design of switching filters as follows [42,48]:

$$\lambda = \sum_{c=\tau}^{\tau+2} d(\mathbf{y}^c_{(r,s)}, \mathbf{x}_{(r,s)}), \tag{9.65}$$

where $\mathbf{y}^c_{(r,s)}$ denotes the vector obtained in Equation (9.4.1.3) with the weights set as $w_{(i,j)} = |\zeta| - 2c + 2$ if $(i,j) = (r,s)$, and $w_{(i,j)} = 1$ otherwise. The smoothing parameter c, for $c = 1, 2, ..., (|\zeta| + 1)/2$, is used to control the amount of smoothing in the

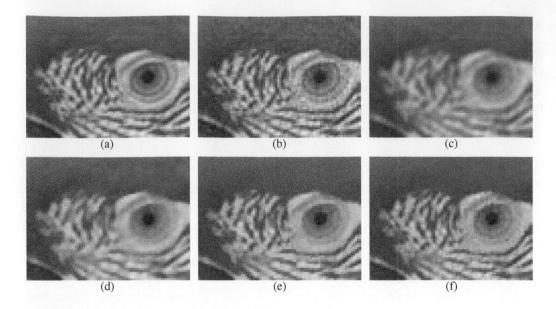

Figure 9.9 Additive Gaussian noise suppression using vector combination methods: (a) noise-free image; (b) noisy image; (c) fuzzy vector filter; (d) data-adaptive filter based on digital paths; (e) standard 4-point anisotropic diffusion filter after 20 iterations; (f) bilateral filter operating using an 11×11 supporting window. Unless otherwise stated, the filtered images are produced in a single iteration using a 3×3 supporting window.

noise detection step, ranging from no smoothing (the identity operation for $c = 1$) to the maximum amount of smoothing ($c = (|\zeta| + 1)/2$).

The methods described above are just examples of switching filters aimed at color images. Recently, various concepts [49–54] have been introduced to further enhance noise detection characteristics and/or computational efficiency of this powerful filter class.

9.4.2   Combination Methods

The averaging-like filters constitute another often-used method of choice for smoothing high-frequency variations and transitions. These filters can be designed to obtain the trade-off between reducing random noise and preserving the desired image content. Here, the underlying idea is to replace each pixel value in an image with the weighted average of pixels inside a localized image area, thus suppressing pixel values that are unrepresentative of their surroundings (Figure 9.9).

9.4.2.1   Fuzzy Vector Filters

The data-adaptive vector filters [12, 55, 56] perform fuzzy weighted averaging inside a localized image area:

$$\mathbf{y}_{(r,s)} = f \left(\sum\nolimits_{(i,j)\in\zeta} w_{(i,j)} \mathbf{x}_{(i,j)} \right), \tag{9.66}$$

where $f(\cdot)$ is a nonlinear function and

$$w_{(i,j)} = \mu_{(i,j)} / \sum_{(g,h)\in\zeta} \mu_{(g,h)} \tag{9.67}$$

is the normalized filter weight associated with the input color vector $\mathbf{x}_{(i,j)} \in \Psi_{(r,s)}$. The filter is constrained via $w_{(i,j)} \geq 0$ and $\sum_{(i,j)\in\zeta} w_{(i,j)} = 1$ to ensure that its output is unbiased.

The weights $w_{(i,j)}$ are calculated using fuzzy membership function terms $\mu_{(i,j)}$, which can have a form of an exponential function [12, 56]:

$$\mu_{(i,j)} = \beta \left(1 + \exp\left\{D_{(i,j)}\right\}\right)^{-r}, \tag{9.68}$$

where r is a parameter adjusting the weighting effect of the membership function, β is a normalizing constant, and $D_{(i,j)}$ is the aggregated distance or similarity measure defined in Equation (9.37). The filter operates without the requirement for fuzzy rules and can adapt to various noise characteristics by tuning the parameters of its membership function.

Improved detail-preserving characteristics of the above filter can be obtained by constraining its output to be a vector from the original input set $\Psi_{(r,s)}$ as follows:

$$w_{(i,j)} = \begin{cases} 1 & \text{if } \mu_{(i,j)} = \mu_{max} \\ 0 & \text{if } \mu_{(i,j)} \neq \mu_{max} \end{cases} \tag{9.69}$$

where $\mu_{max} \in \{\mu_{(i,j)}; (i,j) \in \zeta\}$ is the maximum fuzzy membership value. If the maximum value occurs at a single point only, the above filter reduces to a selection filtering operation $\mathbf{y}_{(r,s)} = \mathbf{x}_{(i,j)}$, for $\mu_{(i,j)} = \mu_{max}$.

9.4.2.2 Data-Adaptive Filters Based on Digital Paths

The data-adaptive filter presented above can be further advanced by exploiting connections between image pixels inside the supporting window using the concept of digital paths [57, 58]. Similar pixels are grouped together, thus forming paths that reveal the underlying structural image content. Note that two distinct pixel locations on the image lattice can be connected by many paths; the number of possible geodesic paths of certain length connecting two distinct points depends on their locations, length of the path, and the neighborhood system employed [31].

A digital path $P^\eta (\rho_0, \rho_1, \dots, \rho_\eta)$ of length η, linking the starting location $\rho_0 = (i,j)$ and the ending location $\rho_\eta = (g,h)$, is associated with the connection cost

$$\Lambda^\eta (\rho_0, \rho_1, \dots, \rho_\eta) = f \left(\mathbf{x}_{\rho_0}, \mathbf{x}_{\rho_1}, \dots, \mathbf{x}_{\rho_\eta}\right) = \sum_{c=1}^\eta \left\| \mathbf{x}_{\rho_c} - \mathbf{x}_{\rho_{c-1}} \right\|, \tag{9.70}$$

which can be seen as a measure of dissimilarity between the pixels forming this path. Thus, the value of $\Lambda^\eta(\cdot) = 0$ for a path consisting of the identical vectors. The appropriateness of the digital paths leading from (i,j) to (g,h) can be evaluated using a similarity function as follows:

$$\mu^{\eta,\Psi} (\rho_0, \rho_\eta) = \sum_{b=1}^\chi f \left(\Lambda^{\eta,b} (\rho_0, \rho_\eta)\right) = \sum_{b=1}^\chi \exp\left[-\beta \cdot \Lambda^{\eta,b} ((i,j), (g,h))\right], \tag{9.71}$$

where χ is the number of all paths connecting (i, j) and (g, h). The term $\Lambda^{\eta,b}\left((i, j), (g, h)\right)$ denotes a dissimilarity value along a specific path b from the set of all χ possible paths leading from (i, j) to (g, h) in the search area $\Psi_{(r,s)}$, $f(\cdot)$ denotes a smooth function of $\Lambda^{\eta,b}$, and β is a design parameter.

The above similarity function can be modified by considering the information on the local image features as follows:

$$\mu^{\eta,\Psi}\left(\rho_0, \rho_1, \eta\right) = \sum_{b=1}^{\chi} \exp\left[-\beta \cdot \Lambda^{\eta,b}(\rho_0, \rho_1, \rho_2^*, \ldots, \rho_\eta^*)\right]. \tag{9.72}$$

Here, the connection costs $\Lambda^{\eta}(\cdot)$ are investigated for digital paths that originate at ρ_0, cross ρ_1, and then pass the successive locations ρ_c, for $c = 2, 3, \ldots, \eta$, until each path reaches length η. Thus, χ denotes the number of the paths $P^{\eta}(\rho_0, \rho_1, p_2^*, \ldots, \rho_\eta^*)$, originating at ρ_0 and crossing ρ_1 to end at ρ_η^*, which are totally included in the search area $\Psi_{(r,s)}$. If the constraint of crossing the location ρ_1 is omitted, then $\Lambda^{\eta,b}(\rho_0, \rho_1, \rho_2^*, \ldots, \rho_\eta^*)$ reduces to $\Lambda^{\eta,b}(\rho_0, \rho_1^*, \rho_2^*, \ldots, \rho_\eta^*)$.

Using the digital path approach, the data-adaptive filter can be defined as follows:

$$\mathbf{y}_{(r,s)} = \sum_{(i,j) \Leftrightarrow (r,s)} w_{(i,j)}^{(r,s)} \mathbf{x}_{(i,j)} = \sum_{(i,j) \Leftrightarrow (r,s)} \frac{\mu^{\eta,\Psi}\left((r, s), (i, j)\right) \mathbf{x}_{(i,j)}}{\sum_{(g,h) \Leftrightarrow (r,s)} \mu^{\eta,\Psi}\left((r, s), (g, h)\right)}, \tag{9.73}$$

where $(i, j) \Leftrightarrow (r, s)$ denotes all points (i, j) connected by digital paths with (r, s) contained in $\Psi_{(r,s)}$. The filter output $\mathbf{y}_{(r,s)}$ is equivalent to the weighted average of all vectors $\mathbf{x}_{(i,j)}$ connected by digital paths with the vector $\mathbf{x}_{(r,s)}$.

By restricting the digital path definition, the output vector $\mathbf{y}_{(r,s)}$ can be expressed as a weighted average of the nearest neighbors of $\mathbf{x}_{(r,s)}$ as follows [57]:

$$\mathbf{y}_{(r,s)} = \sum_{\rho_1^* \sim \rho_0} w(\rho_0, \rho_1^*)\mathbf{x}_{\rho_1^*}, \tag{9.74}$$

where the normalized weights are obtained by exploring all digital paths starting from the central pixel of $\Psi_{(r,s)}$ and crossing its neighbors. To produce more efficient smoothing, the filtering function can be defined as a weighted average of all pixels contained in the supporting window as follows [57]:

$$\mathbf{y}_{(r,s)} = \sum_{\rho_\eta^*} w(\rho_0, \rho_\eta^*)\mathbf{x}_{\rho_\eta^*}, \tag{9.75}$$

where the weights $w(\rho_0, \rho_\eta^*)$ are obtained by exploring all digital paths leading from the pixel $\mathbf{x}_{(r,s)}$, for $\rho_0 = (r, s)$, to any of the pixels in the supporting window.

9.4.2.3 Anisotropic Diffusion Filters

The concept of anisotropic diffusion [59] has been derived from its isotropic predecessor by introducing the conduction coefficients to inhibit the smoothing of the image edges. In the case of color images, the anisotropic diffusion filter can be implemented as follows:

$$\mathbf{y}_{(r,s)} = \mathbf{x}_{(r,s)} + \lambda \sum_{(i,j)\in\zeta} f(\mathbf{x}_{(i,j)} - \mathbf{x}_{(r,s)})(\mathbf{x}_{(i,j)} - \mathbf{x}_{(r,s)}), \tag{9.76}$$

where λ is the parameter used to control the rate of the diffusion process; the term $\zeta = \{(r-1,s), (r,s-1), (r,s+1), (r+1,s)\}$ denotes the neighboring pixels located above, left, right, and below with respect to the pixel location (r,s) under consideration. The term $f(\cdot)$ denotes the nonlinear conduction function that determines the behavior of the diffusion process. Note that diffusion filtering is an iterative process and depending on the filter settings and the characteristics of the image being processed, a number of iterations may be needed to obtain the desired result.

The conduction function should produce large values in homogenous regions to encourage image smoothing and small in edge regions to preserve the structural content. Ideally, this function should satisfy four properties: $f(0) = M$ for $0 < M < \infty$, which ensures isotropic smoothing in regions of similar intensity; $f(a) = 0$ for $a \to \infty$, which ensures edge preservation; $f(a) \geq 0$ to ensure a forward diffusion process; and that $af(a)$ is strictly decreasing function to avoid numerical instability. However, many functions successfully used in practice accomplish just the first three properties.

The conduction function is defined to perform an edge-sensing operation based on image gradients by producing the diffusion coefficients that can be seen as the weights associated with the neighboring vectors $\mathbf{x}_{(i,j)}$, for $(i,j) \in \zeta$. The two most common forms of the conduction function are

$$f(a) = \exp(-(c(a)/\tau)^2) \tag{9.77}$$

$$f(a) = \frac{1}{1 + (c(a)/\tau)^2} \tag{9.78}$$

where τ states the diffusion rate and $c(\cdot)$ denotes a color diffusivity operator. In its simplest form, this operator can be defined to use the gradients from each color channel separately, resulting in componentwise diffusion filtering that may form edges at different locations for each color channel. To avoid this drawback, one should use a function that combines information from all three color channels. This can be accomplished, for example, by applying the conduction function to the luminance gradients or the value produced by some distance or similarity measure presented earlier in this chapter. Various diffusion approaches suitable for color images can be found in [60–62].

9.4.2.4 Bilateral Filters

Bilateral filtering [63–65] aims at performing anisotropic image smoothing using a non-iterative formulation. This filter is well-known for its ability to reduce image noise and simultaneously preserve the desired structural content, such as edges, textures, and details. Its weighted averaging formulation is obtained using the weights that indicate range and spatial closeness between the pixel located in the window center and its neighbors:

$$\mathbf{y}_{(r,s)} = \sum_{(i,j)\in\zeta} w_{(i,j)} u_{(i,j)} \mathbf{x}_{(i,j)} \Big/ \sum_{(i,j)\in\zeta} w_{(i,j)} u_{(i,j)}, \tag{9.79}$$

where ζ denotes pixel locations inside the filter window.

The range closeness is a function of differences in the range of the signal, often referred to as a photometric space. To avoid using different weights in different color channels and thus producing color artifacts, range weights $w_{(i,j)}$ can be defined to take advantage of color correlation as follows:

$$w_{(i,j)} = \exp\left(-\sum_{k=1}^{3} \frac{(x_{(r,s)k} - x_{(i,j)k})^2}{\sigma_k^2/\lambda_k}\right),\qquad(9.80)$$

where λ_k is a scaling factor and σ_k^2 denotes the noise variance in the kth color channel.

The range weights are combined with the spatial weights

$$u_{(i,j)} = \frac{1}{2\pi\sigma_d} \exp\left(\frac{-\|(r,s) - (i,j)\|^2}{2\sigma_d^2}\right)\qquad(9.81)$$

obtained using a traditional Gaussian-weighting function defined in the spatial domain to reduce the contribution of pixels that reside farther from the pixel being filtered. Note that unlike the range weights, which are not known a priori because they depend on actual samples inside the filter window, the spatial weights can be precomputed.

Parameters σ_d, σ_k, and λ_k control the amount of image smoothing. As the range parameter σ_k^2/λ_k increases, the bilateral filter approximates Gaussian convolution more closely, because the range weights become more constant over the intensity interval of the image. Increasing the spatial parameter σ_d smoothes larger features, as more averaging is allowed.

Bilateral filtering is considered expensive, and therefore various implementation strategies have been proposed in the past, such as separable kernels [66], local histograms [67], layered approximation [68], bilateral grid [69], and bilateral pyramid [70]. The bilateral filter has a solid theoretical foundation; it can be shown [71, 72] that it has some relation to robust filtering, anisotropic diffusion, and local mean filtering.

9.5 Color Image Quality Evaluation

The performance of color image filtering methods is usually evaluated through the closeness of the filtered image to the reference image. The two basic approaches used for assessing such an image closeness are the subjective evaluation approach and the objective evaluation approach. The subjective approach is usually required in applications where the original images are unavailable, and thus standard objective measures that are based on the difference in the statistical distributions of the pixel values cannot be employed.

9.5.1 Subjective Assessment

Most images are typically intended for human inspection; therefore, human opinion on visual quality is considered representative of the palatability of visual signals [73]. Statistically meaningful results are usually obtained by conducting a large-scale study where human observers view and rate images, presented either in specific or random order and under identical viewing conditions, using a predetermined rating scale. In order to simulate a realistic situation where viewing of images is done by ordinary citizens and not image processing experts, the observers should be unaware of the specifics of the experiments.

Because edges indicate the boundaries and shape of the objects in the image and the human visual system is sensitive to changes in color appearance, subjective quality assessment of filtered color images usually focuses on the presence of edge/details, residual noise, and color artifacts in the image. In a typical five-point rating scale [20], the score ranges from 1 to 5, reflecting poor, fair, good, very good, and excellent noise removal, respectively. Using the same scoring scale, the presence of any distortion can be evaluated as very disruptive, disruptive, destructive but not disruptive, perceivable but not destructive, and imperceivable.

9.5.2 Objective Assessment

Subjective assessment of image quality is usually time-consuming and often impractical. Therefore, whenever possible, automatic assessment of visual quality should be performed instead. There exist various methods that are capable of assessing the perceptual quality of images and some of these methods, comprehensively surveyed in [73], can produce the scores that correlate well with subjective opinion.

9.5.2.1 Perceptual Measures

In the CIELuv or CIELab color space, perceptual differences between two colors can be measured using the Euclidean distance. If the image data is in the RGB format, the procedure includes the conversion of each pixel from the RGB values to the corresponding XYZ values using Equation (9.27), and then from these XYZ values to the target Luv or Lab values using the conversion formulas described in Section 9.3.4. Operating on the pixel level, the resulting error between the two color vectors can be expressed as follows [4]:

$$\Delta E_{uv}^* = \sqrt{(L_1^* - L_2^*)^2 + (u_1^* - u_2^*)^2 + (v_1^* - v_2^*)^2}, \quad (9.82)$$

$$\Delta E_{ab}^* = \sqrt{(L_1^* - L_2^*)^2 + (a_1^* - a_2^*)^2 + (b_1^* - b_2^*)^2}, \quad (9.83)$$

The resulting perceptual difference between the two images is usually calculated as an average of ΔE errors obtained in each pixel location. Psychovisual experiments have shown that the value of ΔE equal to unity represents a just noticeable difference (JND) in either of these two color models [74], although higher JND values may be needed to account for the complexity of visual information in pictorial scenes.

Because any color can be described in terms of its lightness, chroma, and hue, the CIELab error can be equivalently expressed as follows [4, 28]:

$$\Delta E_{ab}^* = \sqrt{(\Delta L_{ab}^*)^2 + (\Delta C_{ab}^*)^2 + (\Delta H_{ab}^*)^2}, \quad (9.84)$$

where $\Delta L_{ab}^* = L_1^* - L_2^*$ denotes the difference in lightness,

$$\Delta C_{ab}^* = C_1^* - C_2^* = \sqrt{(a_1^*)^2 + (b_1^*)^2} - \sqrt{(a_2^*)^2 + (b_2^*)^2} \quad (9.85)$$

denotes the difference in chroma, and

$$\Delta H_{ab}^* = \sqrt{(a_1^* - a_2^*)^2 + (b_1^* - b_2^*)^2 - (\Delta C_{ab}^*)^2} \qquad (9.86)$$

denotes a measure of hue difference. Note that the hue angle can be expressed as $h_{ab}^* = \arctan(b/a)$.

A more recent ΔE formulation weights the chroma and hue components by a function of chroma [75]:

$$\Delta E_{94}^* = \sqrt{\left(\frac{\Delta L_{ab}^*}{k_L S_L}\right)^2 + \left(\frac{\Delta C_{ab}^*}{k_C S_C}\right)^2 + \left(\frac{\Delta H_{ab}^*}{k_H S_H}\right)^2}, \qquad (9.87)$$

where $S_L = 1$, $S_H = 1 + 0.015\sqrt{C_1^* C_2^*}$, $S_C = 1 + 0.045\sqrt{C_1^* C_2^*}$, and $k_L = k_H = k_C = 1$ for reference conditions. Even more sophisticated formulations of the ΔE measure can be found in [76] and [77]. An improved measure for pictorial scenes, referred to as S-CIELab [78], employs spatial filtering. The spatiotemporal nature of digital video is accounted for in ST-CIELab [79].

9.5.2.2 Standard Measures

Because the RGB color space is widely used to acquire, process, store, and display color images, there is a need for objective criteria that can assess image quality by operating directly on the RGB data. The mean absolute error (MAE) and the mean square error (MSE), defined as

$$\text{MAE} = \frac{1}{3K_1 K_2} \sum_{r=1}^{K_1} \sum_{s=1}^{K_2} \sum_{k=1}^{3} \left| o_{(r,s)k} - y_{(r,s)k} \right| \qquad (9.88)$$

$$\text{MSE} = \frac{1}{3K_1 K_2} \sum_{r=1}^{K_1} \sum_{s=1}^{K_2} \sum_{k=1}^{3} \left(o_{(r,s)k} - y_{(r,s)k} \right)^2, \qquad (9.89)$$

well address this need; these simple measures are often considered an indication of signal-detail preservation and noise suppression, respectively. As their names suggest, the resulting error is calculated as the average of errors calculated in individual pixel locations. In the above equations, $\mathbf{o}_{(r,s)} = [o_{(r,s)1}, o_{(r,s)2}, o_{(r,s)3}]$ denotes the reference RGB pixel (usually taken from the original noise-free image), which is compared with its actual (filtered or noisy) version $\mathbf{y}_{(r,s)} = [y_{(r,s)1}, y_{(r,s)2}, y_{(r,s)3}]$ occupying the same pixel location (r, s) in an image with K_1 rows and K_2 columns. Similar to ΔE formulations, lower MAE and MSE values correspond to better image quality.

The mean square error concept is adopted in the signal-to-noise ratio (SNR) and the peak signal-to-noise ratio (PSNR) measures, which are defined as follows:

$$\text{SNR} = 10 \log_{10} \left[\frac{\sum_{r=1}^{K_1} \sum_{s=1}^{K_2} \sum_{k=1}^{3} \left(o_{(r,s)k} \right)^2}{\sum_{r=1}^{K_1} \sum_{s=1}^{K_2} \sum_{k=1}^{3} \left(o_{(r,s)k} - y_{(r,s)k} \right)^2} \right] \qquad (9.90)$$

$$\text{PSNR} = 10 \log_{10} \left[\frac{255^2}{\text{MSE}} \right], \tag{9.91}$$

where 255 corresponds to the maximum signal level in a standard eight bits per RGB color component representation. Because many signals have a wide dynamic range, both these measures are expressed in terms of the logarithmic decibel scale. The PSNR is the ratio between the maximum possible power of a signal and the power of corrupting noise that affects the fidelity of its representation. Both SNR and PSNR values increase with the improvement in image quality.

9.6 Conclusion

This chapter surveyed popular filtering methods for restoring the color image data from the noisy measurements by operating inside the localized image area using the supporting window sliding over the entire image. These methods take advantage of color correlation-driven robust estimation and data-adaptive image processing, which make them effective in removing image noise while preserving desired structural and color features. Various vector median filters, vector directional filters, selection weighted vector filters, and switching vector filters were presented as example methods based on order statistics, which are particularly useful for color images corrupted with impulsive noise. For images with the presence of additive Gaussian noise, example methods described in this chapter included fuzzy vector filters, digital path-based filters, anisotropic diffusion-based filters, and bilateral filters, which all suppress noise by combining the input data.

Knowing the process of image formation is crucial in noise modeling, which is needed to simulate the noise observed in real-life digital images and to study the effect of the filters on both the noise and the desired image features. Filtering performance greatly depends on the measure employed in the filter design to discriminate color vectors and judge their similarities by taking into account the length and/or the orientation of color vectors. The variety of such measures and efficient filter design methodology make the presented vector processing framework very valuable in modern digital imaging and multimedia systems that attempt to mimic the human visual perception and use color as a cue for better image understanding and improved processing performance.

Bibliography

[1] R. Gonzalez and R.E. Woods, *Digital Image Processing*. Reading, MA: Prentice Hall, 3rd edition, 2007.

[2] G. Sharma, "Color fundamentals for digital imaging," in *Digital Color Imaging Handbook*, G. Sharma (Ed.), Boca Raton, FL: CRC Press / Taylor & Francis, 2002, pp. 1–113.

[3] G. Wyszecki and W.S. Stiles, *Color Science: Concepts and Methods, Quantitative Data and Formulas*. New York: Wiley-Interscience, 2nd edition, 2000.

[4] H.J. Trussell, E. Saber, and M. Vrhel, "Color image processing," *IEEE Signal Processing Magazine*, vol. 22, no. 1, pp. 14–22, January 2005.

[5] M. Stokes, M. Anderson, S. Chandrasekar, and R. Motta, "A standard default color space for the Internet—sRGB," *Technical Report*, available online, http://www.w3.org/Graphics/Color/sRGB.html.

[6] R. Lukac, B. Smolka, K. Martin, K.N. Plataniotis, and A.N. Venetsanopulos, "Vector filtering for color imaging," *IEEE Signal Processing Magazine*, vol. 22, no. 1, pp. 74–86, January 2005.

[7] J. Gomes and L. Velho, *Image Processing for Computer Graphics*. New York: Springer-Verlag, 1997.

[8] P. Milanfar, *Super Resolution Imaging*. Boca Raton, FL: CRC Press / Taylor & Francis, September 2010.

[9] S.T. McHugh, "Digital camera image noise." Available online, http://www.cambridgeincolour.com/tutorials/noise.htm.

[10] R. Lukac, "Single-sensor digital color imaging fundamentals," in *Single-Sensor Imaging: Methods and Applications for Digital Cameras*, R. Lukac (Ed.), Boca Raton, FL: CRC Press / Taylor & Francis, September 2008, pp. 1–29.

[11] J. Astola, P. Haavisto, and Y. Neuvo, "Vector median filters," *Proceedings of the IEEE*, vol. 78, no. 4, pp. 678–689, April 1990.

[12] K.N. Plataniotis, D. Androutsos, and A.N. Venetsanopoulos, "Adaptive fuzzy systems for multichannel signal processing," *Proceedings of the IEEE*, vol. 87, no. 9, pp. 1601–1622, September 1999.

[13] V. Kayargadde and J.B. Martens, "An objective measure for perceived noise," *Signal Processing*, vol. 49, no. 3, pp. 187–206, March 1996.

[14] J. Zheng, K.P. Valavanis, and J.M. Gauch, "Noise removal from color images," *Journal of Intelligent and Robotic Systems*, vol. 7, no. 3, pp. 257–285, 1993.

[15] K.K. Sung, *A Vector Signal Processing Approach to Color*. M.S. thesis, Massachusetts Institute of Technology, 1992.

[16] R. Lukac, K.N. Plataniotis, B. Smolka, and A.N. Venetsanopoulos, "Generalized selection weighted vector filters," *EURASIP Journal on Applied Signal Processing*, vol. 2004, no. 12, pp. 1870–1885, September 2004.

[17] R. Lukac and K.N. Plataniotis, "A taxonomy of color image filtering and enhancement solutions," in *Advances in Imaging and Electron Physics*, P.W. Hawkes (Ed.), Elsevier/Academic Press, vol. 140, June 2006, pp. 187–264.

[18] K. Tang, J. Astola, and Y. Neuvo, "Nonlinear multivariate image filtering techniques," *IEEE Transactions on Image Processing*, vol. 4, no. 6, pp. 788–798, June 1995.

[19] R.M. Nosovsky, "Choice, similarity and the context theory of classification," *Journal of Experimental Psychology: Learning, Memory, and Cognition*, vol. 10, no. 1, pp. 104–114, January 1984.

[20] K.N. Plataniotis and A.N. Venetsanopoulos, *Color Image Processing and Applications*, New York: Springer Verlag, 2000.

[21] B. Smolka, R. Lukac, A. Chydzinski, K.N. Plataniotis, and W. Wojciechowski, "Fast adaptive similarity based impulsive noise reduction filter," *Real-Time Imaging*, vol. 9, no. 4, pp. 261–276, August 2003.

[22] R.W.G. Hunt, *Measuring Colour*. Kingston-upon-Thames, England: Fountain Press, 3rd edition, 1998.

[23] S. Susstrunk, R. Buckley, and S. Swen, "Standard RGB color spaces," in *Proceedings of the Seventh Color Imaging Conference: Color Science, Systems, and Applications*, Scottsdale, AZ, November 1999, pp. 127–134.

[24] D. Alleysson D, B.C. de Lavarene, S. Susstrunk S, and J. Herault, "Linear minimum mean square error demosaicking," in *Single-Sensor Imaging: Methods and Applications for Digital Cameras*, R. Lukac (Ed.), Boca Raton, FL: CRC Press / Taylor & Francis, September 2008, pp. 213–237.

[25] S. Argyropoulos, N.V. Boulgouris, N. Thomos, Y. Kompatsiaris, and M.G. Strintzis, "Coding of two-dimensional and three-dimensional color image sequences," in *Color Image Processing: Methods and Applications*, R. Lukac and K.N. Plataniotis (Eds.), Boca Raton, FL: CRC Press / Taylor & Francis, October 2006, pp. 503–523.

[26] L. Guan, S.Y. Kung, and J. Larsen, *Multimedia Image and Video Processing*. Boca Raton, FL: CRC Press, 2001.

[27] C.A. Poynton, *A Technical Introduction to Digital Video*. Toronto, ON, Canada: Prentice Hall, 1996.

[28] S. Susstrunk, "Colorimetry," in *Focal Encyclopedia of Photography*, M.R. Peres (Ed.), Burlington, MA: Focal Press / Elsevier, 4th edition, 2007, pp. 388–393.

[29] R.G. Kuehni, *Color Space and Its Divisions: Color Order from Antiquity to the Present*. Hoboken, NJ: Wiley-Interscience, 2003.

[30] I. Pitas and A.N. Venetsanopoulos, *Nonlinear digital Filters, Principles and Applications*. Dordrecht: Kluwer Academic Publishers, 1990.

[31] B. Smolka, K.N. Plataniotis, and A.N. Venetsanopoulos, "Nonlinear techniques for color image processing," in *Nonlinear Signal and Image Processing: Theory, Methods, and Applications*, K.E. Barner and G.R. Arce (Eds.), Boca Raton, FL: CRC Press, 2004, pp. 445–505.

[32] M. Barni, V. Cappelini, and A. Mecocci, "Fast vector median filter based on Euclidean norm approximation," *IEEE Signal Processing Letters*, vol. 1, no. 6, pp. 92–94, June 1994.

[33] T. Viero, K. Oistamo, and Y. Neuvo, "Three-dimensional median related filters for color image sequence filtering," *IEEE Transactions on Circuits, Systems and Video Technology*, vol. 4, no. 2, pp. 129–142, April 1994.

[34] R. Lukac, B. Smolka, K.N. Plataniotis, and A.N. Venetsanopoulos, "Selection weighted vector directional filters," *Computer Vision and Image Understanding*, vol. 94, no. 1–3, pp. 140–167, April–June 2004.

[35] Y. Shen and K.E. Barner, "Fast adaptive optimization of weighted vector median filters," *IEEE Transactions on Signal Processing*, vol. 54, no. 7, pp. 2497–2510, July 2006.

[36] C. Kotropoulos and I. Pitas, *Nonlinear Model-Based Image/Video Processing and Analysis*, New York: J. Wiley, 2001.

[37] N. Nikolaidis and I. Pitas, "Multichannel L filters based on reduced ordering," *IEEE Transactions on Circuits and Systems for Video Technology*, vol. 6, no. 5, pp. 470–482, October 1996.

[38] P.E. Trahanias and A.N. Venetsanopoulos, "Vector directional filters: A new class of multichannel image processing filters," *IEEE Transactions on Image Processing*, vol. 2, no. 4, pp. 528–534, October 1993.

[39] P.E. Trahanias, D. Karakos, and A.N. Venetsanopoulos, "Directional processing of color images: Theory and experimental results," *IEEE Transactions on Image Processing*, vol. 5, no. 6, pp. 868–881, June 1996.

[40] M.E. Celebi, H.A. Kingravi, R. Lukac, and F. Celiker, "Cost-effective implementation of order-statistics-based vector filters using minimax approximations," *Journal of the Optical Society of America A*, vol. 26, no. 6, pp. 1518–1524, June 2009.

[41] R. Lukac and K.N. Plataniotis, "cDNA microarray image segmentation using root signals," *International Journal of Imaging Systems and Technology*, vol. 16, no. 2, pp. 51–64, April 2006.

[42] R. Lukac, "Adaptive color image filtering based on center-weighted vector directional filters," *Multidimensional Systems and Signal Processing*, vol. 15, no. 2, pp. 169–196, April 2004.

[43] D.G. Karakos and P.E. Trahanias, "Generalized multichannel image-filtering structure," *IEEE Transactions on Image Processing*, vol. 6, no. 7, pp. 1038–1045, July 1997.

[44] M. Gabbouj and A. Cheickh, Vector median-vector directional hybrid filter for color image restoration. *Proceedings of the European Signal Processing Conference*, Trieste, Italy, September 1996, pp. 879-881.

[45] R. Lukac, "Adaptive vector median filtering," *Pattern Recognition Letters*, vol. 24, no. 12, pp. 1889–1899, August 2003.

[46] R. Lukac, K.N. Plataniotis, A.N. Venetsanopoulos, and B. Smolka, "A statistically-switched adaptive vector median filter," *Journal of Intelligent and Robotic Systems*, vol. 42, no. 4, pp. 361–391, April 2005.

[47] R. Lukac, B. Smolka, K.N. Plataniotis, and A.N. Venetsanopoulos, "Vector sigma filters for noise detection and removal in color images," *Journal of Visual Communication and Image Representation*, vol. 17, no. 1, pp. 1–26, February 2006.

[48] R. Lukac, V. Fischer, G. Motyl, and M. Drutarovsky, "Adaptive video filtering framework," *International Journal of Imaging Systems and Technology*, vol. 14, no. 6, pp. 223–237, December 2004.

[49] L. Jin and D. Li, "An efficient color impulse detector and its application to color images," *IEEE Signal Processing Letters*, vol. 14, no. 6, pp. 397–400, June 2007.

[50] L. Jin and D. Li, "A switching vector median filter based on CIELAB color spaces for color image processing," *Signal Processing*, vol. 87, no. 6, pp. 1345–1354, June 2007.

[51] J.G. Camarena, V. Gregori, S. Morillas, and A. Sapena, "Fast detection and removal of impulsive noise using peer groups and fuzzy metrics," *Journal Visual Communication and Image Representation*, vol. 19, no. 1, pp. 20–29, January 2008.

[52] M.E. Celebi and A. Aslandogan, "Robust switching vector median filter for impulsive noise removal," *Journal of Electronic Imaging*, vol. 17, no. 4, pp. 43006, October 2008.

[53] Z. Xu, H.R. Wu, B. Qiu and X. Yu, "Geometric features-based filtering for suppression of impulse noise in color images," *IEEE Transactions on Image Processing*, vol. 18, no. 8, pp. 1742–1759, August 2009.

[54] B. Smolka, "Peer froup switching filter for impulse noise reduction in color images," *Pattern Recognition Letters*, vol. 31, no. 6, pp. 484–495, April 2010.

[55] K.N. Plataniotis, D. Androutsos, and A.N. Venetsanopulos, "Fuzzy adaptive filters for multichannel image processing," *Signal Processing*, vol. 55, no. 1, pp. 93–106, January 1996.

[56] R. Lukac, K.N. Plataniotis, B. Smolka, and A.N. Venetsanopoulos, "cDNA microarray image processing using fuzzy vector filtering framework," *Fuzzy Sets and Systems*, vol. 152, no. 1, pp. 17–35, May 2005.

[57] M. Szczepanski, B. Smolka, K.N. Plataniotis, and A.N. Venetsanopoulos, "On the geodesic paths approach to color image filtering," *Signal Processing*, vol. 83, no. 6, pp. 1309–1342, June 2003.

[58] M. Szczepanski, B. Smolka, K.N. Plataniotis, and A.N. Venetsanopoulos, "On the distance function approach to color image enhancement," *Discrete Applied Mathematics*, vol. 139, no. 1–3, pp. 283–305, April 2004.

[59] P. Perona and J. Malik, "Scale space and edge detection using anisotropic diffusion," *IEEE Transactions on Pattern Analysis and Machine Intelligence*, vol. 12, no. 7, pp. 629–639, July 1990.

[60] G. Sapiro and D.L. Ringach, "Anisotropic diffusion of multivalued images with applications to color filtering," *IEEE Transactions on Image Processing*, vol. 5, no. 11, pp. 1582–1586, November 1996.

[61] B. Coll, J.L. Lisani, and C. Shert, "Color images filtering by anisotropic diffusion," in *Proceedings of the 12th International Workshop on Systems, Signals, and Image Processing*, Chalkida, Greece, September 2005.

[62] B. Smolka, R. Lukac, K.N. Plataniotis, and A.N. Venetsanopoulos, "Modied anisotropic diffusion framework," *Proceedings of SPIE*, vol. 5150, pp. 1657–1666, June 2003.

[63] V. Aurich and J. Weule, "Non-linear Gaussian filters performing edge preserving diffusion," in *Proceedings of the DAGM Symposium*, pp. 538–545, 1995.

[64] S.M. Smith and J.M. Brady, "SUSAN A new approach to low level image processing," *International Journal of Computer Vision*, vol. 23, no. 1, pp. 45–78, May 1997.

[65] C. Tomasi and R. Manduchi, "Bilateral filtering for gray and color images," in *Proceedings of the IEEE International Conference on Computer Vision*, Bombay, India, pp. 839–846, January 1998.

[66] T.Q. Pham and L.J. van Vliet, "Separable bilateral filtering for fast video preprocessing," in *Proceedings of the IEEE International Conference on Multimedia and Expo*, Amsterdam, The Netherlands, July 2005.

[67] B. Weiss, "Fast median and bilateral filtering," *ACM Transactions on Graphics*, vol. 25, no. 3, pp. 519–526, July 2006.

[68] F. Durand and J. Dorsey, "Fast bilateral filtering for the display of high dynamic-range images," *ACM Transactions on Graphics*, vol. 21, no. 3, pp. 257–266, July 2002.

[69] S. Paris and F. Durand, "A fast approximation of the bilateral filter using a signal processing approach," *International Journal of Computer Vision*, vol. 81, no. 1, pp. 24–52, January 2009.

[70] R. Fattal, M. Agrawala, and S. Rusinkiewicz, "Multiscale shape and detail enhancement from multi-light image collections," *ACM Transactions on Graphics*, vol. 26, no. 3, August 2007.

[71] S. Paris, P. Kornprobst, J. Tumblin and F. Durand, "Bilateral filtering: Theory and applications," *Foundations and Trends in Computer Graphics and Vision*, vol. 4, no. 1, pp. 1–73, 2008.

[72] B.K. Gunturk, "Bilateral filter: Theory and applications," in *Computational Photography: Methods and Applications*, R. Lukac (ed.), Boca Raton, FL, USA: CRC Press / Taylor & Francis, October 2010, pp. 339–366.

[73] A.K. Moorthy, K. Seshadrinathan, and A.C. Bovik, "Image and video quality assessment: Perception, psychophysical models, and algorithms," in *Perceptual Imaging: Methods and Applications*, R. Lukac (Ed.), Boca Raton, FL: CRC Press / Taylor & Francis, 2012.

[74] D.F. Rogers and R.E. Earnshaw, *Computer Graphics Techniques: Theory and Practice*, New York: Springer-Verlag, 2001.

[75] CIE publication No. 116 (1995), Industrial colour difference evaluation, Central Bureau of the CIE.

[76] M.R. Luo, G. Cui, and B. Rigg, "The development of the CIE 2000 colour difference formula: CIEDE2000," *Color Research and Applications*, vol. 26, no. 5, pp. 340–350, 2001.

[77] M.R. Luo, G. Cui, and B. Rigg, "Further comments on CIEDE2000," *Color Research and Applications*, vol. 27, pp. 127–128, 2002.

[78] B. Wandell, "S-CIELAB: A spatial extension of the CIE L*a*b* DeltaE color difference metric," Available online, http://white.stanford.edu/~brian/scielab/.

[79] X. Tong, D.J. Heeger, L. van den Branden, and J. Christian, "Video quality evaluation using ST-CIELAB," *Proceedings of SPIE*, vol. 3644, pp. 185–196, 1999.

Chapter 10

Document Image Restoration and Analysis as Separation of Mixtures of Patterns: From Linear to Nonlinear Models

ANNA TONAZZINI
Istituto di Scienza e Tecnologie dell'Informazione

IVAN GERACE
Istituto di Scienza e Tecnologie dell'Informazione
Università degli Studi di Perugia

FRANCESCA MARTINELLI
Istituto di Scienza e Tecnologie dell'Informazione

10.1   Introduction

Conservation, readability, and content analysis of ancient documents are often compromised by several and different damages that they have undertaken over time, and that continue to cause a progressive decay. Natural aging, usage, poor storage conditions, humidity, molds, insect infestations, and fires are the most diffuse degradation factors. In addition, the materials used in the original production of the documents, that is, paper or parchment and inks, are usually highly variable in consistency and characteristics. All these factors concur to cause ink diffusion and fading, seeping of ink from the reverse side (bleed-through distortion), transparency from either the reverse side or from subsequent pages (show-through distortion), stains, noise, low contrast, unfocused, faint, fragmented, or joined characters. Furthermore, these defects are usually varying across the document. These problems are common to the majority of governmental, historical, ecclesiastic and commercial archives in Europe, so that seeking a remedy would have an enormous social and technological impact. Digital imaging can play a fundamental role in this respect. Indeed, it is an essential tool for generating digital archives, in order to ensure the documents' accessibility and conservation, especially for those rare or very important historical documents, whose fragility prevents direct access by scholars and historians. Moreover, OCR (optical character recognition) processing for automatic transcription and indexing facilitates access to digital archives and the retrieval of information. Finally, a very common need is the improvement of the readability by interested scholars.

Often, the digital images of documents are acquired only in grayscale or, at best, in the visible range of the spectrum, due to the larger diffusion of related acquisition equipment. However, owing to specific damages, some documents may be very difficult to read by the naked eye. This particularly concerns documents produced during the sixteen and seventeenth centuries, due to the corrosion, fading, seeping, and diffusion of the ink used (iron-gall mostly), and those produced even more recently, due to the bad quality of the paper that started being used after the nineteenth century. Furthermore, interesting features are often barely detectable in the original color document, while revealing the whole contents is an important aid to scholars who are interested in dating or establishing the origin of the document itself, or reading the hidden text it may contain. As an example, in the case of palimpsests, usually ancient manuscripts that have been erased and then rewritten, what is desired is to enhance and let "emerge" traces of the original underwriting. Thus, additional information can sometimes be obtained from images taken at nonvisible wavelengths, for instance, in the near-infrared and ultraviolet ranges. Alternatively, or in conjunction with multispectral/hyperspectral acquisitions, digital image processing techniques can be used for enhancing the readability of the document contents and seeking out new information.

10.1.1   Related Work

One of the most frequent degradations considered in the literature regarding document restoration is the presence of background artifacts, which must be removed prior (e.g., binarization). This is mainly caused by back-to-front interferences. Bleed-through is intrinsic to the document, and occurs when some patterns interfere with the main text due to seeping of ink from the reverse page side, caused, for example, by humidity. Show-through may appear also in scans of modern, well-preserved documents when the paper is not completely opaque. Removing the bleed-through or show-through patterns from a digital image of a document is not trivial, especially with ancient originals, where interferences of this kind are usually very strong. Indeed, dealing with heavy bleed-through degradation is practically impossible by any simple thresholding technique, because the intensities of the unwanted background can be very close to those of the main text. Under these conditions, thresholding either does not remove bleed-through, or also eliminates part of the information in the front side. Thus, adaptive and/or structural approaches be adopted [1]. In [2], the front side of a color image is processed via multiscale analysis, employing adaptive binarization and edge magnitude thresholding. Adaptive thresholding is also used in [3]. Other authors propose segmentation of the different color clusters in the image via adaptations of the k-means algorithm [4, 5]. In [6], for the grayscale scan of a single-sided document, a classification approach based on a double binary Markov Random Field, one for the front side and the other for the back side, is proposed.

Most methods are based on the exploitation of information from the front and back sides, usually referred to as recto and verso, and mainly treat grayscale images. Although these methods require a preliminary registration of the two sides, they usually outperform the methods based on a single-sided scan, as exploiting the information contained in the two sides allows for better discriminating the foreground text from interferences, especially when they are almost as strong as the main text itself. In [7], the nonlinear physical model of the show-through in modern scanners is first simplified for deriving a linear mathematical model, and then an adaptive linear filter is exploited that uses scans of both sides of the document. In [8], the two sides of a graylevel manuscript are compared at each pixel, and, basically, a thresholding is applied. In [9], a wavelet technique is applied for iteratively enhancing the foreground strokes and smearing the interfering strokes. In [10], steps of segmentation, to identify the bleed-through areas, are followed by inpainting of estimated pure background areas. Segmentation-classification is the basis also for the methods derived in [11] and [12]. More recently, variational approaches, based on nonlinear diffusion, have been proposed to model and then remove this kind of degradation [13, 14].

A common drawback of all the approaches above is that they often remove all the structured background, and, in addition to canceling the strokes coming from the reverse side, they can also remove other patterns originally belonging to the front side. This may be undesirable when these patterns are signs of the document history and authenticity. Furthermore, specific situations in which the interfering texts, or some of them, can be of interest themselves have received little attention. Consider, for instance, the already-mentioned case of underwritings in palimpsests, or the presence of stamps, or paper watermarks. Often, these patterns represent the most significant information from a cultural and historical point of view, whereas they are usually barely perceivable in the originals. Hence, the goal of document restoration should be their enhancement and recovery. In other cases, even the recovery of the bleed-through/show-through pattern can be of interest per se, for instance when the verso scan is not available.

10.1.2 Blind Source Separation Approach

Blind source separation (BSS) [15] techniques have recently attracted great interest for restoring degraded documents affected by back-to-front interferences. Indeed, the most physically reliable data models consider the appearance of a degraded document as the overlapping of several individual layers of information, or patterns, or sources. BSS, based on the availability of multispectral or multiview (e.g., recto–verso) scans, enables, at least in principle, to disentangle these patterns and then to analyze the overall content of the document itself. With this approach, artifact removal and text readability improvement could be achieved preserving the genuine features of the document (e.g., paper texture and color) and simultaneously enhancing and extracting other salient structures it might contain.

One of the first works in this direction is the already-cited paper [7], although the concepts of pattern mixtures and blind separation are not explicitly stated. Furthermore, the model is limited to grayscale recto–verso scans, and to mild show-through distortions. In [16], for the first time, we proposed a linear instantaneous mixing model for the data, and, assuming statistical independence of the sources, an Independent Component Analysis (ICA) strategy to analyze multispectral single-sided scans with several overlapping information layers. In [17] and [18], we extended this approach to the reduction of the show-through/bleed-through interference in registered double-sided grayscale and RGB scans, respectively, employing data decorrelation.

Nevertheless, as highlighted in [7]and [13], the physical model underlying text overlapping is very complicated because it derives from complex chemical processes of ink diffusion, paper absorption, and light transmission, which are expected to be nonlinear and nonstationary across the document page. As a fundamental aspect, it is apparent that, in the pixels where two texts are superimposed on each other, the resulting intensity is not the vector sum of the intensities of the two components, but it is likely some nonlinear combination of them. Furthermore, in the recto–verso case, it is also apparent that the pattern interfering from the back side is always blurred, whereas the same pattern is not, or much less, on the side where it was originally written. Hence, the inclusion of blur kernels, or Point Spread Functions (PSFs), into the model is necessary to permit the two versions of a same pattern to match each other. Also in the multispectral single-side case, due to the different filters used, the various overlapped patterns can be affected by different blur kernels, and, in addition, these blur kernels can be different from one channel to the other, thus making clearly inconsistent the instantaneous assumption.

Despite that, our initial linear instantaneous model, solved with BSS algorithms, proved to give satisfactory results in many cases of ancient documents affected by show-through or bleed-through, and even to remove other kinds of degradations, such as spots, or enhance features that are hidden in the document. As a step toward a mathematical model that is likely to be more adhering to the physics of the problem, we also attempted to model the observations as linear convolutional mixtures of the source images. In this formulation, the value of the generic pixel in the observations results from a weighted average of the source values in the supports of the blur kernels. In [19], we proposed this extended model for both the multispectral single-side case and the multispectral double-side case. We solved the fully blind estimation problem through the use of a regularizing edge-preserving model for the source images, and a mixed deterministic-stochastic minimization of the related energy function, with respect to the images and the model parameters. Along the line of including a blur kernel, for the grayscale double-side case alone, [20] proposes a convolutional BSS formulation, which accounts also for a nonlinearity assumed known and derived from [7]. The solution is based on the total variation stabilizer for the source images. In [21], a maximum likelihood approach is proposed for two nonlinear mixtures of two sources, where the nonlinearity, derived experimentally, is approximated as quadratic, and a blur kernel on the interfering pattern is accounted for.

10.1.3   Chapter Outline

In this chapter we summarize our activity on document analysis and restoration, carried out in the framework of the blind source separation approach. We describe the various mixing models adopted and the related separation algorithms. Starting with a description of our very initial linear and instantaneous mixing model, solved through statistical methods such as Independent Component Analysis (ICA) and data decorrelation, we move toward its convolutional extension, solved through regularization techniques, exploiting edge-preserving stabilizers for the images. In both cases, our approach is fully blind, that is, null or little information is assumed on the parameters of the model, which must be estimated along with the restored images. The two models/methods were developed for both the multispectral single-side case and the multispectral double-side case, with the aim of restoring the document from interferences that compromise its readability, but also at analyzing its contents, that is, to separately extract its components and/or enhance hidden patterns. Comparisons between the performances of the two different models and the related solution techniques are provided in real examples. Finally, we describe our more recent work based on the adoption of a nonlinear mixing model, inspired by the one proposed in [7]. While the original model in [7] is specifically designed for restoring grayscale double-sided documents affected by mild show-through, we generalized it to cope also with the more invasive bleed-through distortion. Differently from the adaptive filtering algorithm proposed in [7], our fully blind solution strategy is based on edge-preserving regularization to achieve the data-driven estimation of both the restored images and the model parameters. The method can be easily extended to the multispectral case by restoring the various pairs of recto–verso channels independently. Preliminary experimental results on real data will be shown and compared with those obtained with the methods based on linear data models.

10.2 Linear Instantaneous Data Model

As mentioned in Section 10.1, in our first approach, the restoration of degraded documents was formulated as the problem of recovering and enhancing all the patterns overlapped in the document, by exploiting multiple observations, modeled as linear instantaneous mixtures of the patterns themselves, and employing statistical BSS algorithms. In this way, we have been able to formulate several different document analysis problems as various instances of the same BSS problem. This unified formulation, although oversimplified, is sufficiently versatile and flexible to treat many real scenarios, as it will be shown in the section devoted to the experimental results.

In the literature, the original BSS framework of linear and instantaneous mixtures has been mainly considered for 1-D signals [22–24], and, more recently, for solving important image processing and computer vision problems as well [15]. In this formulation, each pixel of the observations solely results from a linear combination of the component values in the corresponding pixel. The linear and instantaneous assumption is often physically grounded, as for instance in the case of radiation sky maps in astrophysics [25]. For the application considered in this chapter, it represents a first approximation of more complex combination phenomena.

10.2.1 Single-Side Document Case

In the multispectral/hyperspectral single-side scenario, we assume that available multiple reflectance maps of the recto side of a document has a vector value $\mathbf{x}(t)$ of N components, where t is the pixel index, and the total number of pixels in the images is T. Similarly, we assume to have M superimposed sources represented, at each pixel t, by the vector $\mathbf{s}(t)$. Because we consider images of documents containing homogeneous texts or drawings, we can also reasonably assume that the color of each source, in its pristine state (i.e., undegraded), is almost uniform. We call A_{ij} the mean reflectance index for the jth source at the ith observation channel. Thus, the source functions $s_i(t)$, $i = 1, 2, ..., M$, denote the "quantity" of M patterns that concur to form the color at point t. In formulas, it is

$$\mathbf{x}(t) = A\mathbf{s}(t), \qquad t = 1, 2, ..., \mathrm{T}, \tag{10.1}$$

where the $N \times M$ matrix A has the meaning of a mixing matrix. We call this model *the interchannel model*. One of its advantages is that, because it uses the reflectance maps of the document at different bands, there is no need for registration. Furthermore, when the observations amount to the RGB scan of the document, the original RGB representation of the individual sources can also be recovered by exploiting the estimated mixing coefficients. For the ith source, it is

$$
\begin{aligned}
s_i^R(t) &= A_{1i}s_i(t), \\
s_i^G(t) &= A_{2i}s_i(t), \\
s_i^B(t) &= A_{3i}s_i(t), \qquad t = 1, 2, ..., \mathrm{T}. \tag{10.2}
\end{aligned}
$$

This possibility, of course, requires that a full separation is achieved for the source we want to colorize.

However, as already highlighted, this model does not perfectly describe the phenomenon of interfering texts in documents because it does not account for color saturation in correspondence of occlusions, nor for the sensible blurring effect due to ink diffusion or light spreading through the support. However, we can reasonably assume that the linear approximation holds when the patterns do not occlude each other, and that the occlusion areas are few. Similarly, we assume that each source pattern is affected by its own blurring operator that is approximately the same for a given source in the different observations. At this stage, we then assume that blurring can be disregarded.

10.2.2 Recto–Verso Document Case

In a second scenario, the available document observations amount to two sets of N multispectral scans of the recto and verso side of a page, affected by either show-through or bleed-through. To build a model for our data, we only consider two distinct patterns, one in the clean recto side and one in the clean verso side of the page, which appear as overlapped in the N scans. If, as already done for the multispectral single-side case, we neglect both the occlusion effect and the blur on the interfering patterns, the recovery of the clean recto and verso images can then be formulated as the solution of a set of N, 2×2 linear instantaneous BSS problems, one for each channel, resulting in the following instance of the model of Equation (10.1):

$$
\begin{aligned}
x_1^k(t) &= A_{11}^k s_1^k(t) + A_{12}^k s_2^k(t), \\
x_2^k(t) &= A_{21}^k s_1^k(t) + A_{22}^k s_2^k(t), \quad t = 1, 2, \ldots, \mathrm{T}, \quad k = 1, 2, \ldots, N,
\end{aligned}
\tag{10.3}
$$

where $x_1^k(t)$ and $x_2^k(t)$ are the recto and the horizontally flipped verso observed reflectances, respectively, at the kth channel; $s_1^k(t)$ and $s_2^k(t)$ are the ideal reflectance maps at the kth channel associated with the recto and verso pattern, respectively; t is a pixel index; and A_{ij}^k are unknown mixing coefficients. Physically, coefficients A_{ij}^k represent the percentage of ink intensity attenuation of the two patterns on the two sides. We call this model the *intrachannel model*.

The general formulation of Equation (10.3) can be simplified. Let us consider that, at least in an idealized setting, the two sides have been written with the same ink, same pressure, and at two close moments. Then, it is reasonable to assume that, at each channel, the attenuation of the bleed-through/show-through pattern on the two sides is the same, that is, $A_{12}^k = A_{21}^k$. For similar considerations, it is also expected that $A_{11}^k = A_{22}^k$. Furthermore, the ink intensity of the front text pattern on the recto side should be higher than that of the bleed-through pattern, that is, $A_{11}^k > A_{12}^k$, with the same relationship holding, reversed, on the back side. It is worth noting that these constraints on the mixing operator, if exploited in a direct manner, would prevent the permutation indeterminacy otherwise inherent in BSS methods. However, here, we will exploit them indirectly, through the choice of a suitable separation algorithm, as explained later on.

Note that here the mutual independence of the overall set of sources cannot be assumed, as the different colors of a same pattern are certainly highly correlated. On the other hand, the overall $2N \times 2N$ system of equations is separable, so that the N problems can be solved independently. Furthermore, the model in Equation (10.3), together with its constraints, makes sense even in the case where the grayscale scans of the recto and verso sides are only available, so that the model reduces to a single pair of equations ($N = 1$). Finally, it is worth mentioning that, in the RGB case ($N = 3$), the recovered pairs of source patterns at the various channels can be recomposed to give the two restored images in their RGB representation.

10.2.3 Solution through Independent Component Analysis

For simplicity's sake, in this subsection and in the following one, we refer to the single-side model of Equation (10.1), although all considerations and techniques directly apply to each pair of equations of the double-side model of Equation (10.3) as well. Because the ICA techniques we are going to describe are mainly derived for the same number of sources and observations, and noiseless data, we restrict the model to the case $M = N$. As will be shown in the experimental result subsection, this implies that when $M < N$, the remaining estimated sources are noise maps, and when $M > N$, some sources remains mixed.

Solving the system in Equation (10.1) with respect to both A and $\mathbf{s} = (\mathbf{s}(1), ..., \mathbf{s}(T))$ would clearly give an undetermined problem, unless more information is exploited. Under the fundamental hypothesis of a nonsingular mixing matrix, which means linear independence of the views and, in our case, of spectral diversity for the different patterns, blind separation can only be achieved by assuming the mutual statistical independence of the sources. This is the kind of information used in the ICA approach to BSS. Assuming to know the prior distribution for each source, the joint prior distribution for $\mathbf{s}$ is thus given by

$$P(\mathbf{s}) = \prod_{i=1}^{N} P_i(\mathbf{s}_i), \tag{10.4}$$

where $\mathbf{s}_i = (s_i(1), s_i(2), ..., s_i(\mathrm{T}))$. The separation problem can be formulated as the maximization of Equation (10.4), subject to the constraint $\mathbf{x} = A\mathbf{s}$. Calling W the unknown matrix A^{-1}, the problem reduces to the search for a W, $W = (\mathbf{w}_1, \mathbf{w}_2, ..., \mathbf{w}_N)^T$, such that, when applied to the data $\mathbf{x} = (\mathbf{x}_1, \mathbf{x}_2, ..., \mathbf{x}_N)$, produces the set of vectors $\hat{s}_i = \mathbf{w}_i^T \mathbf{x}$ that are maximally independent, and whose distributions are given by the P_i. By taking the logarithm of Equation (10.4), the problem solved by ICA algorithms is then [15]

$$\hat{W} = \arg\max_W \sum_t \sum_i \log P_i(\mathbf{w}_i^T \mathbf{x}(t)) + \mathrm{T} \log |\det(W)|. \qquad (10.5)$$

Once we have estimated W, the vectors $W\mathbf{x}$ are copies of the original sources $\mathbf{s}$, apart from unavoidable scale and permutation indeterminacies. It has been shown that, in addition to independence, to make separation possible, a necessary extra condition for the sources is that they all, but at most one, must be non-Gaussian. To enforce non-Gaussianity, generic super-Gaussian or sub-Gaussian distributions can be used as priors for the sources, and have been proven to give very good estimates for the mixing matrix and for the sources as well, no matter the true source distributions, which, on the other hand, are usually unknown.

The FastICA algorithm gives the possibility of choosing among a number of "nonlinearities," to be used in place of the derivatives of the log-distributions. It solves Equation (10.5) and returns the estimated sources using a fixed-point iteration scheme [26] that has been found in independent experiments to be 10 to 100 times faster than conventional gradient descent methods.

10.2.4 Solution through Data Decorrelation

To enforce statistical independence, the cross-central moments of all orders between each pair of estimated sources must be constrained to zero. Thus, in principle, no source separation can be obtained by only constraining second-order statistics, that is, by enforcing uncorrelation alone. However, decorrelation has the advantage of being always less computationally expensive than most ICA algorithms, with no parameters to be set. Furthermore, it has been proven that, under some conditions, it can be equivalent to ICA [15], and, when the main aim is not full separation but interference reduction, this can be achieved with channel decorrelation if the overlapped unwanted pattern and the main text are actually uncorrelated.

To enforce statistical uncorrelation on the basis of the available color channels, which are normally highly correlated, we must estimate the data covariance matrix and diagonalize it. This is equivalent to orthogonalizing the different color channels. The result of orthogonalization is, of course, not unique. We experimentally tested the performances of two different strategies, namely Principal Component Analysis (PCA) and Symmetric Whitening (SW) [27]. More formally, given the $N \times N$ data covariance matrix

$$R_{\mathbf{xx}} = <\mathbf{x}\mathbf{x}^T>, \qquad (10.6)$$

approximated from the available multispectral observations, its eigenvalue decomposition is

$$R_{\mathbf{xx}} = V_{\mathbf{x}}\Lambda_{\mathbf{x}}V_{\mathbf{x}}^T, \qquad (10.7)$$

where $V_\mathbf{x}$ is the matrix of the eigenvectors of $R_\mathbf{xx}$ and $\Lambda_\mathbf{x}$ is the diagonal matrix of its eigenvalues, in decreasing order. It is easy to verify that the two following choices for W yield a diagonal $R_{\hat{\mathbf{s}}\hat{\mathbf{s}}}$:

$$W_{PCA} = V_\mathbf{x}^T \tag{10.8}$$

$$W_{SW} = V_\mathbf{x}\Lambda_\mathbf{x}^{-\frac{1}{2}}V_\mathbf{x}^T. \tag{10.9}$$

Matrix W_{PCA} produces a set of vectors $\mathbf{s}_i$ that are orthogonal to each other, and whose Euclidean norms are equal to the eigenvalues of the data covariance matrix. Using matrix W_{SW}, vectors $\mathbf{s}_i$ are orthonormal. It is worth noting that matrix W_{SW} has the further property of being symmetric, and then its application is equivalent to ICA when the mixing matrix A is symmetric as well [15]. This is just the property that we exploit to equivalently apply symmetric whitening in place of ICA to solve each pair of equations of the double-side model of Equation (10.3).

We experimentally observed that PCA never performs full separation on our documents, as the component corresponding to the largest eigenvalue is a weighted sum of all the observations, with positive weights. When the document is acquired in RGB, this PCA component can be seen as an optimal graylevel representation of the color image, in the sense that it carries on the maximum of information contained in the original color image. However, sometimes, PCA produces at least one component where some of the interfering patterns are reduced. Conversely, it is interesting to note that symmetric whitening almost always performs show-through/bleed-through reduction, sometimes even outperforming ICA, and often achieves a full separation of the superimposed patterns.

10.2.5 Discussion of the Experimental Results

In this section we give some examples of the potentiality of the linear instantaneous data model and the related BSS algorithms, for the restoration and the analysis of the content of ancient degraded documents. In a first example, shown in Figure 10.1, all three algorithms described above have been tested on the front side of an RGB manuscript affected by heavy show-through. With SW, we were able to recover a map of the main text, cleansed from all interferences, and a second map where the main text appears overlapped with a printed text, unnoticeable in the original RGB manuscript (Figures 10.1(b) and (e)). This demonstrates the capability of these techniques to remove interferences while enhancing hidden patterns. The cleansed main text has been colorized according to Equation (10.2). Note, in this map, the paper texture including a watermark as well. PCA produced a cleansed map of the hidden printed text alone (Figure 10.1(c)), while with ICA, we obtained a cleansed map of the show-through pattern (Figure 10.1(d)). Figure 10.1(f) shows a pseudocolor image built using the grayscale version of the image in Figure 10.1(b) and the image in Figure 10.1(e), where the handwritten and printed texts appear in different colors, then perceptually well separated. It is to be said that, with three observations, we are able, at best, to separate three of the overlapped patterns that the document might contain. For this manuscript, we obtained the individual recovery of three patterns, out of the four patterns actually contained in it. Indeed, the paper texture remained mixed with the main handwritten text. The availability of a fourth observation, for instance in a nonvisible band, might have permitted the individual recovery of the paper texture as well. As a final consideration, in this case

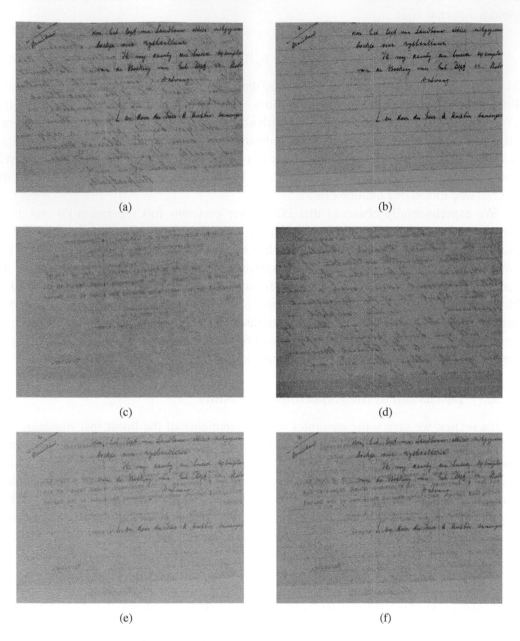

(a)

(b)

(c)

(d)

(e)

(f)

Figure 10.1 Example of removing show-through while enhancing a hidden pattern: (a) the original RGB document; (b) the document cleansed from interferences and colorized; (c) a cleansed map of the hidden printed text alone; (d) a cleansed map of the show-through pattern alone; (e) the handwritten main text plus the hidden printed text; (f) a pseudocolor image built using images in (b) (grayscale version) and (e).

(a) (b) (c)

Figure 10.2 Example of stain attenuation via symmetric whitening: (a) the original RGB document; (b) the grayscale version of the document; (c) first output of SW (note the enhancement of the bleed-through pattern).

none of the three employed BSS algorithms was able, alone, to produce the three separated patterns.

In a second experiment, we applied SW to another RGB manuscript, degraded by a large stain due to humidity. In Figures 10.2(b) and (c) we compare the grayscale version of the color manuscript in Figure 10.2(a) with one of the SW outputs, to can better appreciate the removal of the spot and the simultaneous enhancement of the underlying bleed-through pattern caused by water.

The two previous experiments regarded the analysis of single-sided documents, according to the model of Equation (10.1). Many other examples can be found in [16,27,28]. In the two following experiments, we show the performance of SW for the reduction of show-through or bleed-through from a pair of recto–verso documents, modeled according to Equation (10.3). As already highlighted, because we assume the symmetry of the mixing matrix for this application, SW is equivalent to ICA, and usually more cheaply. Because the system of equations in Equation (10.3) is separable, for simplicity sake we limited our experimentation to the case of grayscale documents, that is, $N = 1$. However, we recall that an interesting application of the technique is to RGB documents, where the separate solution of the three pairs of problems, one for each of the R, G, and B channels of the images, permits, in a straightforward manner, to recover the color appearance of the restored recto and verso images. Some examples in this respect can be found in [18]. On the other hand, the grayscale setting has application to the large databases of digitized documents that already exist in many archives and libraries, where the documents have been acquired in grayscale only or as black- and-white microfilms.

In Figure 10.3 we show the application of SW to a recto–verso pair drawn from the Google Book Search dataset [29]. The results of Figures 10.3(b) and (e), although satisfactory, present a number of defects. First of all, the apparent nonstationarity of the bleed-through patterns prevents their full removal when using a stationary model (see, for example, the mid-gray residual strokes in the center of the recto side). Second, even where removed, the bleed-through strokes leave white halos. This is probably due to the fact that a pattern in one observation does not exactly match the corresponding pattern in the other observation. Indeed, we already highlighted that the bleed-through/show-through pattern is always blurred. However, the pixel-by-pixel processing performed by SW presents also has some advantages, for example with respect to standard thresholding, even when successfully applicable. First of all, the background is not flat, and preserves the genuine features of the side they belong to. Consider, for example, the handwritten annotation in the top-left corner of the recto side, or the word "ECONOMIA" in the center of the same side. The annotation is intrinsic to the recto side, whereas the word "ECONOMIA" is likely to be a show-through effect from a third page. In both cases, these marks cannot be removed with our techniques because there are no corresponding patterns on the verso side. Nevertheless, we deem this a positive characteristic of the technique. Indeed, indiscriminately removing all the textured background could also be negative, at least when the aim of the processing is to restore the appearance of the document in its pristine state. Furthermore, simple post-processing techniques can be applied to try to remove the white halos, for example by cutting the accumulated histogram at a certain percentage of the whole histogram, and the original color of the background can be recovered by rescaling the image in the range of the original data, rather than to full range. The results of these post-processings are shown in Figures 10.3(c) and (f).

In Figures 10.4(c) and (d) the results of SW and subsequent post-processing are shown for the recto–verso pair of Figures 10.4(a) and (b). As is apparent, in this case the post-processing was not able to mitigate the effect of the heavy nonstationarity of the back-to-front interferences. Moreover, because of the large superposition between the two over-lapped patterns, it can be clearly seen that, in correspondence with the occlusions, the recovered main text is lighter. This defect is due to the linearity of the model adopted. In the next section we show that extending the model to be linear convolutional rather than instantaneous, and adopting a model of regularity for the images to be recovered can help in solving the drawbacks related to the nonstationarity of the degradation. However, the problem of occlusions can only be solved if a nonlinear model is adopted, as will be shown in Section 10.4.

10.3   Linear Convolutional Data Model

For the single-side multispectral case, the previous linear instantaneous data generation model (the interchannel model) of Equation (10.1) can be extended to be linear convolutional as follows:

$$x_i(t) = \sum_{j=1}^{M} A_{ij} h_{ij}(t) \otimes s_j(t) + n_i(t), \quad t = 1, 2, \ldots, \mathrm{T}$$

$$i = 1, 2, \ldots, N, \qquad (10.10)$$

where quantity $h_{ij}(t) \otimes s_j(t)$ is the blurred version of source $\mathbf{s}_j$ that contributes to the observed reflectance map $\mathbf{x}_i$, at pixel t, and h_{ij} is a blur kernel. Owing to the presence, in the model, of the mixing scalar coefficients A_{ij}, we assume a unitary gain for each blur kernel. Below, the set of all blur kernels indicated with $\mathbf{H}$. Note that, because we are going to solve the system through regularization, the constraint that $M = N$ is not necessary anymore, and we can allow for a noise term in each observation.

According to Equation (10.10), the data images are noisy linear instantaneous mixtures of the blurred sources through matrix A, but are convolutional mixtures of the ideal, un-

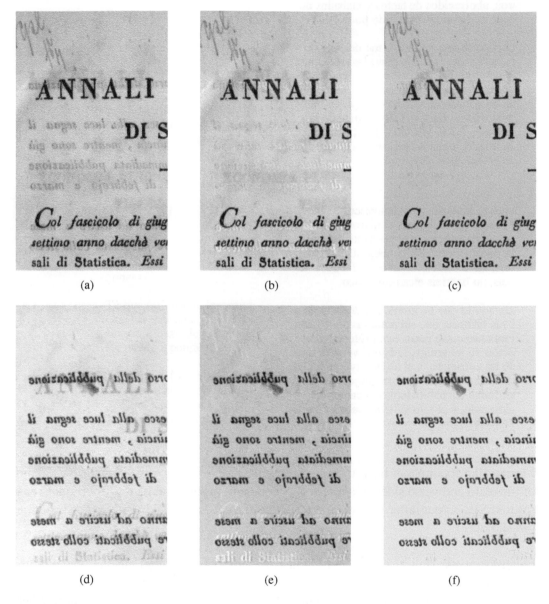

Figure 10.3 Example of bleed-through removal through symmetric whitening of the recto and verso sides of a grayscale document: (a) original recto; (b) full range first SW output; (c) the result in (b) after histogram thresholding; (d) original verso (horizontally flipped); (e) full-range second SW output; (f) result in (e) after histogram thresholding.

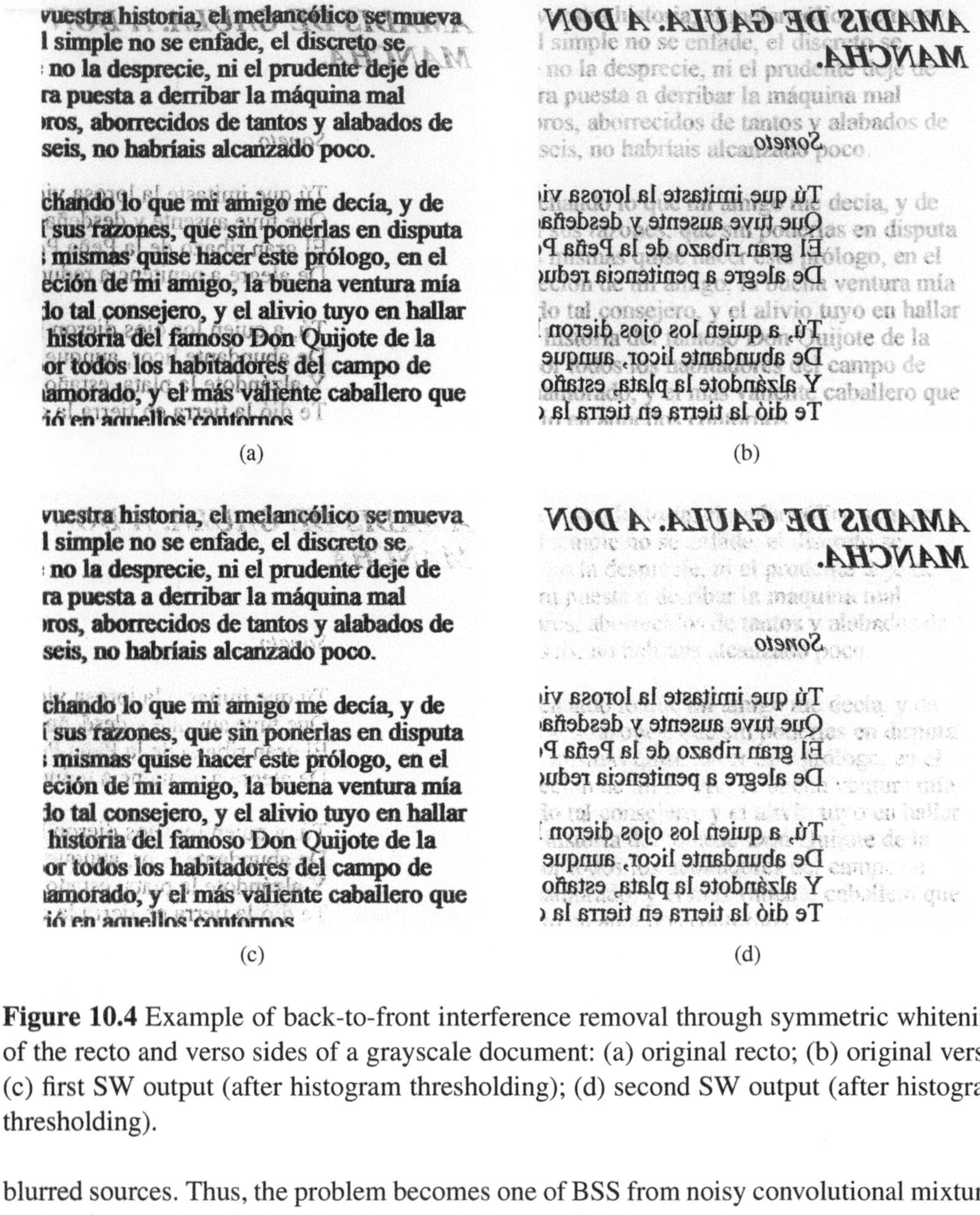

vuestra historia, el melancólico se mueva
l simple no se enfade, el discreto se
: no la desprecie, ni el prudente deje de
ra puesta a derribar la máquina mal
ros, aborrecidos de tantos y alabados de
seis, no habríais alcanzado poco.

chando lo que mi amigo me decía, y de
l sus razones, que sin ponerlas en disputa
s mismas quise hacer este prólogo, en el
eción de mi amigo, la buena ventura mía
lo tal consejero, y el alivio tuyo en hallar
historia del famoso Don Quijote de la
or todos los habitadores del campo de
amorado, y el más valiente caballero que
ió en aquellos contornos

(a)

vuestra historia, el melancólico se mueva
l simple no se enfade, el discreto se
: no la desprecie, ni el prudente deje de
ra puesta a derribar la máquina mal
ros, aborrecidos de tantos y alabados de
seis, no habríais alcanzado poco.

chando lo que mi amigo me decía, y de
l sus razones, que sin ponerlas en disputa
s mismas quise hacer este prólogo, en el
eción de mi amigo, la buena ventura mía
lo tal consejero, y el alivio tuyo en hallar
historia del famoso Don Quijote de la
or todos los habitadores del campo de
amorado, y el más valiente caballero que
ió en aquellos contornos

(c)

(b)

(d)

Figure 10.4 Example of back-to-front interference removal through symmetric whitening of the recto and verso sides of a grayscale document: (a) original recto; (b) original verso; (c) first SW output (after histogram thresholding); (d) second SW output (after histogram thresholding).

blurred sources. Thus, the problem becomes one of BSS from noisy convolutional mixtures or, in other words, of joint blind deconvolution and blind separation.

Research on blind separation methods from convolutional mixtures has largely been addressed for one dimensional signals, with applications, for example, in speech and audio processing and in biomedical signal processing [30–33]. Some specific works on blind separation of convolutional mixtures of images have regarded, for instance, photography of semireflections, or microscopy and tomography [34–36].

Note that classical ICA for instantaneous mixtures can still be applied to the models of Equation (10.10) only when each source is affected by the same blur operator in every mixture, that is, $h_{ij} = h_j \ \forall j, \ \forall i$. However, the recovered sources can be, at best, the blurred versions of the original ones, with amplified noise. More likely, because blurring can introduce cross-correlation among the sources, separation could not be achieved. On the other hand, available knowledge on the sources, the blur kernels, and the mixing cannot be enforced within the ICA approach, which solely relies on source independence.

The linear convolutional data model for the multispectral double-side case (the intra-channel model) becomes

$$
\begin{aligned}
x_1^k(t) &= A_{11}^k h_{11}^k(t) \otimes s_1^k(t) + A_{12}^k h_{12}^k(t) \otimes s_2^k(t) + n_1^k(t), \\
x_2^k(t) &= A_{21}^k h_{21}^k(t) \otimes s_1^k(t) + A_{22}^k h_{22}^k(t) \otimes s_2^k(t) + n_2^k(t), \quad t = 1, 2, \ldots, \mathrm{T} \\
&\hspace{9cm} k = 1, 2, \ldots, N. \ (10.11)
\end{aligned}
$$

For the sake of generality, kernels h_{ii}^k are included to model the possible blurring effect due to the digital acquisition process, ink smearing and fading, and other degradations which might affect also the pattern on the side where it was originally written. Conversely, kernels $h_{ij}^k, i \neq j$, are filters modeling the blurring effect proper of the back-to-front interferences.

The same properties and symmetries of the attenuation coefficients A_{ij}^k assumed in the linear instantaneous model are kept for this model too, and extend to the blur kernels. Thus, we assume $h_{12}^k = h_{21}^k$ and $h_{11}^k = h_{22}^k$. Also, the blur of the bleed-through pattern can be reasonably assumed stronger than that affecting the main text, on both sides. These constraints will be explicitly enforced in the solution strategy, as they allow us to regularize the problem and reduce its dimension. Actually, a further dimension reduction could arise by assuming $A_{11}^k = A_{22}^k = 1$, which means that the main text on the two sides is supposed non-attenuated, apart from the blurring effect. However, we found that considering a possible attenuation for these patterns also helps in increasing the contrast in the reconstructions.

10.3.1 Solution through Regularization

The problem of estimating matrix A, the blur kernels $\mathbf{H}$, and the deblurred sources $\mathbf{s}$ can be stated as the optimization of a regularized energy function $E(\mathbf{s}, A, \mathbf{H})$, composed of a data fidelity term plus stabilizers for the sources, subjected to the model constraints for A and $\mathbf{H}$. In formulas, it is [19]

$$
\begin{aligned}
(\hat{\mathbf{s}}, \hat{A}, \hat{\mathbf{H}}) &= \arg\min_{\mathbf{s}, A, \mathbf{H}} E(\mathbf{s}, A, \mathbf{H}) = \arg\min_{\mathbf{s}, A, \mathbf{H}} \left(\sum_t (A\mathbf{H}\mathbf{s}(t) - \mathbf{x}(t))^T \Sigma^{-1}(A\mathbf{H}\mathbf{s}(t) \right. \\
&\hspace{2cm} \left. -\mathbf{x}(t)) + \sum_i U_i(\mathbf{s}_i) \right), \quad\quad\quad\quad (10.12)
\end{aligned}
$$

where Σ is the covariance matrix of the noise, assumed to be uncorrelated and location independent, and $\mathbf{H}\mathbf{s}$ indicates the $N \times \mathrm{T}$ matrix of the degraded sources.

For each source, we adopt an exhaustive autocorrelation model, which includes regularity of both intensity and edges in the images. This kind of model is particularly suitable for describing images of texts, both printed and handwritten. In our case, we express the regularity of edges by penalizing parallel, adjacent edges, and chose $U_i(\mathbf{s}_i)$ as

$$U_i(\mathbf{s}_i) = \sum_t \sum_{(r,z)\in N_t} \psi_i\left((s_i(t) - s_i(r)), (s_i(r) - s_i(z))\right), \qquad (10.13)$$

where N_t is the set of the two couples of adjacent locations (r, z), $z < r$, which, in the 2D grid of pixels, precede location t in horizontal and in vertical. As functions ψ_i, we chose the following [19, 37]:

$$\psi_i(\xi_1, \xi_2) = \begin{cases} \begin{cases} \lambda_i \xi_1^2 & \text{if } |\xi_1| < \theta \\ \\ \alpha_i & \text{if } |\xi_1| \geq \theta \end{cases} & \text{if } |\xi_2| < \theta, \\ \\ \begin{cases} \lambda_i \xi_1^2 & \text{if } |\xi_1| < \bar{\theta} \\ \\ \alpha_i + \varepsilon_i & \text{if } |\xi_1| \geq \bar{\theta} \end{cases} & \text{if } |\xi_2| \geq \theta, \end{cases} \qquad (10.14)$$

where λ_i is the positive regularization parameter, $\theta = \sqrt{\alpha_i/\lambda_i}$ has the meaning of a threshold on the intensity gradient above which a discontinuity is expected, and $\bar{\theta} = \sqrt{(\alpha_i + \varepsilon_i)/\lambda_i}$ is a suprathreshold, higher than the threshold, to lower the expectation of an edge when a parallel, close edge is likely to be present. Functions ψ_i are equal for all sources, but have possible different hyperparameters, in order to graduate the constraint strength in dependence of the specific source considered.

Problems like those of Equation (10.12) are usually approached by means of alternating componentwise minimization with respect to the three sets of variables in turn. We propose the following theoretical scheme [19]:

$$\left(\hat{A}, \hat{\mathbf{H}}\right) = \underset{A,\mathbf{H}}{\arg\min} \, E(\mathbf{s}(A, \mathbf{H}), A, \mathbf{H}), \qquad (10.15)$$

$$\mathbf{s}(A, \mathbf{H}) = \underset{\mathbf{s}}{\arg\min} \, E(\mathbf{s}, A, \mathbf{H}), \qquad (10.16)$$

subject to the proper constraints on A and $\mathbf{H}$.

In this formulation, the original joint minimization problem is restated as the constrained minimization of the energy function with respect to the whole mixing operator alone, while the sources are kept clamped to the minimizer of the energy function, for any given status of the mixing.

Solving the problem of Equations (10.15) and (10.16) presents some computational difficulties. Indeed, in general, it is not possible to derive analytical formulas for the sources viewed as functions of A and $\mathbf{H}$, and the stabilizers are not convex. Thus, a simulated annealing (SA) algorithm must be adopted for the updating of A and $\mathbf{H}$, and, for each proposal of a new status for each of them, the sources must be computed through numerical estimation. If a correct SA schedule was used, and for each proposal of a new status for A and $\mathbf{H}$ the corresponding sources were estimated, this scheme would ensure convergence to the global optimum. Nevertheless, its computational complexity would be prohibitive. However, some reasonable approximations can be adopted to reduce the complexity of the original problem, while keeping the effectiveness of the approach. First of all, due to the usual small number of mixing and blur coefficients, SA is not particularly cumbersome in this case. On the other hand, based on the feasible assumption that small changes in either A or $\mathbf{H}$ do not significantly affect the sources, these can be updated only after significant modifications—for example, at the end of a complete visitation of all the mixing parameters. Furthermore, although the energy function is nonconvex with respect to the sources, the image models we exploit allow for adopting efficient deterministic nonconvex optimization algorithms, such as the Graduated Non-Convexity (GNC) algorithm [38]. The GNC algorithm is based on the minimization, in sequence, of a set of approximations of the original energy function, where the first approximation is a convex function and the last approximation coincides with the original function. By observing that, in our case, the data consistency term of the energy is quadratic in s, whereas the stabilizer is not, in order to find a first convex approximation for the energy function we only need to properly approximate the specific stabilizer in Equation (10.14) [39]. The whole blind separation algorithm thus reduces to an alternating scheme governed by an external simulated annealing for the estimation of A and $\mathbf{H}$, according to Equation (10.15), interrupted at the end of each Metropolis cycle, to perform an update of the sources s, according to Equation (10.16) and via GNC [19]. In order to avoid premature freezing of the solution, at each temperature of SA we employ only ten approximations for the energy function and minimize each of them via ten steps of gradient descent. This further reduces the computational costs. In all the experiments performed using this scheme, stabilization of the solutions was always observed.

10.3.2 Discussion of the Experimental Results

The regularization method described above to separate overlapped patterns in documents, based on a linear convolutional data model, was conceived for application to multispectral scans of both single-sided and double-sided documents.

In Figure 10.5(b) we show the result obtained when the goal of the analysis of a single-sided RGB manuscript (Figure 10.5(a)) was the extraction of the main text, cleansed from all interferences. This result is compared with the best one we were able to obtain with BSS algorithms. It is apparent, here, the effectiveness of enforcing a local smoothness constraint for the pattern we want to extract.

In a second example, we applied the same method to the grayscale recto–verso pair of Figures 10.4(a) and (b). A comparison with the results obtained through SW (see Figures 10.4(c) and (d)) clearly highlights the superiority of the present approach. As expected, the inclusion of a convolutional kernel into the data model permits a better match between a same pattern in the two views, and then a better removal of the interferences, without residual halos. At the same time, the adoption of a local autocorrelation model for the patterns themselves allows to mitigate, at least to some extent, the effects of the nonstationarity of the back-to-front interferences, even when a stationary data model is considered. However, note that the attenuation of the ink intensity in correspondence with the occlusions remains

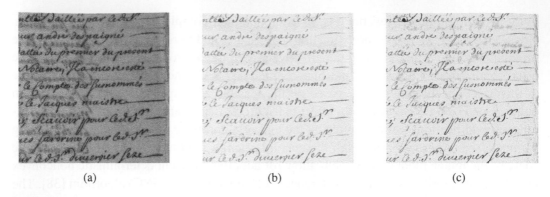

(a)   (b)   (c)

Figure 10.5 Example of interference removal and recovery of the main text from an RGB manuscript: (a) the original image; (b) the clean main text obtained with the regularization technique based on a linear convolutional data model; (c) the best extraction of the main text obtainable with ICA (after histogram thresholding).

and is even more apparent. In the following section we show that this drawback can be avoided only within a nonlinear convolutional data model.

10.4   Nonlinear Convolutional Data Model for the Recto–Verso Case

To derive a nonlinear convolutional data model, we drew inspiration from the show-through model for modern scanners proposed in [7], and restrict ourselves to the case of grayscale recto–verso documents affected by back-to-front interferences.

The show-through distortion in modern scanners is driven by a very complex physical model, depending on the paper scattering and transmittance parameters, the spreading of light in the paper, and the reflectance of the backing. In [7], a simplified, yet complex, mathematical model was first derived and then further approximated to make it tractable. In particular, the assumption that the fraction of light transmitted is much smaller than the fraction of light scattered permitted to "linearize" the model in such a way that the observed optical density on the recto side is described as the sum between the ideal recto optical density and the absorptance of the verso written pattern convolved with an unknown blur kernel, the so-called "show-through point spread function (PSF)." A specular equation describes the observed optical density of the verso side in terms of the absorptance of the printed pattern on the recto side. Note that although the model is linear convolutional in density and absorptance, it results are nonlinear when written in terms of all densities or all reflectances, due to the logarithmic relationship between optical density and reflectance. The model, in all its instances, is symmetric and stationary; that is, the same show-through PSF is assumed for both sides, and a single parameter is taken for the reflectance of white paper, unprinted on either sides.

To derive the show-through cancelation algorithm of [7], the model is further simplified by approximating the unknown show-through corrected absorptances with the corresponding absorbances from the observed scans, so that the two sides can be processed indepen-

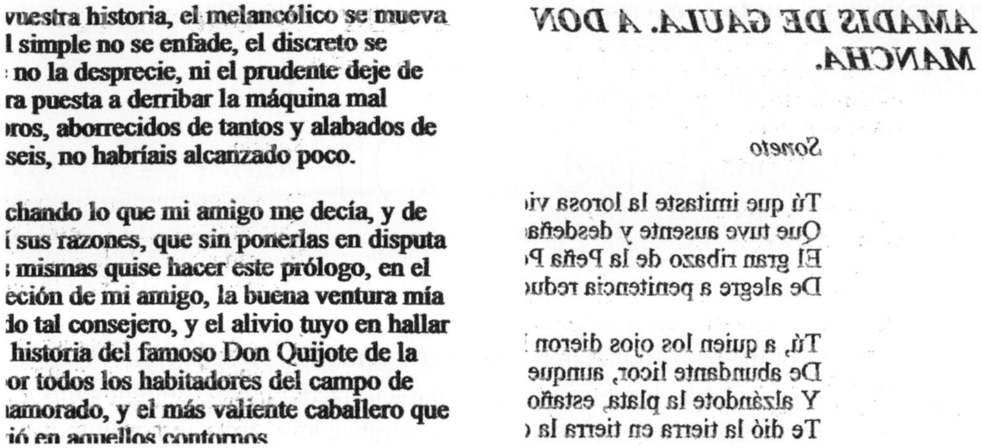

vuestra historia, el melancólico se mueva
l simple no se enfade, el discreto se
: no la desprecie, ni el prudente deje de
ra puesta a derribar la máquina mal
)ros, aborrecidos de tantos y alabados de
seis, no habríais alcanzado poco.

chando lo que mi amigo me decía, y de
í sus razones, que sin ponerlas en disputa
; mismas quise hacer este prólogo, en el
eción de mi amigo, la buena ventura mía
lo tal consejero, y el alivio tuyo en hallar
historia del famoso Don Quijote de la
or todos los habitadores del campo de
iamorado, y el más valiente caballero que
ió en aquellos contornos

(a)

(mirrored text, right column)

AMADIS DE GAULA, A DON
MANCHA.

Soneto

Tú que imitaste la llorosa vi
Que tuve ausente y desdeña
El gran ribazo de la Peña P
De alegre a penitencia redu

Tú, a quien los ojos dieron
De abundante licor, aunque
Y alzándote la plata, estaño
Te dió la tierra en tierra la

(b)

Figure 10.6 Example of back-to-front interference removal from the recto–verso pair of Figures 10.4(a) and (b), using the regularization technique based on a linear convolutional data model: (a) the restored recto side; (b) the restored verso side.

dently and in a pixel-by-pixel fashion. This approximation allows us to relax the symmetry and stationarity of the original model, so that a possibly different PSF is estimated on each side, and tracked across the areas where the currently processed side has no printing and the other has printing, through a least-mean square adaptive filtering algorithm. The resulting algorithm is very simple and effective; however, it requires a number of parameters to be set manually; for example, thresholds to segment the images in printed or nonprinted areas, and the show-through PSF is assumed small in comparison to unity. This means that the algorithm is specifically designed for modern scanned documents, affected by a relatively mild show-through distortion.

Unfortunately, the situation is typically worse when bleed-through is considered, in that it is caused by ink seeping through the paper fiber and chemical transformations of the materials, rather than by pure light transmission. In addition, the bleed-through pattern is usually an interference that is stronger than show-through, and it is likely to be highly nonstationary, due to unpredictable causes such as accidental humidity undergone by some parts of the sheet, inhomogeneity of the support, or higher pressure during the writing process on the back side. From all the considerations above, it is clear that a general and comprehensive mathematical model for generic back-to-front interferences in ancient documents, accounting for these large variability of degradation effects, cannot be easily formulated. We then attempted to generalize the original show-through model in [7] to account for higher levels of interferences, and for nonsymmetries between the two sides. Furthermore, without decoupling the two equations, we propose to solve the data model jointly in both images and the data model parameters through regularization, including a stabilizer for the solutions that, as shown in the previous section, can also help in managing the nonstationarity.

Assuming, for simplicity's sake, a single observation channel (or a grayscale observation), the nonlinear recto–verso model we considered is the following:

$$x_1(t) = s_1(t) \cdot \exp\left\{-q_2 \left[h_2(t) \otimes \left(1 - \frac{s_2(t)}{R_2}\right)\right]\right\} + n_1(t),$$

$$x_2(t) = s_2(t) \cdot \exp\left\{-q_1 \left[h_1(t) \otimes \left(1 - \frac{s_1(t)}{R_1}\right)\right]\right\} + n_2(t), \quad t = 1, 2, \ldots, T,$$

$$(10.17)$$

where, again, $x_1(t)$ and $x_2(t)$ are the observed reflectances; $s_1(t)$ and $s_2(t)$ are the ideal reflectances, of the front and back side, respectively, at pixel t; and R_1 and R_2 are the mean reflectance values of the background in the recto and verso side, respectively. We assumed two different PSFs, h_1 and h_2 for the two sides, characterized by two different gains q_1 and q_2, possibly higher than 1, which represent the interference level from the front to the back and from the back to the front, respectively. With this notation, h_1 and h_2 are intended to be of unitary sum.

Let $\mathbf{q} = \{q_1, q_2\}$ and $\mathbf{H} = \{h_1, h_2\}$; then the data model of Equation (10.17) can be rewritten in compact matrix form as:

$$\mathbf{x}(t) = B(\mathbf{s}, \mathbf{q}, \mathbf{H}; t)\mathbf{s}(t) + \mathbf{n}(t), \quad t = 1, 2, \ldots, T, \quad (10.18)$$

where $B(\mathbf{s}, \mathbf{q}, \mathbf{H}; t)$ is a 2×2 diagonal matrix, whose diagonal elements are

$$B_{11}(t) = \exp\left\{-q_2 \left[h_2(t) \otimes \left(1 - \frac{s_2(t)}{R_2}\right)\right]\right\},$$

and

$$B_{22}(t) = \exp\left\{-q_1 \left[h_1(t) \otimes \left(1 - \frac{s_1(t)}{R_1}\right)\right]\right\},$$

respectively.

10.4.1 Solution through Regularization

We propose to solve the system of Equation (10.17) with the same regularization strategy already adopted for the linear convolutional data model, employing an edge-preserving autocorrelation model for the sources. The fully blind problem we must solve then becomes:

$$(\hat{\mathbf{s}}, \hat{\mathbf{q}}, \hat{\mathbf{H}}) = \arg\min_{\mathbf{s}, \mathbf{q}, \mathbf{H}} E(\mathbf{s}, \mathbf{q}, \mathbf{H}) \quad (10.19)$$

$$= \arg\min_{\mathbf{s}, \mathbf{q}, \mathbf{H}} \left(\sum_t (B(\mathbf{s}, \mathbf{q}, \mathbf{H}; t)\mathbf{s}(t) - \mathbf{x}(t))^T \Sigma^{-1} (B(\mathbf{s}, \mathbf{q}, \mathbf{H}; t)\mathbf{s}(t) - \mathbf{x}(t)) \right.$$

$$\left. + \sum_i U_i(\mathbf{s}_i) \right),$$

where the stabilizers $U_i(\mathbf{s}_i)$ are given by Equations (10.13) and (10.14). Nevertheless, the problem of Equation (10.19) presents an extra difficulty with respect to the homologous problem of Equation (10.12). Indeed, while in that case the cost energy, seen as a function of $\mathbf{s}$, was nonconvex only for the stabilizer part, this time also the data fidelity term is nonconvex. Thus, in order to use a GNC strategy to minimize the energy function in s, we need to built a first convex approximation for the data term as well. Unfortunately, this is not an easy task, if both s_1 and s_2 are considered as variables, as, basically, one variable multiplies the exponential of the other. Conversely, things becomes tractable when we only linearize the exponential, by considering fixed the multiplicative variable. This strategy leads us to modify the estimation scheme of Equations (10.15)-(10.16) as follows:

$$\left(\hat{q}, \hat{H}\right) = \arg\min_{q,H} E(s(q,H), q, H), \tag{10.20}$$

$$s_1(s_2, q, H) = \arg\min_{s_1} E(s, q, H), \tag{10.21}$$

$$s_2(s_1, q, H) = \arg\min_{s_2} E(s, q, H), \tag{10.22}$$

subject to the proper constraints on q and H.

In [40], the derivation of a convex approximation for the data term is discussed in detail. Basically, this is obtained by approximating the exponential function through the polynomial of best approximation, which must be of degree 1 in this case, and exploiting the Gram–Schmidt orthogonalization. In particular, we are interested in a projection onto a subspace in which the approximation function vanishes in zero.

The whole blind separation algorithm thus reduces to an alternating scheme governed by an external simulated annealing for the estimation of q and H, according to Equation (10.20), interrupted at the end of each Metropolis cycle, to perform an update of the sources s, according to Equations (10.21) and (10.22) via GNC. The same implementation solutions already adopted in the case of the linear convolutional data model are adopted here as well, in order to reduce the computational costs and achieve stabilization of the estimates. Briefly, these consist of terminating SA when the reconstructions stabilize and employing, at each temperature, only ten approximations for the energy function. In addition, each approximation is minimized via ten steps of gradient descent only.

10.4.2 Discussion of the Experimental Results

As a preliminary example of the results obtainable with this regularized nonlinear convolutional data model, we applied the method described above to the same recto–verso pair shown in Figures 10.4(a) and (b). The results, shown in Figure 10.7, clearly outperform both those obtained with SW (Figures 10.4(c) and (d)) and those obtained with the regularized linear convolutional data model (Figures 10.6(a) and (b)). In particular, the inclusion of nonlinearity allows a perfect recovery of the main text in correspondence with the occlusions, and the use of a local smoothness constraint for the images allows us to remedy the nonstationarity of the degradation. Further experimentation to test the validity of the approach on a large dataset of documents affected by a strong bleed-through distortion is currently underway.

10.5 Conclusions and Future Prospects

In this chapter we surveyed our activity on document analysis and restoration, formulating the problem as one of blind source separation (BSS). Indeed, ancient documents are of-

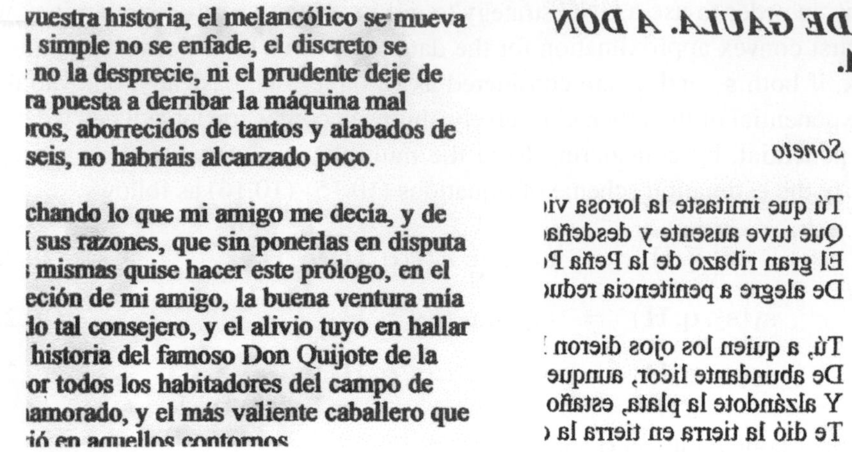

vuestra historia, el melancólico se mueva
l simple no se enfade, el discreto se
no la desprecie, ni el prudente deje de
ra puesta a derribar la máquina mal
ros, aborrecidos de tantos y alabados de
seis, no habríais alcanzado poco.

chando lo que mi amigo me decía, y de
l sus razones, que sin ponerlas en disputa
mismas quise hacer este prólogo, en el
eción de mi amigo, la buena ventura mía
lo tal consejero, y el alivio tuyo en hallar
historia del famoso Don Quijote de la
or todos los habitadores del campo de
amorado, y el más valiente caballero que
ió en aquellos contornos

(a)

(b)

Figure 10.7 Example of back-to-front interference removal from the recto–verso pair of Figure 10.4(a) and (b), using the regularization technique based on a nonlinear convolutional data model: (a) the restored recto side; (b) the restored verso side.

ten affected by several types of artifacts, such as back-to-front interferences, which can be viewed as the overlapping of several individual layers of information, or patterns. Although a general, comprehensive, and reliable physical model accounting for the large variability of degradation effects that cause the phenomenon of pattern superposition in documents is not known, some mathematical models have recently been proposed to approximate specific situations. For example, in [7], an efficient mathematical model for the show-through distortion in grayscale recto–verso scans of modern document has been proposed; and in [13], still for grayscale recto–verso scans, the bleed-through effect has been modeled via nonlinear diffusion.

Our experience in this field began with assuming multispectral views of single-sided documents as linear instantaneous mixtures, with unknown coefficients, of, possibly, all the individual patterns overlapped in the document appearance. Indeed, these patterns can be the main text plus back-to-front interferences, but also other document features, such as stamps, paper watermarks, underwritings, annotations, drawings, and so on. Relying on the spectral diversity of the various patterns and on their statistical independence, interesting results have been obtained using very fast BSS algorithms, namely, ICA and channel decorrelation, to separate and individually extract the various patterns. Similar results have been obtained when the data amounts to multiview observations, for example, multispectral recto–verso scans. Some improvements have been obtained by accounting for different blur operators affecting the various patterns in the various observation channels. Although this extended linear convolutional data model requires the adoption of more expensive regularization techniques exploiting edge-preserving stabilizers for the images, its use has proven very useful in permitting the matching of the different patterns in the different observations, and when the overlapped patterns are nonstationary. Further improvements can be achieved by adopting, within the same regularization approach, a nonlinear convolutional mixing model, extending the one proposed in [7] to treat high levels of interferences and to account for nonsymmetries.

In this chapter we described the three mixing models above, as well as the related separation algorithms. In all cases, our approach was fully blind, that is, null or little information is assumed on the parameters of the model, which must be estimated along with the restored images. Advantages and drawbacks of the three models/methods were discussed, also through comparisons of their performances on real examples. In particular, we found that the method based on the nonlinear convolutional mixing model is promising for overcoming the drawbacks of the methods based on linear models, that is the poor reconstructions in correspondence of the occlusions between different patterns.

However, some relevant problems remain to be explored. These mainly regard the high computational cost of the regularization methods, caused by the adoption of an SA scheme for the estimation of the mixing parameters, and the presence, in the regularizing image model, of some hyperparameters that, at the moment, are selected in a heuristic way, based on the visual inspection of the mixtures, or by trial and error. Another limitation of our approach is the stationarity of the data models, which is clearly invalid in many real cases, especially when the document is affected by invasive bleed-through distortion. Finally, because the nonlinear data model is presently devised for the multispectral recto–verso case alone, and for two overlapped patterns only, we plan to extend it to the multispectral single-side case, thus accounting for the overlapping of several layers of information.

Acknowledgment

The work of Anna Tonazzini and Francesca Martinelli has been supported by Project N. 1220000119, Call PIA 2008, funded by the European Program POR CALABRIA FESR 2007-2013. Partners: TEA sas di Elena Console & C., Istituto di Scienza e Tecnologie dell'Informazione CNR, Dipartimento di Meccanica Università della Calabria. The work of Ivan Gerace has been supported by PRIN 2008 N. 20083KLJEZ.

Bibliography

[1] G. Leedham, S. Varma, A. Patankar, and V. Govindaraju, "Separating text and background in degraded document images — A comparison of global thresholding techniques for multi-stage thresholding," in *Proceedings of the Eighth International Workshop on Frontiers in Handwriting Recognition*, 2002, pp. 244–249.

[2] H. Nishida and T. Suzuki, "A multiscale approach to restoring scanned color document images with show-through effects," in *Proceedings of the International Conference on Document Analysis and Recognition ICDAR*, 2003.

[3] Q. Wang, T. Xia, L. Li, and C. L. Tan, "Document image enhancement using directional wavelet," in *Proceedings of the IEEE Conference on Computer Vision Pattern Recognition*, vol. 2, pp. 534–539, 2003.

[4] Y. Leydier, F. LeBourgeois, and H. Emptoz, "Serialized unsupervised classifier for adaptive color image segmentation: Application to digitized ancient manuscripts," in *Proceedings of the International Conference on Pattern Recognition*, 2004, pp. 494–497.

[5] F. Drida, F. LeBourgeois, and H. Emptoz, "Restoring ink bleed-through degraded document images using a recursive unsupervised classification technique," in *Proceedings of the 7th Workshop on Document Analysis Systems*, 2006, pp. 38–49.

[6] C. Wolf, "Document ink bleed-through removal with two hidden markov random fields and a single observation field," Laboratoire d'Informatique en Images et Systémes d'Information, INSA de Lyon, Tech. Rep. RR-LIRIS-2006-019, November 2006.

[7] G. Sharma, "Show-through cancellation in scans of duplex printed documents," *IEEE Transactions on Image Processing*, vol. 10, no. 5, pp. 736–754, 2001.

[8] E. Dubois and A. Pathak, "Reduction of bleed-through in scanned manuscript documents," in *Proceedings of the IS&T Image Processing, Image Quality, Image Capture Systems Conference*, 2001, pp. 177–180.

[9] C. L. Tan, R. Cao, and P. Shen, "Restoration of archival documents using a wavelet technique," *IEEE Transactions on Pattern Analysis and Machine Intelligence*, vol. 24, no. 10, pp. 1399–1404, 2002.

[10] P. Dano, "Joint restoration and compression of document images with bleed-through distortion," Master's thesis, Ottawa-Carleton Institute for Electrical and Computer Engineering, School of Information Technology and Engineering, University of Ottawa, June 2003.

[11] K. Knox, "Show-through correction for two-sided documents," United States Patent 5832137, November 1998.

[12] Q. Wang and C. L. Tan, "Matching of double-sided document images to remove interference," in *IEEE Conference on Computer Vision and Pattern Recognition (CVPR)*, 2001, p. 1084.

[13] R. F. Moghaddam and M. Cheriet, "Low quality document image modeling and enhancement," *International Journal on Document Analysis and Recognition*, vol. 11, no. 4, pp. 183–201, March 2009.

[14] ——, "A variational approach to degraded document enhancement," *IEEE Transactions on Pattern Analysis and Machine Intelligence*, vol. 32, no. 8, pp. 1347–1361, August 2010.

[15] A. Cichocki and S. Amari, *Adaptive Blind Signal and Image Processing*. New York: Wiley, 2002.

[16] A. Tonazzini, L. Bedini, and E. Salerno, "Independent component analysis for document restoration," *International Journal on Document Analysis and Recognition*, vol. 7, pp. 17–27, 2004.

[17] A. Tonazzini, E. Salerno, and L. Bedini, "Fast correction of bleed-through distortion in grayscale documents by a blind source separation technique," *International Journal on Document Analysis and Recognition*, vol. 10, pp. 17–25, June 2007.

[18] A. Tonazzini, G. Bianco, and E. Salerno, "Registration and enhancement of double-sided degraded manuscripts acquired in multispectral modality," in *Proceedings of the 10th International Conference on Document Analysis and Recognition ICDAR 2009*, 2009, pp. 546 – 550.

[19] A. Tonazzini, I. Gerace, and F. Martinelli, "Multichannel blind separation and deconvolution of images for document analysis," *IEEE Transactions on Image Processing*, vol. 19, no. 4, pp. 912–925, April 2010.

[20] B. Ophir and D. Malah, "Show-through cancellation in scanned images using blind source separation techniques," in *Proceedings of the International Conference on Image Processing (ICIP)*, vol. III, 2007, pp. 233–236.

[21] F. Merrikh-Bayat, M. Babaie-Zadeh, and C. Jutten, "Linear-quadratic blind source separating structure for removing show-through in scanned documents," *International Journal on Document Analysis and Recognition*, vol. 14, 319–333, 2011.

[22] S. Amari and A. Cichocki, "Adaptive blind signal processing-neural network approaches," *Proceedings of the IEEE*, vol. 86, no. 10, pp. 2026–2048, October 1998.

[23] T. Lee, M. Lewicki, and T. Sejnowski, "Independent component analysis using an extended infomax algorithm for mixed sub-Gaussian and super-Gaussian sources," *Neural Computation*, vol. 11, pp. 409–433, 1999.

[24] A. Hyvärinen, J. Karhunen, and E. Oja, *Independent Component Analysis*. New York: Wiley, 2001.

[25] E. Kuruoglu, L. Bedini, M. T. Paratore, E. Salerno, and A. Tonazzini, "Source separation in astrophysical maps using independent factor analysis," *Neural Networks*, vol. 16, pp. 479–491, 2003.

[26] A. Hyvärinen, "Fast and robust fixed-point algorithms for independent component analysis," *IEEE Transactions on Neural Networks*, vol. 10, no. 3, pp. 626–634, 1999.

[27] A. Tonazzini, E. Salerno, M. Mochi, and L. Bedini, "Bleed-through removal from degraded documents using a color decorrelation method," in *Document Analysis Systems VI*, Series Lecture Notes in Computer Science, S. Marinai and A. Dengel, Eds. Berlin: Springer, 2004, vol. 3163, pp. 250–261.

[28] A. Tonazzini, "Color space transformations for analysis and enhancement of ancient degraded manuscripts," *Pattern Recognition and Image Analysis*, vol. 20, no. 3, pp. 404–417, 2010.

[29] Google, "Book Search Dataset," 2007.

[30] K. Kokkinakis and A.K. Nandi, "Multichannel blind deconvolution for source separation in convolutive mixtures of speech," *IEEE Transactions on Audio, Speech, and Language Processing*, vol. 14, no. 1, pp. 202–212, 2006.

[31] H. Buchner, R. Aichner, and W. Kellermann, "A generalization of blind source separation algorithms for convolutive mixtures based on second-order statistics," *IEEE Transactions on Speech and Audio Processing*, vol. 13, no. 1, pp. 120–134, January 2005.

[32] C. Simon, P. Loubaton, and C. Jutten, "Separation of a class of convolutive mixtures: A contrast function approach," *Signal Processing*, vol. 81, pp. 883–887, 2001.

[33] S. Douglas, H. Sawada, and S. Makino, "Natural gradient multichannel blind deconvolution and speech separation using causal FIR filters," *IEEE Transactions on Speech and Audio Processing*, vol. 13, no. 1, pp. 92–104, 2005.

[34] S. Shwarts, Y. Schechner, and M. Zibulevsky, "Blind separation of convolutive image mixtures," *Neurocomputing*, vol. 71, no. 10-12, pp. 2164–2179, 2008.

[35] M. Castella and J.-C. Pesquet, "An iterative blind source separation method for convolutive mixtures of images," *Lecture Notes in Computer Science*, vol. 3195, pp. 922–929, 2004.

[36] E. Be'ery and A. Yeredor, "Blind separation of superimposed shifted images using parameterized joint diagonalization," *IEEE Transactions on Image Processing*, vol. 17, no. 3, pp. 340–353, March 2008.

[37] A. Tonazzini and I. Gerace, "Bayesian MRF-based blind source separation of convolutive mixtures of images," in *Proceedings of the 13th European Signal Processing Conference (EUSIPCO)*, September 2005.

[38] A. Blake and A. Zissermann, *Visual Reconstruction*. Cambridge, MA: MIT Press, 1987.

[39] A. Boccuto, M. Discepoli, I. Gerace, R. Pandolfi, and P. Pucci, "A GNC algorithm for deblurring images with interacting discontinuitiess," in *Proceedings of the VI SIMAI*, July 2002, pp. 296–310.

[40] I. Gerace, F. Martinelli, and A. Tonazzini, "See-through correction in recto-verso documents via a regularized nonlinear model," CNR-ISTI, Pisa, Tech. Rep. TR-001, May 2011.

Chapter 11

Correction of Spatially Varying Image and Video Motion Blur Using a Hybrid Camera

YU-WING TAI
Korea Advanced Institute of Science and Technology

MICHAEL S. BROWN
National University of Singapore

11.1 Introduction

This chapter focuses on an approach to reduce spatially varying motion blur in image and video footage using a hybrid camera system [1,2]. The hybrid camera was first proposed by Ben-Ezra and Nayar [3,4] and is an imaging system that couples a standard video camera with an auxiliary low-resolution camera sharing the same optical path but capturing at a significantly higher frame rate. The auxiliary video is temporally sharper but at a lower resolution, while the lower-frame-rate video has higher spatial resolution but is susceptible to motion blur. Using the information in these two videos, our method has two aims: (1) to deblur the frames in the high-resolution video, and (2) to estimate new high-resolution video frames at a higher temporal sampling.

While high-resolution, high-frame-rate digital cameras are becoming increasingly more affordable, the hybrid-camera design remains promising. Even at 60 fps, high-speed photography/videography is susceptible to motion blur artifacts. In addition, as the frame-rate of high-resolution cameras increase, low-resolution camera frame-rate speeds increase accordingly, with cameras available now with over 1,000 fps at lower resolution. Thus, our approach has application to ever-increasing temporal imaging. In addition, the use of hybrid cameras, and hybrid-camera-like designs, have been demonstrated to offer other advantages over single-view cameras including object segmentation and matting [4–6], depth estima-

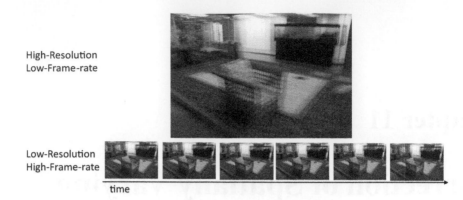

High-Resolution
Low-Frame-rate

Low-Resolution
High-Frame-rate

time

Figure 11.1 Trade-off between resolution and frame rates. Top: Image from a high-resolution, low-frame-rate camera. Bottom: Images from a low-resolution, high-frame-rate camera. (© 2010 IEEE.)

tion [7], and high dynamic range imaging [8]. The ability to perform object segmentation is key in deblurring moving objects, as demonstrated by [4] and our own work in Section 11.5.

Unlike the previous work in [3,4] that focused on static images suffering from globally invariant motion blur, we address the broader problem of correcting spatially varying motion blur and aim to deblur temporal sequences. In addition, our work achieves improved deblurring performance by exploiting the available information acquired in the hybrid camera system, including optical flow, back-projection constraints between low-resolution and high-resolution images, and temporal coherence along image sequences. Another benefit of our approach is that it can be used to increase the frame rate of the high-resolution camera by estimating the missing frames.

The central idea in our formulation is to combine the benefits of both deconvolution and super resolution. Deconvolution of motion-blurred, high-resolution images yields high-frequency details, but with ringing artifacts due to the lack of low-frequency components. In contrast, super resolution-based reconstruction from low-resolution images recovers artifact-free low-frequency results that lack high-frequency detail. We show that the deblurring information from deconvolution and super resolution are complementary to each other and can be used together to improve deblurring performance. In video deblurring applications, our method further capitalizes on additional deconvolution constraints that can be derived from consecutive video frames. We demonstrate that this approach produces excellent results in reducing spatially varying motion blur. In addition, the availability of the low-resolution imagery and subsequently derived motion vectors further allows us to perform estimate new temporal frames in the high-resolution video, which we also demonstrate.

The processing pipeline of our approach is shown in Figure 11.2, which also relates process components to their corresponding section in the chapter. The remainder of the chapter is organized as follows: Section 11.2 discusses related work, Section 11.3 describes the hybrid camera setup and the constraints on deblurring available in this system, Section 11.4 describes our overall deconvolution formulation expressed in a maximum a posteriori (MAP) framework, Section 11.5 discusses how to extend our framework to han-

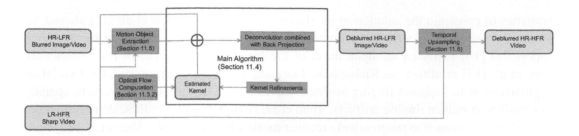

Figure 11.2 The processing pipeline of our system. Optical flows are first calculated from the Low-Resolution, High-Frame-Rate (LR-HFR) video. From the optical flows, spatially varying motion blur kernels are estimated (Section 11.3.2). Then the main algorithm performs an iterative optimization procedure that simultaneously deblurs the High-Resolution, Low-Frame-Rate (HR-LFR) image/video and refines the estimated kernels (Section 11.4). The output is a deblurred HR-LFR image/video. For the case of deblurring a moving object, the object is separated from the background prior to processing (Section 11.5). In the deblurring of video, we can additionally enhance the frame rate of the deblurred video to produce a High-Resolution, High-Frame-Rate (HR-HFR) video result (Section 11.6). (© 2010 IEEE.)

dle moving objects, Section 11.6 describes how to perform temporal upsampling with our framework, and Section 11.7 provides results and comparisons with other current work, followed by a discussion and summary in Section 11.8.

11.2 Related Work

Motion deblurring can be cast as the deconvolution of an image that has been convolved with either a global motion point spread function (PSF) or a spatially varying PSF. The problem is inherently ill-posed as there are a number of unblurred images that can produce the same blurred image after convolution. Nonetheless, this problem is well studied given its utility in photography and video capture. The following describes several related works.

11.2.1 Traditional Deblurring

The majority of related work involves traditional blind deconvolution, which simultaneously estimates a global motion PSF and the deblurred image. These methods include well-known algorithms such as Richardson–Lucy [9, 10] and Wiener deconvolution [11]. For a survey on blind deconvolution, readers are referred to [12,13]. These traditional approaches often produce less than desirable results that include artifacts such as ringing.

11.2.2 PSF Estimation and Priors

A recent trend in motion deblurring is to either constrain the solution of the deblurred image or to use auxiliary information to aid in either the PSF estimation or the deconvolution itself (or both). Examples include work by Fergus et al. [14], which used natural image statistics to constrain the solution to the deconvolved image. Raskar et al. [15] altered the shuttering sequence of a traditional camera to make the PSF more suitable for deconvolution. Jia [16] extracted an alpha mask of the blurred region to aid in PSF estimation. Dey et al. [17] modified the Richardson–Lucy algorithm by incorporating total variation regularization to suppress ringing artifacts. Levin et al. [18] introduced gradient sparsity constraints to reduce ringing artifacts. Yuan et al. [19] proposed a multiscale nonblind deconvolution approach to progressively recover motion-blurred details. Shan et al. [20] studied the relationship between estimation errors and ringing artifacts, and also proposed the use of a spatial distribution model of image noise together with a local prior that suppresses ringing to jointly improve global motion deblurring. An evaluation to the performance of recent blind deconvolution algorithms can be found in [21].

Other recent approaches use more than one image to aid in the deconvolution process. Bascle et al. [22] processed a blurry image sequence to generate a single unblurred image. Yuan et al. [23] used a pair of images—one noisy and one blurred. Rav-Acha and Peleg [24] considered images that have been blurred in orthogonal directions to help estimate the PSF and constrain the resulting image. Chen and Tang [25] extended the work of [24] to remove the assumption of orthogonal blur directions. Bhat et al. [26] proposed a method that uses high-resolution photographs to enhance low-quality video, but this approach is limited to static scenes. Most closely related to ours is the work of Ben-Ezra and Nayar [3,4], which used an additional imaging sensor to capture high-frame-rate imagery for the purpose of computing optical flow and estimating a global PSF. Li et al. [7] extended the work of Ben-Ezra and Nayar [3,4] by using parallel cameras with different frame rates and resolutions, but their work targets depth map estimation—not deblurring.

The previously mentioned approaches assume that the blur arises from a global PSF. Recent work addressing spatially varying motion blur includes that of Levin [27], which used image statistics to correct a single-motion blur on a stable background. Bardsley et al. [28] segmented an image into regions exhibiting similar blur, while Cho et al. [29] used two blurred images to simultaneously estimate local PSFs as well as deconvolve the two images. Ben-Ezra and Nayar [4] demonstrated how the auxiliary camera could be used to separate a moving object from the scene and apply deconvolution to this extracted layer. These approaches [4, 27–29], however, assume that the motion blur is globally invariant within each separated layer. Work by Shan et al. [30] allows the PSF to be spatially varying; however, the blur is constrained to that from rotational motion. Levin et al. [31] proposed a parabolic camera designed for deblurring images with one-dimensional object motion. During exposure, the camera moves in a manner that allows the resulting image to be deblurred using a single deconvolution kernel. Tai et al. [32, 33] proposed a projective motion blur model that uses a sequence of homographies to describe motion PSF instead of a conventional patch-based approach.

11.2.3 Super Resolution and Upsampling

The problem of super resolution can be considered a special case of motion deblurring in which the blur kernel is a low-pass filter that is uniform in all motion directions. High-frequency details of a sharp image are therefore completely lost in the observed low-resolution image. There are two main approaches to super resolution: (1) image hallucination based on training data and (2) image super resolution computed from multiple low-resolution images. Our work is closely related to the latter approach, which is reviewed here. The most common technique for multiple image super resolution is the back-projection algorithm proposed by Irani and Peleg [34, 35]. The back-projection algorithm is an iterative refinement procedure that minimizes the reconstruction errors of an estimated high-resolution image through a process of convolution, downsampling, and upsampling. A brief review that includes other early work on multiple image super resolution is given in [36]. More recently, Patti et al. [37] proposed a method to align low-resolution video frames with arbitrary sampling lattices to reconstruct a high-resolution video. Their approach also uses optical flow for alignment and PSF estimation. These estimates, however, are global and do not consider local object motion. This work was extended by Elad and Feuer [38] to use adaptive filtering techniques. Zhao and Sawhney [39] studied the performance of multiple image super resolution against the accuracy of optical flow alignment and concluded that the optical flows must be reasonably accurate in order to avoid ghosting effects in super resolution results. Shechtman et al. [40] proposed space-time super resolution, in which multiple video cameras with different resolutions and frame rates are aligned using homographies to produce outputs of either higher temporal and/or spatial sampling. When only two cameras are used, this approach can be considered a demonstration of a hybrid camera; however, this work did not address the scenario where severe motion blur is present in the high-resolution, low-frame-rate camera. Sroubek et al. [41] proposed a regularization framework for solving the multiple image super resolution problem. This approach also does not consider local motion blur effects. Recently, Agrawal and Raskar [42] proposed a method to increase the resolution of images that have been deblurred using a coded exposure system. Their approach can also be considered a combination of motion deblurring and super resolution, but is limited to translational motion.

While various previous works are related in part, our work is unique in its focus on spatially varying blur with no assumption on global or local motion paths. Moreover, our approach takes full advantage of the information available from the hybrid-camera system, using techniques from both deblurring and super resolution together in a single MAP framework. Specifically, our approach incorporates spatially varying deconvolution, together with back-projection against the low-resolution frames. This combined strategy produces deblurred images with less ringing than traditional deconvolution, but with more detail than approaches using regularization and prior constraints. As with other deconvolution methods, we cannot recover frequencies that have been completely lost due to motion blur and downsampling. A more detailed discussion of our approach is provided in Section 11.4.4.

11.3 Hybrid Camera System

The advantages of a hybrid camera system are derived from the additional data acquired by the low-resolution, high-frame-rate (LR-HFR) camera. While the spatial resolution of this camera is too low for many practical applications, the high-speed imagery is reasonably blur-free and is therefore suitable for optical flow computation. Figure 11.1 illustrates an example. Because the cameras are assumed to be synchronized temporally and observing

Figure 11.3 Our hybrid camera combines a Point Grey Dragonfly II camera, which captures images of 1024×768 resolution at 25 fps (6.25 fps for image deblurring examples), and a Mikrotron MC1311 camera that captures images of 128×96 resolution at 100 fps. A beamsplitter is employed to align their optical axes and respective images. Video synchronization is achieved using a 8051 microcontroller. (© 2010 IEEE.)

the same scene, the optical flow corresponds to the motion of the scene observed by the high-resolution, low-frame-rate (HR-LFR) camera, whose images are corrupted with motion blur due to the slower temporal sampling. This ability to directly observe fast-moving objects in the scene with the auxiliary camera allows us to handle a larger class of object motions without the use of prior motion models, because optical flow can be computed.

In the following, we discuss the construction of a hybrid camera, the optical flow and motion blur estimation, and the use of the low-resolution images as reconstruction constraints on high-resolution images.

11.3.1 Camera Construction

Three conceptual designs of the hybrid camera system were discussed by Ben-Ezra and Nayar [3]. In their work, they implemented a simple design in which the two cameras are placed side-by-side, such that their viewpoints can be considered the same when viewing a distant scene. A second design avoids the distant scene requirement by using a beam splitter to share between two sensing devices the light rays that pass through a single aperture, as demonstrated by McGuire et al. [6] for the studio matting problem. A promising third design is to capture both the HR-LFR and LR-HFR video on a single sensor chip. According to [43], this can readily be achieved using a programmable CMOS sensing device.

In our work, we constructed a hand-held hybrid camera system based on the second design as shown in Figure 11.3. The two cameras were positioned such that their optical axes and pixel arrays were well aligned. Video synchronization was achieved using an 8051

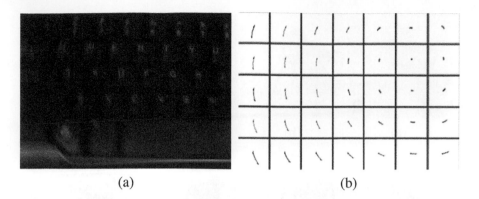

(a)   (b)

Figure 11.4 Spatially varying blur kernel estimation using optical flows: (a) motion blur image; (b) estimated blur kernels of (a) from optical flows. (© 2010 IEEE.)

microcontroller. To match the color responses of the two devices we employed histogram equalization. In our implemented system, the exposure levels of the two devices were set to be equal, and the signal-to-noise ratios in the HR-LFR and LR-HFR images were approximately the same.

11.3.2 Blur Kernel Approximation Using Optical Flows

In the absence of occlusion, disocclusion, and out-of-plane rotation, a blur kernel can be assumed to represent the motion of a camera relative to objects in the scene. In [3], this relative motion is assumed to be constant throughout an image. In such a scenario, the globally invariant blur kernel is obtained through the integration of global motion vectors over a spline curve. However, because optical flow is in fact a local estimation of motions, we can calculate spatially varying blur kernels from optical flows. We use the multiscale Lucas-Kanade algorithm [44] to calculate the optical flow at each pixel location.

Following the brightness constancy assumption of optical flow estimation, we assume that our motion-blurred images are captured under constant illumination, such that the change of pixel color of moving scene/object points over the exposure period can be neglected. The per-pixel motion vectors are then integrated to form spatially varying blur kernels, $K(x, y)$, one per pixel. This integration is performed as described in [3] for global motion.

We use a C1 continuity spline curve fit to the path of optical flow at position (x, y). The number of frames used to fit the spline curve is sixteen for image examples and four for video examples (Figure 11.3). Figure 11.4 shows an example of spatially varying blur kernels estimated from optical flows.

The optical flows estimated with the multiscale Lucas–Kanade algorithm [44] may contain noise that degrades blur kernel estimation. We found such noisy estimates to occur mainly in smooth or homogeneous regions that lack features for correspondence, while regions with sharp features tend to have accurate optical flows. Because deblurring artifacts are evident primarily around such features, the Lucas-Kanade optical flows are effective for our purposes. Optical flow noise in relatively featureless regions has little effect on deblurring results, as these areas are relatively unaffected by errors in the deblurring kernel. As a

(a)   (b)   (c)   (d)

(e)   (f)   (g)   (h)

Figure 11.5 Performance comparisons for different deconvolution algorithms on a synthetic example. The ground truth motion blur kernel is used to facilitate comparison. The signal-to-noise ratio (SNR) of each result is reported. (a) A motion-blurred image [SNR(dB) = 25.62] with the corresponding motion blur kernel shown in the inset. Deconvolution results using (b) Wiener filter [SNR(dB) = 37.0]; (c) Richardson–Lucy [SNR(dB) = 33.89]; (d) Total Variation Regularization [SNR(dB) = 36.13]; (e) Gradient Sparsity Prior [SNR(dB) = 46.37]; (f) our approach [SNR(dB) = 50.26], which combines constraints from both deconvolution and super resolution. The low-resolution image in (g) is 8x downsampled from the original image, shown in (h). (© 2010 IEEE.)

measure to heighten the accuracy and consistency of estimated optical flows, we use local smoothing [45] as an enhancement of the multiscale Lucas-Kanade algorithm [44].

The estimated blur kernels contain quantization errors due to low resolution of the optical flows. Additionally, motion vector integration may provide an imprecise temporal interpolation of the flow observations. Our MAP optimization framework addresses these issues by refining the estimated blur kernels in addition to deblurring the video frames or images. Details of this kernel refinement are discussed fully in Section 11.4.

11.3.3   Back-Projection Constraints

The capture of low-resolution frames in addition to the high-resolution images not only facilitates optical flow computation, but also provides super resolution-based reconstruction constraints [34–38, 40, 46] on the high-resolution deblurring solution. The back-projection algorithm [34, 35] is a common iterative technique for minimizing the reconstruction error, and can be formulated as follows:

$$I^{t+1} = I^t + \sum_{j=1}^{M}(u(W(I_{l_j}) - d(I^t \otimes h))) \otimes p, \qquad (11.1)$$

where M represents the number of corresponding low-resolution observations, t is an iteration index, I_{l_j} refers to the jth low-resolution image, $W(\cdot)$ denotes a warp function that aligns I_{l_j} to a reference image, $\otimes$ is the convolution operation, h is the convolution filter before downsampling, p is a filter representing the back-projection process, and $d(\cdot)$ and $u(\cdot)$ are the downsampling and upsampling processes respectively. Equation (11.1) assumes that each observation carries the same weight. In the absence of a prior, h is chosen to be a Gaussian filter with a size proportionate to the downsampling factor, and p is set equal to h.

In the hybrid camera system, a number of low-resolution frames are captured in conjunction with each high-resolution image. To exploit this available data, we align these frames according to the computed optical flows, and use them as back-projection constraints in Equation (11.1). The number of low-resolution image constraints M is determined by the relative frame rates of the cameras. In our implementation, we choose the first low-resolution frame as the reference frame to which the estimated blur kernel and other low-resolution frames are aligned. Choosing a different low-resolution frame as the reference frame would lead to a different deblurred result, which is a property that can be used to increase the temporal samples of the deblurred video as later discussed in Section 11.6.

The benefits of using the back-projection constraint, and multiple such back-projection constraints, is illustrated in Figure 11.5. Each of the low-resolution frames presents a physical constraint on the high-resolution solution in a manner that resembles how each offset image is used in a super resolution technique. The effectiveness of incorporating the back-projection constraint to suppress ringing artifacts is demonstrated in Figure 11.5 in comparison to several other deconvolution algorithms.

11.4 Optimization Framework

Before presenting our deblurring framework, we briefly review the Richardson–Lucy deconvolution algorithm, as our approach is fashioned in a similar manner.

11.4.1 Richardson–Lucy Image Deconvolution

The Richardson–Lucy algorithm [9, 10] is an iterative maximum likelihood deconvolution algorithm derived from Bayes' theorem that minimizes the following estimation error:

$$\arg\min_{I} n(||I_b - I \otimes K||^2), \tag{11.2}$$

where I is the deblurred image, K is the blur kernel, I_b is the observed blur image, and $n(\cdot)$ is the image noise distribution. A solution can be obtained using the iterative update algorithm defined as follows:

$$I^{t+1} = I^t \times K * \frac{I_b}{I^t \otimes K}, \tag{11.3}$$

where $*$ is the correlation operation. A blind deconvolution method using the Richardson–Lucy algorithm was proposed by Fish et al. [47], which iteratively optimizes I and K in alternation using Equation (11.3) with the positions of I and K switched during optimization iterations for K. The Richardson–Lucy algorithm assumes that image noise $n(\cdot)$ follows a

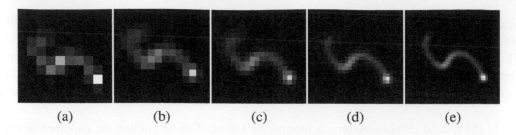

(a)   (b)   (c)   (d)   (e)

Figure 11.6 Multiscale refinement of a motion blur kernel for the image in Figure 11.11. (a) to (e) exhibit refined kernels at progressively finer scales. Our kernel refinement starts from the coarsest level. The result of each coarser level is then upsampled and used as an initial kernel estimate for the next level of refinement. (© 2010 IEEE.)

Poisson distribution. If we assume that image noise follows a Gaussian distribution, then a least squares method can be employed [34]:

$$I^{t+1} = I^t + K * (I_b - I^t \otimes K), \tag{11.4}$$

which shares the same iterative back-projection update rule as Equation (11.1).

From video input with computed optical flows, multiple blurred images I_b and blur kernels K may be acquired by reversing the optical flows of neighboring high-resolution frames. These multiple observation constraints can be jointly applied in Equation (11.4) [24] as

$$I^{t+1} = I^t + \sum_{i=1}^{N} w_i K_i * (I_{b_i} - I^t \otimes K_i), \tag{11.5}$$

where N is the number of aligned observations. That image restoration can be improved with additional observations under different motion blurs is an important property that we exploit in this work. The use of neighboring frames in this manner may also serve to enhance the temporal consistency of the deblurred video frames.

11.4.2   Optimization for Global Kernels

In solving for the deblurred images, our method jointly employs the multiple deconvolution and back-projection constraints available from the hybrid camera input. For simplicity, we assume in this subsection that the blur kernels are spatially invariant. Our approach can be formulated into a MAP estimation framework as follows:

$$
\begin{aligned}
& \arg\max_{I,K} P(I, K | I_b, K_o, I_l) \\
= \; & \arg\max_{I,K} P(I_b|I,K)P(K_o|I,K)P(I_l|I)P(I)P(K) \\
= \; & \arg\min_{I,K} L(I_b|I,K) + L(K_o|I,K) + L(I_l|I) + L(I) + L(K) \tag{11.6}
\end{aligned}
$$

where I and K denote the sharp images and the blur kernels we want to estimate; I_b, K_o, and I_l are the observed blur images, estimated blur kernels from optical flows, and the low-resolution, high-frame-rate images, respectively; and $L(\cdot) = -log(P(\cdot))$. In our formulation, the priors $P(I)$ and $P(K)$ are taken to be uniformly distributed. Assuming that $P(K_o|I, K)$ is conditionally independent of I, that the estimation errors of likelihood probabilities $P(I_b|I, K)$, $P(K_o|I, K)$, and $P(I_l|I)$ follow Gaussian distributions, and that each observation of I_b, K_o, and I_l is independent and identically distributed, we can then rewrite Equation (11.6) as

$$\arg\min_{I,K} \quad \sum_i^N ||I_{b_i} - I \otimes K_i||^2 + \lambda_B \sum_j^M ||I_{l_j} - d(I \otimes h)||^2$$

$$+ \lambda_K \sum_i^N ||K_i - K_{o_i}||^2, \tag{11.7}$$

where λ_K and λ_B are the relative weights of the error terms. To optimize the above equation for I and K, we employ alternating minimization. Combining Equations (11.1) and (11.5) yields our iterative update rules:

1. Update $I^{t+1} = I^t + \sum_{i=1}^N K_i^t * (I_{b_i} - I^t \otimes K_i^t) + \lambda_B \sum_{j=1}^M h \otimes (u(W(I_{l_j}) - d(I^t \otimes h)))$,

2. Update $K_i^{t+1} = K_i^t + \tilde{I}^{t+1} * (I_{b_i} - I^{t+1} \otimes K_i^t) + \lambda_K (K_{o_i} - K_i^t)$,

where $\tilde{I} = I / \sum_{(x,y)} I(x, y)$, $I(x, y) \geq 0$, $K_i(u, v) \geq 0$, and $\sum_{(u,v)} K_i(u, v) = 1$. The two update steps are processed in alternation until the change in I falls below a specified level or until a maximum number of iterations is reached. The $W(I_{l_j})$ term is the warped aligned observations. The reference frame to which these are aligned to can be any of the M low-resolution images. Thus, for each deblurred high-resolution frame, we have up to M possible solutions. This will later be used in the temporal upsampling described in Section 11.6. In our implementation, we set $N = 3$ in correspondence to the current, previous, and next frames, and M is set according to the relative camera settings (4/16 for video/image deblurring in our implementation). We also initialize I^0 as the currently observed blur image I_b, K_i^0 as the estimated blur kernel K_{o_i} from optical flows, and set $\lambda_B = \lambda_K = 0.5$.

For more stable and flexible kernel refinement, we refine the kernel in a multiscale fashion as done in [14, 23]. Figure 11.6 illustrates the kernel refinement process. We estimate PSFs from optical flows of the observed low-resolution images and then downsample to the coarsest level. After refinement at a coarser level, kernels are then upsampled and refined again. The multiscale pyramid is constructed using a downsampling factor of $1/\sqrt{2}$ with five levels. The likelihood $P(K_o|K)$ is applied at each level of the pyramid with a decreasing weight, so as to allow more flexibility in refinement at finer levels. We note that starting at a level coarser than the low-resolution images allows our method to recover from some error in PSF estimation from optical flows.

11.4.3 Spatially Varying Kernels

A spatially varying blur kernel can be expressed as $K(x, y, u, v)$, where (x, y) is the image coordinate and (u, v) is the kernel coordinate. For large-sized kernels (e.g., 65×65), this representation is impractical due to enormous storage requirements. Recent work has

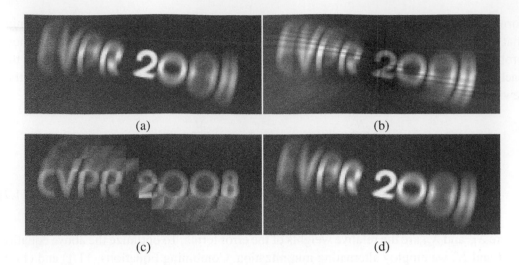

(a)   (b)

(c)   (d)

Figure 11.7 Convolution with kernel decomposition: (a) convolution result without kernel decomposition, where full blur kernels are generated on-the-fly per pixel using optical flow integration; (b) convolution using thirty PCA-decomposed kernels; (c) convolution using patch-based decomposition; (d) convolution using delta function decomposition of kernels, with at most thirty delta functions per pixel. (© 2010 IEEE.)

suggested ways to reduce the storage size, such as by constraining the motion path [30]; however, our approach places no constraints on possible motion. Instead, we decompose the spatially varying kernels into a set of P basis kernels k_l whose mixture weights a_l are a function of image location:

$$K(x, y, u, v) = \sum_{l=1}^{P} a_l(x, y) k_l(u, v).$$ (11.8)

The convolution equation then becomes

$$I(x, y) \otimes K(x, y, u, v) = \sum_{l=1}^{P} a_l(x, y)(I(x, y) \otimes k_l(u, v)).$$ (11.9)

In related work [48], principal components analysis (PCA) is used to determine the basis kernels. PCA, however, does not guarantee positive kernel values, and we have found in our experiments that PCA-decomposed kernels often lead to unacceptable ringing artifacts, exemplified in Figure 11.7(b). The ringing artifacts in the convolution result resemble the patterns of basis kernels. Another method is to use a patch representation that segments images into many small patches such that the local motion blur kernel is the same within each small patch. This method was used in [49], but their blur kernels are defocus kernels with very small variations within local areas. For large object motion, blur kernels in the patch-based method would not be accurate, leading to discontinuity artifacts as shown in Figure 11.7(c). We instead choose to use a delta function representation, where each delta function represents a position (u, v) within a kernel. Because a motion blur kernel is typically sparse, we store only thirty to forty delta functions for each image pixel, where the

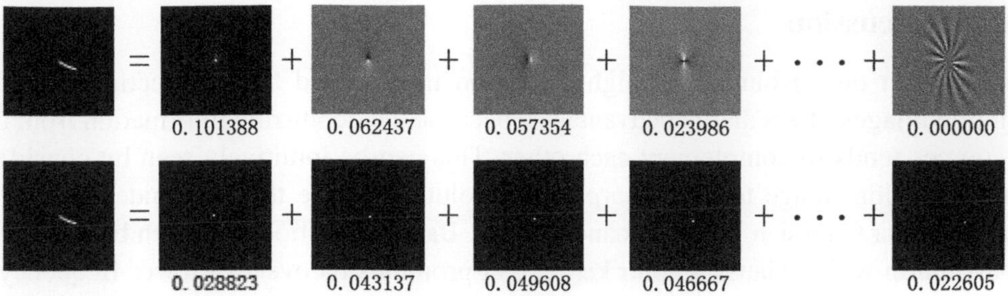

Figure 11.8 PCA versus the delta function representation for kernel decomposition. The top row illustrates the kernel decomposition using PCA, and the bottom row shows the decomposition using the delta function representation. The example kernel is taken from among the spatially varying kernels of Figure 11.7, from which the basis kernels are derived. Weights are displayed below each of the basis kernels. The delta function representation not only guarantees positive values of basis kernels, but also provides more flexibility in kernel refinement. (© 2010 IEEE.)

delta function positions are determined by the initial optical flows. From the total 65×65 possible delta-function in the spatial kernel at each pixel in the image, we find in practice that we only use about 500 to 600 distinct delta functions to provide a sufficient approximation of the spatially varying blur kernels in the convolution process. Examples of basis kernel decomposition using PCA and the delta function representation are shown in Figure 11.8. The delta function representation also offers more flexibility in kernel refinement, while refinements using the PCA representation are limited to the PCA subspace.

Combining Equations (11.9) and (11.7), our optimization function becomes

$$\arg\min_{I,K} \quad \sum_i^N ||I_{b_i} - \sum_l^P a_{il}(I \otimes k_{il})||^2 + \lambda_B \sum_j^M ||I_{l_j} - d(I \otimes h)||^2$$

$$+ \lambda_K \sum_i^N \sum_l^P ||a_{il}k_{il} - a_{o_{il}}k_{il}||^2. \tag{11.10}$$

The corresponding iterative update rules are then

1. Update $I^{t+1} = I^t + \sum_{i=1}^N \sum_l^P a_{il}^t k_{il} * (I_{b_i} - \sum_l^P a_{il}^t (I^t \otimes k_{il})) + \lambda_B \sum_{j=1}^M h \otimes (u(W(I_{l_j}) - d(I^t \otimes h)))$

2. Update $a_{il}^{t+1} = a_{il}^t + (\widetilde{I'}^{t+1} * (I'_{b_i} - \sum_l^P a_{il}^t (I'^{t+1} \otimes k_{il}))) \cdot k_{il} + \lambda_K (a_{o_{il}} - a_{il}^t)$

where I' and I'_b are local windows in the estimated result and the blur image, respectively. This kernel refinement can be implemented in a multiscale framework for greater flexibility and stability. The number of delta functions k_{il} stored at each pixel position may be reduced when an updated value of a_{il} becomes insignificant. For greater stability, we process each update rule five times before switching to the other.

11.4.4 Discussion

Utilizing both deconvolution of high-resolution images and back-projection from low-resolution images offers distinct advantages, because the deblurring information from these two sources tends to complement each other. This can be intuitively seen by considering a low-resolution image to be a sharp high-resolution image that has undergone motion blurring with a Gaussian PSF and bandlimiting. Back-projection may then be viewed as a deconvolution with a Gaussian blur kernel that promotes recovery of lower-frequency image features without artifacts. On the other hand, deconvolution of high-resolution images with the high-frequency PSFs typically associated with camera and object motion generally supports reconstruction of higher-frequency details, especially those orthogonal to the motion direction. While some low-frequency content can also be restored from motion blur deconvolution, there is often significant loss due to the large support regions for motion blur kernels, and this results in ringing artifacts. As discussed in [24], the joint use of images having such different blur functions and deconvolution information favors a better deblurring solution.

Multiple motion blur deconvolutions and multiple back-projections can further help to generate high-quality results. Differences in motion blur kernels among neighboring frames provide different frequency information; and multiple back-projection constraints help to reduce quantization and the effects of noise in low-resolution images. In some circumstances, there exists redundancy in information from a given source, such as when high-resolution images contain identical motion blur, or when low-resolution images are offset by integer pixel amounts. This makes it particularly important to utilize as much deblurring information as can be obtained.

We note that our current approach does not utilize priors on the deblurred image or the kernels. With constraints from the low-resolution images, we have found that these priors are not needed. Figure 11.5 compares our approach with other deconvolution algorithms. Specifically, we compare our approach with Total Variation regularization [17] and Sparsity Priors [18], which have recently been shown to produce better results than traditional Wiener filtering [11] and the Richardson–Lucy [9, 10] algorithm. Both Total Variation regularization and Sparsity Priors produce results with less ringing artifacts. There are almost no ringing artifacts with Sparsity Priors, but many fine details are lost. In our approach, most medium- to large-scale ringing artifacts are removed using back-projection constraints, while fine details are recovered through deconvolution.

Although our approach can acquire and utilize a greater amount of data, high-frequency details that have been lost by both motion blur and downsampling cannot be recovered. This is a fundamental limitation of any deconvolution algorithm that does not hallucinate detail. We also note that reliability in optical flow cannot be assumed beyond a small time interval. This places a restriction on the number of motion blur deconvolution constraints that can be employed to deblur a given frame.

Finally, we note that an iterative back-projection technique incorporated into our framework is known to have convergence problems. Empirically, we have found that stopping after no more than fifty iterations of our algorithm produces acceptable results.

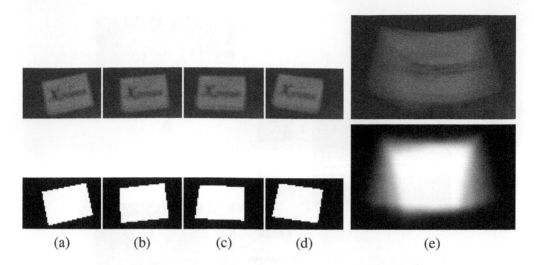

Figure 11.9 Layer separation using a hybrid camera: (a)–(d) low-resolution frames and their corresponding binary segmentation masks; (e) high-resolution frame and the matte estimated by compositing the low-resolution segmentation masks with smoothing. (© 2010 IEEE.)

11.5 Deblurring of Moving Objects

To deblur a moving object, a high-resolution image must be segmented into different layers because pixels on the blended boundaries of moving objects contain both foreground and background components, each with different relative motion to the camera. This layer separation is inherently a matting problem that can be expressed as

$$I = \alpha F + (1 - \alpha)B, \qquad (11.11)$$

where I is the observed image intensity; and F, B and α are the foreground color, background color, and alpha value of the fractional occupancy of the foreground. The matting problem is an ill-posed problem because the number of unknown variables is greater than the number of observations. Single-image approaches require user assistance to provide a trimap [50–52] or scribbles [53–55] for collecting samples of the foreground and background colors. Fully automatic approaches, however, have required either a blue background [56], multiple cameras with different focus [5], polarized illumination [6], or a camera array [57]. In this section we propose a simple solution to the layer separation problem that takes advantage of the hybrid camera system.

Our approach assumes that object motion does not cause motion blur in the high-frame-rate camera, such that the object appears sharp. To extract the alpha matte of a moving object, we perform binary segmentation of the moving object in the low-resolution images, and then compose the binary segmentation masks with smoothing to approximate the alpha matte in the high-resolution image. We note that Ben-Ezra and Nayar [4] used a similar strategy to perform layer segmentation in their hybrid camera system. In Figure 11.9, an example of this matte extraction is demonstrated, together with the moving object separation method of Zhang et al. [58]. The foreground color F must also be estimated for deblurring.

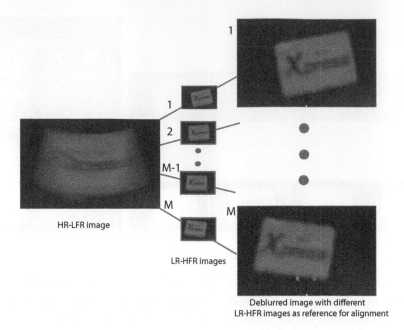

Deblurred image with different
LR-HFR images as reference for alignment

Figure 11.10 Relationship of high-resolution deblurred result to corresponding low-resolution frame. Any of the low-resolution frame can be selected as a reference frame for the deblurred result. This allows up to M deblurred solutions to be obtained. (© 2010 IEEE)

This can be done by assuming a local color smoothness prior on F and B, and then solving for their values with Bayesian matting [50]:

$$\begin{bmatrix} \Sigma_F^{-1} + \mathbf{I}\alpha^2/\sigma_I^2 & \mathbf{I}\alpha(1-\alpha)/\sigma_I^2 \\ \mathbf{I}\alpha(1-\alpha)/\sigma_I^2 & \Sigma_B^{-1} + \mathbf{I}(1-\alpha)^2/\sigma_I^2 \end{bmatrix} \begin{bmatrix} F \\ B \end{bmatrix} = \begin{bmatrix} \Sigma_F^{-1}\mu_F + I\alpha/\sigma_I^2 \\ \Sigma_B^{-1}\mu_B + I(1-\alpha)/\sigma_I^2 \end{bmatrix},$$
(11.12)

where (μ_F, Σ_F) and (μ_B, Σ_B) are the local color mean and covariance matrix (Gaussian distribution) of the foreground and background colors, respectively; $\mathbf{I}$ is a 3×3 identity matrix; and σ_I is the standard derivation of I, which models estimation errors of Equation (11.11). Given the solution of F and B, the α solution can be refined by solving Equation (11.11) in closed form. Refinements of F, B, and α can be done in alternation to further improve the result. A recent regularization-based method targeting for motion-blurred motion object [59] can also be included to enhance the matting results.

Once moving objects are separated, we deblur each layer separately using our framework. The alpha mattes are also deblurred for compositing, and the occluded background areas revealed after alpha mask deblurring can then be filled in either by back-projection from the low-resolution images or by the motion inpainting method of [60].

11.6   Temporal Upsampling

Unlike deblurring of images, videos require deblurring of multiple consecutive frames in a manner that preserves temporal consistency. As described in Section 11.4.2, we can jointly

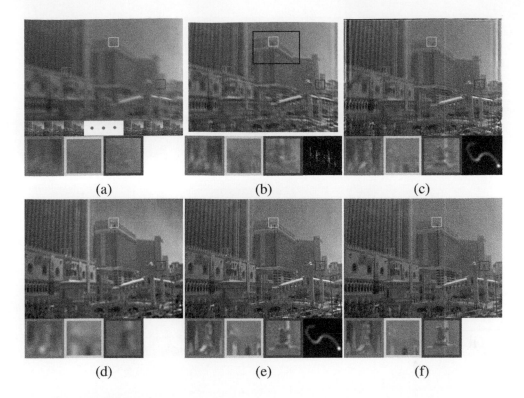

Figure 11.11 Image deblurring using globally invariant kernels: (a) input; (b) result generated with the method of [14], where the user-selected region is indicated by a black box; (c) result generated by [3]; (d) result generated by back-projection [34]; (e) our results; (f) the ground truth sharp image. Close-up views and the estimated global blur kernels are also shown. (© 2010 IEEE)

use the current, previous, and subsequent frames to deblur the current frame in a temporally consistent way. However, after sharpening each individual frame, temporal discontinuities in the deblurred high-resolution, low-frame-rate video may become evident through some jumpiness in the sequence. In this section we describe how our method can alleviate this problem by increasing the temporal sampling rate to produce a deblurred high-resolution, high-frame-rate video.

As discussed by Shechtman et al. [40], temporal super resolution results when an algorithm can generate an output with a temporal rate that surpasses the temporal sampling of any of the input devices. While our approach generates a high-resolution video at greater temporal rate than the input high-resolution, low-frame-rate video, its temporal rate is bounded by the frame rate of the low-resolution, high-frame-rate camera. We therefore refrain from the term *super resolution* and refer to this as *temporal upsampling*.

Our solution to temporal upsampling derives directly from our deblurring algorithm. The deblurring problem is a well-known under-constrained problem because there exist many solutions that can correspond to a given motion-blurred image. In our scenario, we have M high-frame-rate, low-resolution frames corresponding to each high-resolution, low-frame-rate motion-blurred image. Figure 11.10 shows an example. With our algorithm,

Figure 11.12 Image deblurring with spatial varying kernels from rotational motion; (a) input; (b) result generated with the method of [30] (obtained courtesy of the authors of [30]); (c) result generated by [3] using spatially varying blur kernels estimated from optical flow; (d) result generated by back-projection [34]; (e) our results; (f) the ground truth sharp image. Close-ups are also shown. (© 2010 IEEE.)

we therefore have the opportunity to estimate M solutions using each one of the M low-resolution frames as the basic reference frame. While the ability to produce multiple de-blurred frames is not a complete solution to temporal upsampling, here the use of these M different reference frames leads to a set of deblurred frames that is consistent with the temporal sequence. This unique feature of our approach is gained through the use of the hybrid camera to capture low-resolution, high-frame-rate video in addition to the standard high-resolution, low-frame-rate video. The low-resolution, high-frame-rate video not only aids in estimating the motion blur kernels and provides back-projection constraints, but can also help to increase the deblurred video frame rate. The result is a high-resolution, high-frame-rate deblurred video.

11.7   Results and Comparisons

We evaluate our deblurring framework using real images and videos. In these experiments, a ground-truth, blur-free image is acquired by mounting the camera on a tripod and captur-ing a static scene. Motion-blurred images are then obtained by moving the camera and/or introducing a dynamic scene object. We show examples of several different cases: *globally*

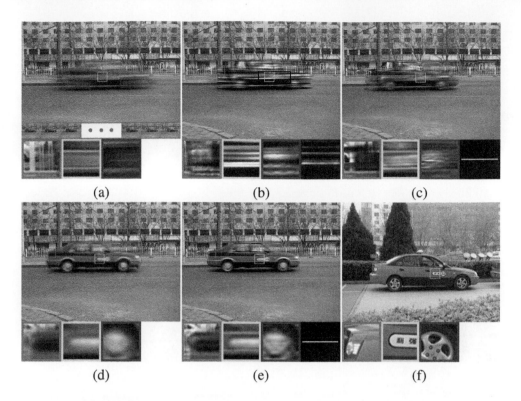

Figure 11.13 Image deblurring with translational motion. In this example, the moving object is a car moving horizontally. We assume that the motion blur within the car is globally invariant. (a) input; (b) result generated by [14], where the user-selected region is indicated by the black box; (c) result generated by [3]; (d) result generated by back-projection [34]; (e) our results; (f) the ground truth sharp image captured from another car of the same model. Close-up views and the estimated global blur kernels within the motion blur layer are also shown. (© 2010 IEEE.)

invariant motion blur caused by handshake, *in-plane rotational motion* of a scene object, *translational motion* of a scene object, *out-of-plane rotational motion* of an object, *zoom-in motion* caused by changing the focal length (i.e., camera's zoom setting), a combination of translational motion and rotational motion with *multiple frames* used as input for deblurring one frame, *video deblurring with out-of-plane rotational motion*, *video deblurring with complex in-plane motion*, and *video deblurring with a combination of translational and zoom-in motion*.

Globally invariant motion blur: In Figure 11.11, we present an image deblurring example with globally invariant motion, where the input is one high-resolution image and several low-resolution images. Our results are compared with those generated by the methods of Fergus et al. [14], Ben-Ezra and Nayar [3], and back-projection [34]. Fergus et al.'s approach is a state-of-the-art blind deconvolution technique that employs a natural image statistics constraint. However, when the blur kernel is not correctly estimated, an unsatisfactory result shown in Figure 11.11(b) will be produced. Ben-Ezra and Nayar use the estimated optical flow as the blur kernel and then perform deconvolution. Their result in Figure

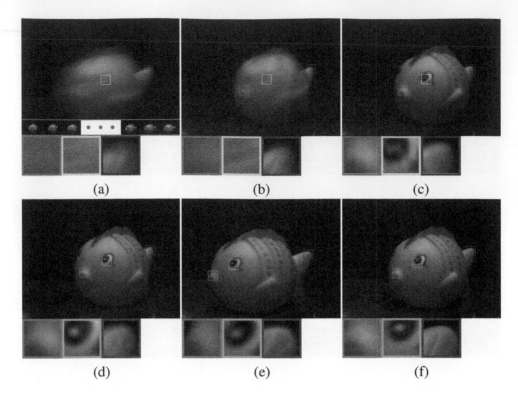

Figure 11.14 Image deblurring with spatially varying kernels. In this example, the moving object contains out-of-plane rotation with both occlusion and disocclusion at the object boundary. (a) input; (b) result generated by [3]; (c) result generated by back-projection [34]; (d) our results using the first low-resolution frame as the reference frame; (e) our results using the last low-resolution frame as the reference frame; (f) the ground truth sharp image. Close-ups are also shown. (© 2010 IEEE.)

11.11(c) is better than that in (b) as the estimated blur kernel is more accurate, but ringing artifacts are still unavoidable. Back-projection produces a super resolution result from a sequence of low-resolution images as shown in Figure 11.11(d). Noting that motion blur removal is not the intended application of back-projection, we can see that its results are blurry because the high-frequency details are not sufficiently captured in the low-resolution images. The result of our method and the refined kernel estimate are displayed in Figure 11.11(e). The ground truth is given in Figure 11.11(f) for comparison.

In-plane rotational motion: Figure 11.12 shows an example with in-plane rotational motion. We compared our results with those by Shan et al. [30], Ben-Ezra and Nayar [3], and back-projection [34]. Shan et al. [30] is a recent technique that targets deblurring of in-plane rotational motion. Our approach is seen to produce less ringing artifacts compared to [30] and [3], and it generates greater detail than [34].

Translational motion: Figure 11.13 shows an example of a car translating horizontally. We assume the motion blur within the car region is globally invariant and thus techniques for removing globally invariant motion blur can be applied after layer separation of the moving object. We use the technique proposed in Section 11.5 to separate the moving car

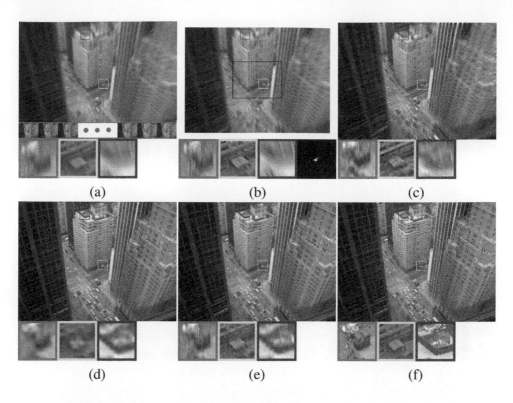

Figure 11.15 Image deblurring with spatially varying kernels. In this example, the camera is zooming into the scene: (a) input; (b) result generated by [14]; (c) result generated by [3]; (d) result generated by back-projection [34]; (e) our results; (f) the ground truth sharp image. Close-ups are also shown. (© 2010 IEEE.)

from the static background. Our results are compared with those generated by Fergus et al. [14], Ben-Ezra and Nayar [3], and back-projection [34]. In this example, the moving car is severely blurred with most of the high-frequency details lost. We demonstrate in Figure 11.13(c) the limitation of using deconvolution alone even with an accurate motion blur kernel. In this example, the super resolution result in Figure 11.13(d) is better than the deconvolution result, but there are some high-frequency details that are not recovered. Our result is shown in Figure 11.13(e), which maintains most low-frequency details recovered by super resolution and also high-frequency details recovered by deconvolution. Some incorrect high-frequency details from the static background are incorrectly retained in our final result because of the presence of some high-frequency background details in the separated moving object layer. We believe that a better layer separation algorithm would lead to improved results. This example also exhibits a basic limitation of our approach. Because there is significant car motion during the exposure time, most high-frequency detail is lost and cannot be recovered by our approach. The ground truth in Figure 11.13(f) shows a similar parked car for comparison.

Out-of-plane rotational motion: Figure 11.14 shows an example of out-of-plane rotation where occlusion/disocclusion occurs at the object boundary. Our result is compared to that of Ben-Ezra and Nayar [3] and back-projection [34]. One major advantage of our

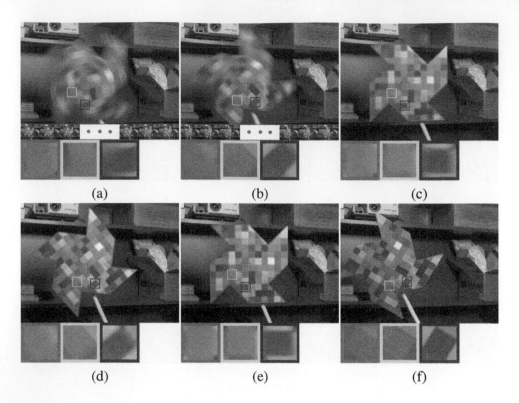

Figure 11.16 Deblurring with and without multiple high-resolution frames: (a)(b) input images containing both translational and rotational motion blur; (c) deblurring using only (a) as input; (d) deblurring using only (b) as input; (e) deblurring of (a) using both (a) and (b) as inputs; (f) ground truth sharp image. Close-ups are also shown. (© 2010 IEEE.)

approach is that we can detect the existence of occlusions/disocclusions of the motion-blurred moving object. This not only helps to estimate the alpha mask for layer separation, but also aids in eliminating irrelevant low-resolution reference frame constraints for back-projection. We show our result by choosing the first frame and the last frame as the reference frame. Both occlusion and disocclusion are contained in this example.

Zoom-in motion: Figure 11.15 shows another example of motion blur from zoom-in effects. Our result is compared to Fergus et al. [14], Ben-Ezra and Nayar [3], and back-projection [34]. We note that the method of Fergus et al. [14] is intended for globally invariant motion blur, and is shown here to demonstrate the effects of using only a single blur kernel to deblur spatially varying motion blur. Again, our approach produces better results with less ringing artifacts and richer detail.

Deblurring with multiple frames: The benefit of using multiple deconvolutions from multiple high-resolution frames is exhibited in Figure 11.16 for a pinwheel with both translational and rotational motion. The deblurring result in Figure 11.16(c) was computed using only Figure 11.16(a) as input. Likewise, Figure 11.16(d) is the deblurred result from only Figure 11.16(b). Using both Figures 11.16(a) and (b) as inputs yields the improved result in Figure 11.16(e). This improvement can be attributed to the difference in high-frequency

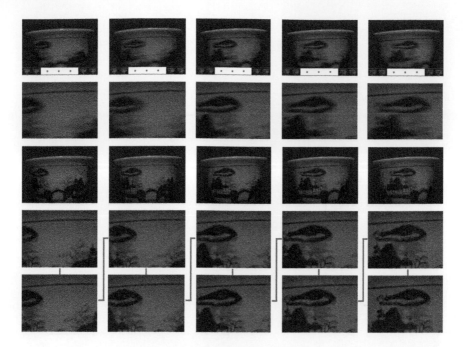

Figure 11.17 Video deblurring with out-of-plane rotational motion. The moving object is a vase with a center of rotation approximately aligned with the image center. First Row (top): Input video frames. Second Row: Close-ups of a motion-blurred region. Third Row: Deblurred video. Fourth Row: Close-ups of deblurred video using the first low-resolution frames as the reference frames. Fifth Row: Close-ups of deblurred video frames using the fifth low-resolution frames as the reference frames. The final video sequence has higher temporal sampling than the original high-resolution video, and is played with frames ordered according to the red lines. (© 2010 IEEE.)

detail that can be recovered from each of the differently blurred images. The ground truth is shown in Figure 11.16(f) for comparison.

Video deblurring with out-of-plane rotational motion: Figure 11.17 demonstrates video deblurring of a vase with out-of-plane rotation. The center of rotation is approximately aligned with the image center. The top row displays five consecutive input frames. The second row shows close-ups of a motion-blurred region. The middle row shows our results with the first low-resolution frames as the reference frames. The fourth and fifth rows show close-ups of our results with respect to the first and fifth low-resolution frames as the reference frames.

This example also demonstrates the ability to produce multiple deblurring solutions as described in Section 11.6. For temporal super resolution, we combine the results together in the order indicated by the red lines in Figure 11.17. With our method, we can increase the frame rate of deblurred high-resolution videos up to the same rate as the low-resolution, high-frame-rate video input.

Video deblurring with complex in-plane motion: Figure 11.18 presents another video deblurring result of a tossed box with complex (in-plane) motion. The top row displays five consecutive input frames. The second row shows close-ups of the motion-blurred moving

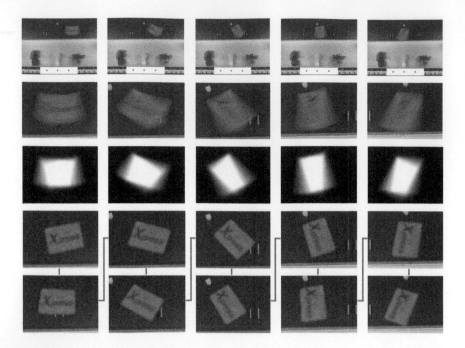

Figure 11.18 Video deblurring with a static background and a moving object. The moving object is a tossed box with arbitrary (in-plane) motion. First Row (top): Input video frames. Second Row: Close-up of the motion-blurred moving object. Third Row: Extracted alpha mattes of the moving-object. Fourth Row: The deblurred video frames using the first low-resolution frames as the reference frames. Fifth Row: The deblurred video frames using the third low-resolution frames as the reference frames. The final video with temporal super resolution is played with frames ordered as indicated by the red lines. (© 2010 IEEE.)

object. The middle row shows our separated mattes for the moving object, and the fourth and the fifth rows present our results with the first and third low-resolution frames as reference. The text on the tossed box is recovered to a certain degree by our video deblurring algorithm. Similar to the previous video deblurring example, our output is a high-resolution, high-frame-rate deblurred video. This result also illustrates a limitation of our method, where the shadow of the moving object is not deblurred and may appear inconsistent. This problem is a direction for future investigation.

Video deblurring with a combination of translational and zoom-in motion: Our final example is shown in Figure 11.19. The moving object of interest is a car driving toward the camera. Both translational effects and zoom-in blur effects exist in this video deblurring example. The top row displays five consecutive frames of input. The second row shows close-ups of the motion-blurred moving object. The middle row shows our extracted mattes for the moving object, and the fourth and the fifth rows present our results with the first and the fifth low-resolution frames as reference.

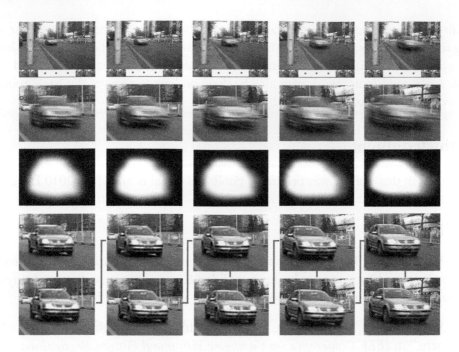

Figure 11.19 Video deblurring in an outdoor scene. The moving object is a car driving toward the camera, which produces both translation and zoom-in blur effects. First Row (top): Input video frames. Second Row (top): The extracted alpha mattes of the moving object. Third Row (top): The deblurred video frames using the first low-resolution frames as the reference frames. Fourth Row: The deblurred video frames using the third low-resolution frames as the reference frames. The final video consists of frames ordered as indicated by the red lines. By combining results from using different low-resolution frames as reference frames, we can increase the frame rate of the deblurred video. (© 2010 IEEE.)

11.8   Conclusion

We have proposed an approach for image/video deblurring using a hybrid camera. Our work has formulated the deblurring process as an iterative method that incorporates optical flow, back-projection, kernel refinement, and frame coherence to effectively combine the benefits of both deconvolution and super resolution. We demonstrate that this approach can produce results that are sharper and cleaner than state-of-the-art techniques.

While our video deblurring algorithm exhibits high-quality results on various scenes, there exist complicated forms of spatially varying motion blur that can be difficult for our method to handle (e.g., motion blur effects caused by object deformation). The performance of our algorithm is also bounded by the performance of several of its components, including optical flow estimation, layer separation and also the deconvolution algorithm. Despite these limitations, we have proposed the first work to handle spatially varying motion blur with arbitrary in-plane/out-of-plane rigid motion. This work is also the first to address video deblurring and to increase video frame rates using a deblurring algorithm.

Future research directions for this work include how to improve deblurring performance by incorporating priors into our framework. Recent deblurring methods have demonstrated the utility of priors, such as the natural image statistics prior and the sparsity prior, for reducing ringing artifacts, and for kernel estimation. Another research direction is to improve layer separation by more fully exploiting the available information in the hybrid camera system. Additional future work may also be done on how to recover the background partially occluded by a motion-blurred object.

Acknowledgment

The material in this chapter was originally published in Tai et al. [1] (© 2010 IEEE.).

Bibliography

[1] Y.-W. Tai, H. Du, M. S. Brown, and S. Lin, "Correction of spatially varying image and video motion blur using a hybrid camera," *IEEE Transactions on Pattern Analysis and Machine Intelligence*, vol. 32, no. 6, pp. 1012–1028, 2010.

[2] Y.-W. Tai, H. Du, M. S. Brown, and S. Lin, "Image/video deblurring using a hybrid camera," in *IEEE Conference on Computer Vision and Pattern Recognition (CVPR)*, 2008.

[3] M. Ben-Ezra and S. Nayar, "Motion deblurring using hybrid imaging," in *IEEE Conference on Computer Vision and Pattern Recognition (CVPR)*, vol. I, pp. 657–664, June 2003.

[4] M. Ben-Ezra and S. Nayar, "Motion-based motion deblurring," *IEEE Transactions on Pattern Analysis and Machine Intelligence*, vol. 26, pp. 689–698, June 2004.

[5] M. McGuire, W. Matusik, H. Pfister, J. F. Hughes, and F. Durand, "Defocus video matting," *ACM Transactions on Graphics*, vol. 24, no. 3, pp. 567–576, 2005.

[6] M. McGuire, W. Matusik, and W. Yerazunis, "Practical, real-time studio matting using dual imagers," in *Eurographics Symposium on Rendering (EGSR)*, 2006.

[7] F. Li, J. Yu, and J. Chai, "A hybrid camera for motion deblurring and depth map super resolution," in *IEEE Conference on Computer Vision and Pattern Recognition (CVPR)*, 2008.

[8] M. Aggarwal and N. Ahuja, "Split aperture imaging for high dynamic range," *International Journal on Computer Vision*, vol. 58, no. 1, pp. 7–17, 2004.

[9] W. Richardson, "Bayesian-based iterative method of image restoration," *Journal of the Optical Society of America*, vol. 62, no. 1, 55–59, 1972.

[10] L. Lucy, "An iterative technique for the rectification of observed distributions," *Astronomical Journal*, vol. 79, no. 6, 745–75, 1974.

[11] Wiener and Norbert, *Extrapolation, Interpolation, and Smoothing of Stationary Time Series*, New York: Wiley, 1949.

[12] P. C. Hansen, J. G. Nagy, and D. P. O'Leary, *Deblurring Images: Matrices, Spectra, and Filtering*, Philadelphia, *Society for Industrial and Applied Mathematic*, 2006.

[13] R. C. Gonzalez and R. E. Woods, *Digital Image Processing (2nd edition).* Englewood Cliffs: Prentice Hall, 2002.

[14] R. Fergus, B. Singh, A. Hertzmann, S. T. Roweis, and W. T. Freeman, "Removing camera shake from a single photograph," *ACM Transactions on Graphics*, vol. 25, no. 3, 2006.

[15] R. Raskar, A. Agrawal, and J. Tumblin, "Coded exposure photography: motion deblurring using fluttered shutter," *ACM Transactions on Graphics*, vol. 25, no. 3, 2006.

[16] J. Jia, "Single image motion deblurring using transparency," in *IEEE Conference on Computer Vision and Pattern Recognition (CVPR)*, 2007.

[17] N. Dey, L. Blanc-Fraud, C. Zimmer, Z. Kam, P. Roux, J. Olivo-Marin, and J. Zerubia, "A deconvolution method for confocal microscopy with total variation regularization," in *IEEE International Symposium on Biomedical Imaging: Nano to Macro*, 2004.

[18] A. Levin, R. Fergus, F. Durand, and W. T. Freeman, "Image and depth from a conventional camera with a coded aperture," *ACM Transactions on Graphics*, 2007.

[19] L. Yuan, J. Sun, L. Quan, and H.-Y. Shum, "Progressive inter-scale and intra-scale non-blind image deconvolution," *ACM Transactions on Graphics*, 2008.

[20] Q. Shan, J. Jia, and A. Agarwala, "High-quality motion deblurring from a single image," *ACM Transactions on Graphics*, 2008.

[21] A. Levin, Y. Weiss, F. Durand, and W. Freeman, "Understanding and evaluating blind deconvolution algorithms," in *IEEE Conference on Computer Vision and Pattern Recognition (CVPR)*, 2009.

[22] B. Bascle, A. Blake, and A. Zisserman, "Motion deblurring and super resolution from an image sequence," in *IEEE European Conference on Computer Vision (ECCV)*, pp. 573–582, 1996.

[23] L. Yuan, J. Sun, L. Quan, and H. Shum, "Image deblurring with blurred/noisy image pairs," in *ACM Transactions on Graphics*, p. 1, 2007.

[24] A. Rav-Acha and S. Peleg, "Two motion blurred images are better than one," *Pattern Recognition Letters*, vol. 26, pp. 311–317, 2005.

[25] J. Chen and C.-K. Tang, "Robust dual motion deblurring," in *IEEE Conference on Computer Vision and Pattern Recognition (CVPR)*, 2008.

[26] P. Bhat, C. L. Zitnick, N. Snavely, A. Agarwala, M. Agrawala, B. Curless, M. Cohen, and S. B. Kang, "Using photographs to enhance videos of a static scene," in *Proceedings Eurographics Symposium on Rendering(EGSR)*, pp. 327–338, 2007.

[27] A. Levin, "Blind motion deblurring using image statistics," in *Advances in Neural Information Processing Systems (NIPS)*, pp. 841–848, 2006.

[28] J. Bardsley, S. Jefferies, J. Nagy, and R. Plemmons, "Blind iterative restoration of images with spatially-varying blur," in *Optics Express*, pp. 1767–1782, 2006.

[29] S. Cho, Y. Matsushita, and S. Lee, "Removing non-uniform motion blur from images," in *IEEE International Conference on Computer Vision (ICCV)*, 2007.

[30] Q. Shan, W. Xiong, and J. Jia, "Rotational motion deblurring of a rigid object from a single image," in *IEEE International Conference on Computer Vision (ICCV)*, 2007.

[31] A. Levin, P. Sand, T. S. Cho, F. Durand, and W. T. Freeman, "Motion-invariant photography," *ACM Transactions on Graphics*, 2008.

[32] Y.-W. Tai, N. Kong, S. Lin, and S. Y. Shin, "Coded exposure imaging for projective motion deblurring," in *IEEE Conference on Computer Vision and Pattern Recognition (CVPR)*, 2010.

[33] Y.-W. Tai, P.Tan, and M. Brown, "Richardson-lucy deblurring for scenes under a projective motion path," *IEEE Transactions on Pattern Analysis and Machine Intelligence*, To appear.

[34] M. Irani and S. Peleg, "Improving resolution by image registration," *Computer Vision, Graphics, and Image Processing (CVGIP)*, vol. 53, no. 3, pp. 231–239, 1991.

[35] M. Irani and S. Peleg, "Motion analysis for image enhancement: Resolution, occlusion and transparency," *Journal of Visual Communication and Image Representation (JVCIR)*, vol. 4, no. 4, 324–335, 1993.

[36] S. Borman and R. Stevenson, "Super resolution from image sequences - a review," *Proceedings of the Midwest Symposium on Circuits and Systems (MWSCAS)*, p. 374, 1998.

[37] A. Patti, M. Sezan, and A.M. Tekalp, "Superresolution video reconstruction with arbitrary sampling lattices and nonzero aperture time," *IEEE Transactions on Image Processing*, vol. 6, no. 8, pp. 1064–1076, 1997.

[38] M. Elad and A. Feuer, "Superresolution restoration of an image sequence: adaptive filtering approach," *IEEE Transactions on Image Processing*, vol. 8, no. 3, pp. 387–395, 1999.

[39] W. Zhao and H. S. Sawhney, "Is super resolution with optical flow feasible?" in *IEEE European Conference on Computer Vision (ECCV)*, pp. 599–613, 2002.

[40] E. Shechtman, Y. Caspi, and M. Irani, "Space-time super resolution," *IEEE Transactions on Pattern Analysis and Machine Intelligence*, vol. 27, no. 4, pp. 531–544, 2005.

[41] F. Sroubek, G. Cristobal, and J. Flusser, "A unified approach to superresolution and multichannel blind deconvolution," *IEEE Transactions on Image Processing*, vol. 16, no. 9, 2322–2332, Sept. 2007.

[42] A. Agrawal and R. Raskar, "Resolving objects at higher resolution from a single motion-blurred image," in *IEEE Conference on Computer Vision and Pattern Recognition (CVPR)*, 2007.

[43] M. Bigas, E. Cabruja, J. Forest, and J. Salvi, "Review of cmos image sensors," *Microelectronics Journal*, vol. 37, no. 5, pp. 433–451, 2006.

[44] B. Lucas and T. Kanade, "An iterative image registration technique with an application to stereo vision," in *Proceedings of Imaging Understanding Workshop*, pp. 121–130, 1981.

[45] J. Xiao, H. Cheng, H. Sawhney, C. Rao, and M. Isnardi, "Bilateral filtering-based optical flow estimation with occlusion detection," in *IEEE European Conference on Computer Vision (ECCV)*, 2006.

[46] S. Baker and T. Kanade, "Limits on super resolution and how to break them," *IEEE Transactions on Pattern Analysis and Machine Intelligence*, vol. 24, no. 9, pp. 1167–1183, 2002.

[47] D. Fish, A. Brinicombe, E. Pike, and J. Walker, "Blind deconvolution by means of the Richardson-Lucy algorithm," *Journal of the Optical Society of America*, vol. 12, 1995.

[48] T. Lauer, "Deconvolution with a spatially-variant PSF," in *Astronomical Data Analysis II*, vol. 4847, pp. 167–173, 2002.

[49] N. Joshi, R. Szeliski, and D. Kriegman, "PSF estimation using sharp edge prediction," in *IEEE Conference on Computer Vision and Pattern Recognition (CVPR)*, 2008.

[50] Y. Chuang, B. Curless, D. H. Salesin, and R. Szeliski, "A Bayesian approach to digital matting," in *IEEE Conference on Computer Vision and Pattern Recognition (CVPR)*, pp. 264–271, 2001.

[51] Y. Chuang, A. Agarwala, B. Curless, D. H. Salesin, and R. Szeliski, "Video matting of complex scenes," *ACM Transactions on Graphics*, pp. 243–248, 2002.

[52] J. Sun, J. Jia, C. Tang, and H. Shum, "Poisson matting," *ACM Transactions on Graphics*, vol. 23, no. 3, 315–321, 2004.

[53] J. Wang and M. Cohen, "An iterative optimization approach for unified image segmentation and matting," in *IEEE International Conference on Computer Vision (ICCV)*, 2005.

[54] A. Levin, D. Lischinski, and Y. Weiss, "A closed form solution to natural image matting," in *IEEE Conference on Computer Vision and Pattern Recognition (CVPR)*, 2006.

[55] J. Wang, M. Agrawala, and M. Cohen, "Soft scissors: An interactive tool for realtime high quality matting," *ACM Transactions on Graphics*, vol. 26, no. 3, 2007.

[56] A. Smith and J. F. Blinn, "Blue screen matting," *SIGGRAPH*, 1996.

[57] N. Joshi, W. Matusik, and S. Avidan, "Natural video matting using camera arrays," *ACM Transactions on Graphics*, vol. 25, no. 3, 779–786, 2006.

[58] G. Zhang, J. Jia, W. Xiong, T. Wong, P. Heng, and H. Bao, "Moving object extraction with a hand-held camera," in *IEEE International Conference on Computer Vision (ICCV)*, 2007.

[59] H. Lin, Y.-W. Tai, and M. Brown, "Motion regularization for matting motion blurred objects," *IEEE Transactions on Pattern Analysis and Machine Intelligence*, vol. 33, no. 11, 2329–2336, 2011.

[60] Y. Matsushita, E. Ofek, W. Ge, X. Tang, and H. Shum, "Full-frame video stabilization with motion inpainting," *IEEE Transactions on Pattern Analysis and Machine Intelligence*, vol. 28, no. 7, 1150–1163, 2006.

Index